MOLECULAR PATHOLOGY LIBRARY SERIES

Philip T. Cagle, MD, Series Editor

AF774098

For other titles published in this series, go to
www.springer.com/series/7723

Basic Concepts of Molecular Pathology

Edited by

Philip T. Cagle, MD
Weill Medical College of Cornell University, New York
The Methodist Hospital, Houston, Texas

and

Timothy Craig Allen, MD, JD
University of Texas Health Science Center at Tyler
Tyler, Texas

Springer

Editors
Philip T. Cagle, MD
Pathology and Laboratory Medicine
Weill Medical College of Cornell University
New York, NY
The Methodist Hospital
Houston, Texas

Timothy Craig Allen, MD, JD
Department of Pathology
University of Texas Health Science
Center at Tyler
Tyler, TX

Series editor
Philip T. Cagle, MD
Pathology and Laboratory Medicine
Weill Medical College of Cornell University
New York, NY
The Methodist Hospital
Houston, Texas

ISSN 1935-987x
ISBN 978-0-387-89626-7 (eBook)
DOI 10.1007/978-0-387-89626-7
Springer Dordrecht Heidelberg London New York

© Springer Science+Business Media, LLC 2009
All rights reserved. This work may not be translated or copied in whole or in part without the written permission of the publisher (Springer Science+Business Media, LLC, 233 Spring Street, New York, NY 10013, USA), except for brief excerpts in connection with reviews or scholarly analysis. Use in connection with any form of information storage and retrieval, electronic adaptation, computer software, or by similar or dissimilar methodology now known or hereafter developed is forbidden.
The use in this publication of trade names, trademarks, service marks, and similar terms, even if they are not identified as such, is not to be taken as an expression of opinion as to whether or not they are subject to proprietary rights.
While the advice and information in this book are believed to be true and accurate at the date of going to press, neither the authors nor the editors nor the publisher can accept any legal responsibility for any errors or omissions that may be made. The publisher makes no warranty, express or implied, with respect to the material contained herein.

Printed on acid-free paper

Springer is part of Springer Science+Business Media (www.springer.com)

To my wife, Kirsten

Philip T. Cagle

To my wife, Fran; and to Caitlin and Erin.

Timothy Craig Allen

Preface

The increasing role of molecular pathology in the daily practice of medicine, including the increasingly routine use of targeted molecular therapy in personalized patient care, has created a need for textbooks that bridge the gap between medical practice and basic molecular science. Each book in the *Molecular Pathology Library Series* by Springer Science+Business Media provides a review of current molecular pathology for a specific organ system. The purpose of *Basic Concepts of Molecular Pathology* is to provide a succinct background of molecular pathology terminology and concepts to serve as a foundation and reference for understanding the medically oriented organ-specific information in the other books in the series. As such, *Basic Concepts of Molecular Pathology* serves as a companion to books in the *Molecular Pathology Library Series* and also stands alone as a quick review of molecular pathology principles.

Philip T. Cagle, MD
Houston, TX
Timothy Craig Allen, MD, JD
Tyler, TX

Series Preface

The past two decades have seen an ever-accelerating growth in knowledge about the molecular pathology of human diseases, which received a large boost with the sequencing of the human genome in 2003. Molecular diagnostics, molecular-targeted therapy, and genetic therapy are now routine in many medical centers. The molecular field now impacts every field in medicine, whether clinical research or routine patient care. There is a great need for basic researchers to understand the potential clinical implications of their research, whereas private practice clinicians of all types (general internal medicine and internal medicine specialists, medical oncologists, radiation oncologists, surgeons, pediatricians, family practitioners), clinical investigators, pathologists, medical laboratory directors, and radiologists require a basic understanding of the fundamentals of molecular pathogenesis, diagnosis, and treatment for their patients.

Traditional textbooks in molecular biology present basic science and are not readily applicable to the medical setting. Most medical textbooks that include a mention of molecular pathology in the clinical setting are limited in scope and assume that the reader already has a working knowledge of the basic science of molecular biology. Other texts emphasize technology and testing procedures without integrating the clinical perspective. There is an urgent need for a text that fills the gap between basic science books and clinical practice.

In the *Molecular Pathology Library* series, basic science and technology are integrated with the medical perspective and clinical application. Each book in the series is divided according to neoplastic and nonneoplastic diseases for each of the organ systems traditionally associated with medical subspecialties.

Each book in the series is organized to provide (1) a succinct background of the essential terminology, concepts, and technology of molecular biology; (2) an overview of the broad application of molecular biology principles to disease; and (3) specific application of molecular pathology to the pathogenesis, diagnosis, and treatment of neoplastic and nonneoplastic diseases specific to each organ system. These broad section topics are divided into succinct chapters, averaging about 15–20 pages each, to cover a very specific disease entity. The chapters are written by established authorities on the specific topic from academic centers around the world. In one book, diverse subjects are included that the reader would have to pursue from multiple sources to have a clear understanding of the molecular pathogenesis, diagnosis, and treatment of specific diseases. Attempting to hunt for the full information from basic concept to specific applications for a disease from the varied sources is time consuming and frustrating. By providing this quick and user-friendly reference, understanding and application of this rapidly growing field are made more accessible to both expert and generalist alike.

As books that bridge the gap between basic science and clinical understanding and practice, the *Molecular Pathology Series* serves the basic scientist, the clinical researcher, and the practicing physician or other health care provider who require more understanding of the application of basic research to patient care, that is, from "bench to bedside." This series is unique and an invaluable resource to those who need to know about molecular pathology from a clinical, disease-oriented perspective. These books will be indispensable to physicians and health care providers in multiple disciplines, as already noted, to residents and fellows in these multiple disciplines as well as their teaching institutions, and to researchers who increasingly must justify the clinical implications of their research.

Philip T. Cagle, MD
Series Editor
Houston, TX

Contents

Contributors

Timothy Craig Allen
Department of Pathology, University of Texas Health Science Center at Tyler, Tyler, TX, USA

Roberto Barrios
Department of Pathology and Laboratory Medicine, Weill Medical College of Cornell University, New York, NY, USA
The Methodist Hospital, Houston, TX, USA

Gregory L. Blakey
Department of Pathology, University of Oklahoma Health Sciences Center, Oklahoma City, OK, USA

Philip T. Cagle
Pathology and Laboratory Medicine, Weill Medical College of Cornell University, New York, NY, USA
The Methodist Hospital, Houston, TX, USA

Chung-Che (Jeff) Chang
Pathology and Laboratory Medicine, Weill Medical College of Cornell University, New York, NY, USA
The Methodist Hospital, Houston, TX, USA

William B. Coleman
Department of Pathology and Laboratory Medicine, UNC Lineberger Comprehensive Cancer Center, University of North Carolina School of Medicine, Chapel Hill, NC, USA

Kumarasen Cooper
Department of Pathology, University of Vermont/Fletcher Allen Healthcare, Burlington, VT, USA

Sanja Dacic
Division of Anatomic Pathology, Department of Pathology, University of Pittsburgh Medical Center, Pittsburgh, PA, USA

Daniel H. Farkas
Center for Molecular Medicine, Grand Rapids, MI, USA

Patrick L. Fitzgibbons
Department of Pathology, St. Jude Medical Center, Fullerton, CA, USA

Larry Fowler
Department of Pathology, University of Texas Health Science Center at San Antonio, San Antonio, TX, USA

Wieslaw Furmaga
Department of Pathology, University of Texas Health Science Center at San Antonio, San Antonio, TX, USA

Thomas J. Goodwin
Department of Biomedical Research and Operations Branch, NASA/Johnson Space Center, Houston, TX, USA

Kevin C. Halling
Molecular Cytology and Imaging Laboratory, Department of Laboratory Medicine and Pathology, Mayo Clinic, Rochester, MN, USA

Louise Hecker
Division of Pulmonary & Critical Care Medicine, University of Michigan Health System, Internal Medicine, Ann Arbor, MI, USA

Jennifer L. Hunt
Surgical Pathology, Head and Neck/Endocrine Pathology, AP Molecular Diagnostics Unit, Department of Anatomic Pathology, Cleveland Clinic, Cleveland, OH, USA

Jeffrey M. Jordan
Department of Pathology, University of Texas Medical Branch, Galveston, TX, USA

Dwight Oliver
Department of Pathology and Laboratory Medicine, University of Texas-Houston Medical School, Houston, TX, USA

Sem H. Phan
Department of Pathology, University of Michigan, Ann Arbor, MI, USA

Sharon C. Presnell
Cell and Tissue Technologies, Becton Dickinson, Research Triangle Park, NC, USA

Ashley G. Rivenbark
UNC Lineberger Comprehensive Cancer Center, Department of Biochemistry and Biophysics, University of North Carolina School of Medicine, Chapel Hill, NC, USA

Pothana Saikumar
Department of Pathology, University of Texas Health Science Center at San Antonio, San Antonio, TX, USA

Munir Shahjahan
Pathology and Laboratory Medicine, The Methodist Hospital, Houston, TX, USA

Travis Solley
Department of Surgery, University of Texas Medical Branch, Galveston, TX, USA

Victor J. Thannickal
Division of Pulmonary & Critical Care Medicine, University of Michigan Health System, Internal Medicine, Ann Arbor, MI, USA

Reinhard Ullmann
Department of Human Molecular Genetics, Max Planck Institute for Molecular Genetics, Berlin, Germany

Manjeri A. Venkatachalam
Department of Pathology, University of Texas Health Science Center, San Antonio, TX, USA

Roger A. Vertrees
Department of Surgery, University of Texas Medical Branch, Galveston, TX, USA

Belinda J. Wagner
Scientific & Technical Operations, Tengion, Inc., Winston-Salem, NC, USA

Amy J. Wendel
Department of Laboratory Medicine and Pathology, Mayo Clinic, Rochester, MN, USA

Youli Zu
Pathology and Laboratory Medicine, Weill Medical College of Cornell University, New York, NY, USA
The Methodist Hospital, Houston, TX, USA

1

Genes, Gene Products, and Transcription Factors

Philip T. Cagle

Introduction

Molecular pathology employs an ever-expanding array of special techniques to study nucleic acids, genes, gene products, receptors, signaling pathways, the cell cycle, and mutations. This chapter and the others in this section provide a quick review of basic terminology and concepts for the understanding of subsequent chapters.

Nucleic Acids, Genes, and Gene Products

Genes are the bits of information that code for the proteins which are necessary for structure and metabolic reactions in living tissues. Genes and the molecules that construct their protein products using the blueprints or genetic code in the genes are composed of nucleic acids. Nucleotides are the building blocks of the nucleic acids, deoxyribonucleic acid (DNA) and ribonucleic acid (RNA), the essential compounds that make up genes and transcribe the genetic code into proteins, respectively. Nucleotides are basic compounds composed of a sugar-phosphate backbone and a nitrogenous base. Nucleotides consist of two types of bases: purines and pyrimidines. In DNA, there are two purines (adenine, abbreviated as A, and guanine, abbreviated as G) and two pyrimidines (thymine, abbreviated as T, and cytosine, abbreviated as C). In RNA, there are also two purines (A and G) and two pyrimidines (uracil, abbreviated as U, replaces T and C). DNA is typically double stranded, with the nucleotide bases paired together as described below, and RNA is typically single stranded. When nucleotides are assembled together in a nucleic acid, formation of covalent phosphodiester bonds results in a free 5′-phosphate at the origin of the nucleic acid and a free 3′-hydroxyl at the end (terminus) of the nucleic acid. For this reason, the synthesis of the nucleic acid is said to occur in a 5′-to 3′-direction (see below).[1–9]

DNA is composed of nucleotides arranged sequentially in a deliberate order that encodes as genes for a matching sequential order of amino acids which will form proteins, as further discussed below. The nucleotides in genes are arranged in a double-stranded right-handed helix in the cell nucleus so that the purine A always binds with the pyrimidine T and the purine G always binds with the pyrimidine C in base pairs. As a result, the DNA in a double helix is arranged in complementary strands: the sequence of nucleotides in one strand of DNA is a "mirror image" of the nucleotide sequence in the other DNA strand. Groups of DNA base pairs are wrapped around small proteins called *histones*, forming arrangements of DNA called *nucleosomes*, allowing the DNA to fit within the cell nucleus. Genes are located on chromosomes that consist of DNA packaged with histone and nonhistone proteins. There are 23 pairs of chromosomes in the human, for a total of 46. Each gene is located at a specific site or locus on a specific pair of chromosomes. Because the chromosomes are in pairs and each chromosome has a locus for each gene, the genes occur as two copies or alleles, one copy or allele on each of the members of the chromosome pair. Chromosomes are ordinarily indistinct in the nuclear chromatin, but are discrete during mitosis or cell division.[1–14]

The genome is the entirety of the DNA sequence or chromosomes of an organism, or its "complete genetic complement." Genomics is the sequencing and study of genomes and cytogenetics is the study of chromosomes, traditionally through visualization of the karyotype or set of chromosomes of an organism. The somatic cells are diploid with a pair of each of the chromosomes and, therefore, two copies or alleles of each gene, one allele at the equivalent locus on each paired chromosome. The gametes are haploid, which means that these cells have only one set of each of the chromosomes with only one allele of each gene. During fertilization, the nuclear material of the two gametes combines, restoring the diploid number of chromosomes and alleles to the diploid number in the fertilized egg.[1–9]

The genotype is the genetic information in an individual's DNA, and the phenotype is how the genotype is manifested or expressed. Genotype and phenotype can differ. If two alleles are the same in one individual, they are said to be homozygous, and if the two alleles are different, they are

P.T. Cagle and T.C. Allen (eds.), *Basic Concepts of Molecular Pathology*, Molecular Pathology Library 2,
DOI: 10.1007/978-0-387-89626-7_1, © Springer Science+Business Media, LLC 2009

said to be heterozygous. If a person is heterozygous for a gene, one allele may be expressed preferentially over the other allele, in which case the former allele is considered dominant and the latter allele is considered recessive. In this situation, the dominant allele codes for features that mask the features coded for by the recessive allele, so that the phenotype is different from the genotype. For example, a person is heterozygous for eye color and has an allele for brown eyes on one chromosome and an allele for blue eyes on the other member of that pair of chromosomes. That person will have brown eyes (the phenotype) because the allele for brown eyes is dominant over the allele for blue eyes. On the other hand, if the person is homozygous and has two recessive alleles for blue eyes, that person will have blue eyes.[1–9]

Single nucleotide polymorphisms, or SNPs (pronounced "snips"), are inherited, naturally occurring variations in one base between the DNA sequences in the same gene in two individuals and account for most of the genetic variation between individuals. Alleles or SNPs that are in close proximity on a chromosome are often inherited together as a haplotype. Polymorphisms are differences in DNA between individuals, and the most simple polymorphism is the SNP.[1–9]

Before cell division, new DNA must be synthesized from an existing strand of DNA, a process called *replication*. The synthesis of new DNA is a tightly controlled phase of the cell cycle (the S phase). The cell cycle is described in greater detail in Chap. 2. Before initiation of DNA replication, a prereplicative complex is constructed. This prereplicative complex is composed of the minichromosome maintenance protein complex (MCM), the origin recognition complex (ORC), and Cdc6/Cdc18. The S-phase kinases Cdc7 and Cdk (cyclin-dependent kinase) activate the prereplicative complex to yield an initiation complex at the origin with binding of Cdc45 to MCM. Replication is initiated at specific points in the DNA, referred to as *origins of replication*, and this creates a Y-shaped replication fork where the parental DNA duplex splits into two daughter DNA duplexes. As further described below, the duplex DNA is unwound with the assistance of special enzymes called topoisomerases, and then replication proteins, including DNA polymerases, bind to the unwound DNA.[15–60]

The synthesis of RNA, including messenger RNA (mRNA), from a strand of DNA (referred to as the *DNA template*) is known as *transcription* and is a fundamental step in the formation of the protein for which the DNA or gene codes. Condensed, inactive DNA at the periphery of the nucleus is called *heterochromatin*, and less-condensed DNA available for transcription is referred to as *euchromatin*, which is generally found in the central part of the nucleus.[1–9]

DNA polymerase is an enzyme that synthesizes DNA using single-stranded DNA as a substrate and requires a small segment of double-stranded DNA to initiate new DNA synthesis. RNA polymerase is an enzyme that synthesizes an RNA transcript from a DNA template during transcription. RNA polymerase first binds to a section of bases on the DNA called the *transcription initiation site* (TIS) or promoter "upstream" of the gene that is being transcribed. RNA polymerase I transcribes genes encoding for ribosomal RNAs (rRNAs), RNA polymerase II transcribes genes encoding for mRNAs (mRNAs), and RNA polymerase III transcribes genes encoding for transfer RNAs (tRNAs).[61–66]

The double-stranded helix of DNA must be unraveled and separated into single strands of DNA before it can undergo either transcription or replication. Topoisomerases are enzymes that break or "nick" a DNA strand, releasing the tension of the coiled helix and allowing the DNA to unwind. Transient DNA single-strand breaks are induced by topoisomerase I, and transient DNA double-strand breaks are induced by topoisomerase II. Once the DNA is separated into single strands, the DNA strand that serves as the template for the mRNA during transcription is referred to as *antisense*, and the complementary DNA strand, which has the identical sequence of bases as the mRNA (except that U replaces T), is referred to as *sense*.[67–78]

During transcription, base pairs are matched with the single strand of antisense DNA template to form a strand of mRNA. The resulting mRNA strand is a "mirror image" of the DNA template, except that uracil replaces thymine such that a DNA template with nucleotide sequence AGTC results in a strand of mRNA nucleotide sequence UCAG.[1–9]

A codon is a series of three base pair nucleotides in a gene that codes for a specific amino acid, and a series of base pair codons codes for a precise sequence of specific amino acids, resulting in the synthesis of a specific protein. The gene product is the final molecule, usually a protein, for which the gene codes that generates the effect of the gene. During translation, the mRNA, derived from the DNA template through transcription, is used as a template for the assembly of the protein product. The assembly of the protein product occurs in association with ribosomes, a component of which are rRNAs, and tRNAs add the amino acids to the protein under assembly. Each tRNA has a specific acceptor arm that attaches a specific amino acid. The tRNA ensures that the amino acid is added to the protein in the correct sequence based on the mRNA template, because the tRNA also has a specific anticodon that binds to the corresponding specific codon in the mRNA. The assembly of the protein product via mRNA is referred to as *gene expression*. Most gene expression is controlled at the level of transcription.[79–81]

Genes are made up of DNA segments called *exons* and *introns*. Exons are translated into the gene product, and introns are intervening DNA segments that are believed to play a regulatory role or serve as "punctuation" in the gene. As such, introns are spliced out of the sequence at the mRNA level, and the splice junction is the site between an exon and an intron where the splicing occurs.[1–9] A short tandem repeat (STR) consists of a sequence of two to five nucleotides that are repeated in tandem, frequently dozens of times, in introns.[82] These STRs are found in microsatellite DNA, which is important in certain types of cancers such as colon cancer, although not significant in lung cancer.[83–99]

The end regions of chromosomes are composed of the nucleotide sequence TTAGGG repeated hundreds of times and are called *telomeres*. Telomere sequences are lost each time that a cell replicates until the cell loses its ability to divide as part of the aging process. Telomerase is a specialized DNA polymerase that replaces the DNA sequences at the telomeres of the chromosomes. Telomerase allows cells to divide indefinitely, a factor that can be important in cancer.[100–115]

Posttranslational Modifications of Gene Products

The specific sequence of amino acids in a protein imparts unique physicochemical properties that cause the polypeptide chain to fold into a tertiary structure which gives the protein its three-dimensional functional form. Domains are compact, spherical units of the three-dimensional tertiary structure.[1–9]

Dimerization is the binding of two proteins together. Binding of proteins to other proteins can enhance or inhibit their function. Dimers are frequently encountered, but trimers (three proteins), tetramers (four proteins), or other combinations can occur. Homodimers consist of two identical proteins bound together, and heterodimers consist of two different proteins bound together.[1–9]

Many proteins present in a cell are inert until they are activated by posttranslational modifications such as proteolytic cleavage or phosphorylation and become functional. The activation and inactivation of proteins by posttranslational modifications is essential in control of receptors, signaling pathways, transcription factors, and the cell cycle.

Phosphorylation and Acetylation

Phosphorylation is the addition of a phosphate group to a protein that is catalyzed by enzymes called *kinases*. Dephosphorylation is removal of a phosphate group from a protein that is catalyzed by enzymes called *phosphatases*. Many of the proteins in signaling pathways, including transcription factors, and the cell cycle are activated or inactivated by kinases and phosphatases. Depending on the domain that is phosphorylated, phosphorylation causes varying effects to a transcription factor. Phosphorylation can cause translocation (movement of a protein from the cytosol into the nucleus) and transactivation of genes, or inhibit binding proteins from binding to DNA.[116–120]

Acetylation is the addition of an acetyl group to a protein that is catalyzed by acetyltransferases, and deacetylation is the removal of an acetyl group from a protein that is catalyzed by deacetylases. Similar to kinases and phosphatases, acetyltransferases and deacetylases activate and inactivate proteins involved in various molecular events.[121]

Protein Degradation and Ubiquitinylation

To limit signaling proteins and remove damaged or abnormal proteins, protein degradation is necessary. Ubiquitinylation or poly ubiquitinylation (the ubiquitin-proteasome pathway) is the rapid degradation of proteins by reversible cross-linkage to a polypeptide called *ubiquitin*. Ubiquitin-activating enzyme (E1) activates ubiquitin, and the activated ubiquitin is transferred to a ubiquitin-conjugating enzyme (E2). The activated ubiquitin is transferred to the specific target protein by ubiquitin ligase (E3). Multiple ubiquitins are added to the protein, resulting in a polyubiquinated protein that is degraded by a large protease complex known as the *proteasome*. During this process, the ubiquitin is released to parti-cipate in more cycles of ubiquitinylation. Ubiquitinylation rapidly removes cell-cycle regulators and signaling proteins, including those involved in cell survival and cell death (apoptosis).[122]

Transcription Factors

Gene expression is primarily controlled at the level of transcription initiation. The transcriptional unit of DNA starts with the 5′-regulatory sequences and ends with the 3′-terminator signal of the gene. Gene-activating proteins are blocked from DNA by the tight binding of histone proteins to the DNA blocks. Histone acetyltransferases acetylate histones, which allows the gene-activating proteins to bind to the DNA. By blocking this process, histone deacetylases silence gene transcription.[123–133]

Transcription factors, also called *trans-acting factors* or *transactivators*, are proteins that bind to DNA and regulate the activity of RNA polymerase. Transcription factors affect gene expression directly by induction or activation of the gene or by reducing transcription levels, causing silencing or inhibition of the gene.[134–141]

Transcription factors that stimulate transcription or the synthesis of an RNA molecule from a DNA template are called *transcriptional activators*. In most cases, transcriptional activators have two domains: the DNA-binding domain recognizes and binds to a specific DNA sequence, and the transactivation domain interacts with the transcriptional machinery to induce transcription. Transcription factors can be categorized into families according to their DNA-binding domains; for example, zinc finger, leucine zipper, copper fist, basic helix-loop-helix, helix-turn-helix, and bZIP.[142–148]

Trans-acting DNA sequences encode for diffusible transcription factors that bind to distant cis-acting DNA regulatory sequences but may sometimes bind to other proteins which, in turn, bind to DNA or the transcription machinery. There are two categories of diffusible transcription factors that bind to DNA. (1) General transcription factors are part of the basic transcription machinery by directly interacting with the RNA polymerase complex. Cis-acting DNA sequences that bind general transcription factors and function in all genes are called *promoters*. (2) Regulatory transcription factors activate or inactivate specific genes. Cis-acting DNA sequences that bind regulatory transcription factors to induce specific genes are called *enhancers*.[134–148]

The initiation of transcription by RNA polymerase requires general transcription factors. A cis-acting DNA regulatory sequence that contains adenine-thymidine-rich nucleotide sequences, referred to as a *TATA box*, is found in the promoters of many genes. The TATA-binding protein (TBP) and TBP-associated factors bind to form the general transcription factor TFIID. This transcription factor, combined with other general transcription factors (TFIIB, TFIIF, TFIIE, and TFIIH), initiate transcription by binding RNA polymerase II to the promoter. A transcription bubble is formed when the transcriptional preinitiation complex binds to a specific sequence of nucleotides and there is separation or melting of the double-stranded DNA in conjunction with histone acetylation. After separation from the preinitiation complex, the transcribing enzyme moves down the DNA template along the reading frame. During transcription elongation, the transcription bubble moves down the DNA template in a 5′- to 3′-direction (as noted earlier). Once transcription is terminated, the resultant mRNA is freed and processed before it is actively transported into the cytoplasm. In the cytoplasm, the mRNA enters the ribosome for translation of the protein product.[149–161]

Loops in the DNA bring enhancers into proximity of the transcription initiation sites even when they are located a distance away in sequence. This proximity allows the enhancers to interact with general transcription factors or RNA polymerase complexes at the promoter, allowing enhancers to stimulate gene transcription above the basal level.

An example of transcription factors is the Myc/Max/Mad network of transcription factors that regulate cell growth and death. The Myc family includes N-myc, c-myc, and L-myc. The Mad family includes Mad1, Mxi1, Mad3, Mad4, Mnt, and Mga. The Mad family functions in part as antagonists of the Myc family. These proteins form heterodimers that determine their effect. Myc/Max heterodimers activate transcription causing cell growth, proliferation, and death. Mad/Max heterodimers competitively inhibit the Myc/Max-induced transcription, causing differentiation, cell survival, and inhibition of growth and proliferation.[162–172]

References

1. Coleman WB, Tsongalis GJ, eds. *The Molecular Basis of Human Cancer*. Totowa, NJ: Humana Press; 2002.
2. Watson JD, Baker TA, Bell SP, Gann A, Levine M, Losick R, eds. *Molecular Biology of the Gene*. 5th ed. Menlo Park, CA: Benjamin Cummings; 2003.
3. Epstein RJ, ed. *Human Molecular Biology: An Introduction to the Molecular Basis of Health and Disease*. Cambridge, UK: Cambridge University Press;2003.
4. Strachan T, Read A, eds. *Human Molecular Genetics*. 3rd ed. New York: Garland Science/Taylor and Francis Group; 2003.
5. Swansbury J, ed. *Cancer Cytogenetics: Methods and Protocols*. Totowa, NJ: Humana Press; 2003.
6. Cooper GM, Hausman RE, eds. *The Cell: A Molecular Approach*. 3rd ed. Washington, DC: ASM Press/Sunderland, MA: Sinauer Associates; 2004.
7. Farkas DH, ed. *DNA from A to Z*. Washington, DC: AACC Press; 2004.
8. Killeen AA, ed. *Principles of Molecular Pathology*. Totowa, NJ: Humana Press; 2004.
9. Leonard DGB, Bagg A, Caliendo A, et al. *Molecular Pathology in Clinical Practice*. New York: Springer; 2005.
10. Watson JD, Crick FH. Molecular structure of nucleic acids: a structure for deoxyribose nucleic acid. *Nature*. 1953;171:737–738.
11. Thoma F, Koller T. Influence of histone H1 on chromatin structure. *Cell*. 1977;12:101–107.
12. Varshavsky AJ, Bakayev VV, Nedospasov SA, Georgiev GP. On the structure of eukaryotic, prokaryotic, and viral chromatin. *Cold Spring Harbor Symp Quant Biol*. 1978;42(pt 1):457–473.
13. Tyler-Smith C, Willard HF. Mammalian chromosome structure. *Curr Opin Genet Dev*. 1993;3:390–397.
14. Lamond AI, Earnshaw WC. Structure and function in the nucleus. *Science*. 1998;280:547–553.
15. Blow JJ, Laskey RA. A role for the nuclear envelope in controlling DNA replication within the cell cycle. *Nature*. 1988;332:546–548.
16. Nishitani H, Nurse P. p65cdc18 plays a major role controlling the initiation of DNA replication in fission yeast. *Cell*. 1995;83: 397–405.
17. Cocker JH, Piatti S, Santocanale C, et al. An essential role for the Cdc6 protein in forming the pre-replicative complexes of budding yeast. *Nature*. 1996;379:180–182.
18. Coleman TR, Carpenter PB, Dunphy WG. The *Xenopus* Cdc6 protein is essential for the initiation of a single round of DNA replication in cell-free extracts. *Cell*. 1996;87:53–63.
19. Muzi Falconi M, Brown GW, Kelly TJ. CDC18$^+$ regulates initiation of DNA replication in *Schizosaccharomyces pombe*. *Proc Natl Acad Sci USA*. 1996;93:1566–1570.
20. Owens JC, Detweiler CS, Li JJ. CDC45 is required in conjunction with CDC7/DBF4 to trigger the initiation of DNA replication. *Proc Natl Acad Sci USA*. 1997;94:12521–12526.
21. Tanaka T, Knapp D, Nasmyth K. Loading of an Mcm protein onto DNA replication origins is regulated by Cdc6p and CDKs. *Cell*. 1997;90:649–660.
22. Williams RS, Shohet RV, Stillman B. A human protein related to yeast Cdc6p. *Proc Natl Acad Sci USA*. 1997;94:142–147.
23. Hateboer G, Wobst A, Petersen BO, et al. Cell cycle-regulated expression of mammalian CDC6 is dependent on E2F. *Mol Cell Biol*. 1998;18:6679–6697.
24. Hua XH, Newport J. Identification of a preinitiation step in DNA replication that is independent of origin recognition complex and cdc6, but dependent on cdk2. *J Cell Biol*. 1998;140:271–281.
25. Leatherwood J. Emerging mechanisms of eukaryotic DNA replication initiation. *Curr Opin Cell Biol*. 1998;10:742–748.

26. McGarry TJ, Kirschner MW. Geminin, an inhibitor of DNA replication, is degraded during mitosis. *Cell.* 1998;93:1043–1053.
27. Mimura S, Takisawa H. *Xenopus* Cdc45-dependent loading of DNA polymerase onto chromatin under the control of S-phase Cdk. *EMBO J.* 1998;17:5699–5707.
28. Saha P, Chen J, Thome KC, et al. Human CDC6/Cdc18 associates with Orc1 and cyclin-cdk and is selectively eliminated from the nucleus at the onset of S phase. *Mol Cell Biol.* 1998;18:2758–2767.
29. Williams GH, Romanowski P, Morris L, et al. Improved cervical smear assessment using antibodies against proteins that regulate DNA replication. *Proc Natl Acad Sci USA.* 1998;95: 14932–14937.
30. Yan Z, DeGregori J, Shohet R, et al. Cdc6 is regulated by E2F and is essential for DNA replication in mammalian cells. *Proc Natl Acad Sci USA.* 1998;95:3603–3608.
31. Zou L, Stillman B. Formation of a preinitiation complex by S-phase cyclin CDK-dependent loading of Cdc45p onto chromatin. *Science.* 1998;280:593–596.
32. Donaldson AD, Blow JJ. The regulation of replication origin activation. *Curr Opin Genet Dev.* 1999;9:62–68.
33. Fujita M, Yamada C, Goto H, et al. Cell cycle regulation of human CDC6 protein. Intracellular localization, interaction with the human mcm complex, and CDC2 kinase-mediated hyperphosphorylation. *J Biol Chem.* 1999;274:25927–25932.
34. Masai H, Sato N, Takeda T, Arai K. CDC7 kinase complex as a molecular switch for DNA replication. *Front Biosci.* 1999;4: D834–D840.
35. Petersen BO, Lukas J, Sorensen CS, et al. Phosphorylation of mammalian CDC6 by cyclin A/CDK2 regulates its subcellular localization. *EMBO J.* 1999;18:396–410.
36. Coverley D, Pelizon C, Trewick S, Laskey RA. Chromatin-bound Cdc6 persists in S and G2 phases in human cells, while soluble Cdc6 is destroyed in a cyclin A-cdk2 dependent process. *J Cell Sci.* 2000;113:1929–1938.
37. Homesley L, Lei M, Kawasaki Y, et al. Mcm10 and the MCM2–7 complex interact to initiate DNA synthesis and to release replication factors from origins. *Genes Dev.* 2000;14:913–926.
38. Maiorano D, Moreau J, Mechali M. XCDT1 is required for the assembly of pre-replicative complexes in *Xenopus laevis*. *Nature.* 2000;404:622–625.
39. Nishitani H, Lygerou Z, Nishimoto T, Nurse P. The Cdt1 protein is required to license DNA for replication in fission yeast. *Nature.* 2000;404:625–628.
40. Petersen BO, Wagener C, Marinoni F, et al. Cell cycle- and cell growth-regulated proteolysis of mammalian CDC6 is dependent on APC-CDH1. *Genes Dev.* 2000;14:2330–2343.
41. Takisawa H, Mimura S, Kubota Y. Eukaryotic DNA replication: from pre-replication complex to initiation complex. *Curr Opin Cell Biol.* 2000;12:690–696.
42. Whittaker AJ, Royzman I, Orr-Weaver TL. *Drosophila* double parked: a conserved, essential replication protein that colocalizes with the origin recognition complex and links DNA replication with mitosis and the down-regulation of S phase transcripts. *Genes Dev.* 2000;14:1765–1776.
43. Wohlschlegel JA, Dwyer BT, Dhar SK, et al. Inhibition of eukaryotic DNA replication by geminin binding to Cdt1. *Science.* 2000;290: 2309–2312.
44. Diffley JF. DNA replication: building the perfect switch. *Curr Biol.* 2001;11:R367–R370.
45. Lei M, Tye BK. Initiating DNA synthesis: from recruiting to activating the MCM complex. *J Cell Sci.* 2001;114:1447–1454.
46. Nishitani H, Taraviras S, Lygerou Z, Nishimoto T. The human licensing factor for DNA replication Cdt1 accumulates in G1 and is destabilized after initiation of S-phase. *J Biol Chem.* 2001;276: 44905–44911.
47. Tada S, Li A, Maiorano D, et al. Repression of origin assembly in metaphase depends on inhibition of RLF-B/Cdt1 by geminin. *Nat Cell Biol.* 2001;3:107–113.
48. Yanow SK, Lygerou Z, Nurse P. Expression of Cdc18/Cdc6 and Cdt1 during G2 phase induces initiation of DNA replication. *EMBO J.* 2001;20:4648–4656.
49. Arentson E, Faloon P, Seo J, et al. Oncogenic potential of the DNA replication licensing protein CDT1. *Oncogene.* 2002;21:1150–1158.
50. Bell SP, Dutta A. DNA replication in eukaryotic cells. *Annu Rev Biochem.* 2002;71:333–374.
51. Bermejo R, Vilaboa N, Cales C. Regulation of CDC6, geminin, and CDT1 in human cells that undergo polyploidization. *Mol Biol Cell.* 2002;13:3989–4000.
52. Bonds L, Baker P, Gup C, Shroyer KR. Immunohistochemical localization of cdc6 in squamous and glandular neoplasia of the uterine cervix. *Arch Pathol Lab Med.* 2002;26:1164–1168.
53. Mihaylov IS, Kondo T, Jones L, et al. Control of DNA replication and chromosome ploidy by geminin and cyclin A. *Mol Cell Biol.* 2002;22:1868–1880.
54. Nishitani H, Lygerou Z. Control of DNA replication licensing in a cell cycle. *Genes Cells.* 2002;7:523–534.
55. Robles LD, Frost AR, Davila M, et al. Down-regulation of Cdc6, a cell cycle regulatory gene, in prostate cancer. *J Biol Chem.* 2002;277: 25431–2538.
56. Shreeram S, Sparks A, Lane DP, Blow JJ. Cell type-specific responses of human cells to inhibition of replication licensing. *Oncogene* 2002; 21:6624–6632.
57. Wohlschlegel JA, Kutok JL, Weng AP, Dutta A. Expression of geminin as a marker of cell proliferation in normal tissues and malignancies. *Am J Pathol.* 2002;161:267–273.
58. Li X, Zhao Q, Liao R, et al. The SCF(Skp2) ubiquitin ligase complex interacts with the human replication licensing factor Cdt1 and regulates Cdt1 degradation. *J Biol Chem.* 2003;278: 30854–30858.
59. Vaziri C, Saxena S, Jeon Y, et al. A p53-dependent checkpoint pathway pre prevents rereplication. *Mol Cell.* 2003;11:997–1008.
60. Yoshida K, Inoue I. Regulation of geminin and Cdt1 expression by E2F transcription factors. *Oncogene.* 2004;23:3802–3812.
61. Krieg PA, Melton DA. In vitro RNA synthesis with SP6 RNA polymerase. *Methods Enzymol.* 1987;155:397–415.
62. Lawyer FC, Stoffel S, Saiki RK, et al. Isolation, characterization, and expression in *Escherichia coli* of the DNA polymerase gene from *Thermus aquaticus*. *J Biol Chem.* 1989;264:6427–6437.
63. Studier FW, Rosenberg AH, Dunn JJ, Dubendorff JW. Use of T7 RNA polymerase to direct expression of cloned genes. *Methods Enzymol.* 1990;185:60–89.
64. Kollmar R, Farnham PJ. Site-specific initiation of transcrip-tion by RNA polymerase II. *Proc Soc Exp Biol Med.* 1993;203:127–139.
65. Chou KC, Kezdy FJ, Reusser F. Kinetics of processive nucleic acid polymerases and nucleases. *Anal Biochem.* 1994;221: 217–230.
66. Tabor S, Richardson CC. A single residue in DNA polymerases of the *Escherichia coli* DNA polymerase I family is critical for

distinguishing between deoxy- and dideoxy-ribonucleotides. *Proc Natl Acad Sci USA*. 1995;92:6339–6343.
67. Goldberg S, Schwartz H, Darnell JE Jr. Evidence from UV transcription mapping in HeLa cells that heterogeneous nuclear RNA is the messenger RNA precursor. *Proc Natl Acad Sci USA*. 1977;74: 4520–4523.
68. Hoffmann-Berling H. DNA unwinding enzymes. *Prog Clin Biol Res*. 1982;102(pt C):89–98.
69. Wang JC. DNA topoisomerases: why so many? *J Biol Chem*. 1991;266:6659–6662.
70. Anderson HJ, Roberge M. DNA topoisomerase II: a review of its involvement in chromosome structure, DNA replication, transcription and mitosis. *Cell Biol Int Rep*. 1992;16:717–724.
71. Gasser SM, Walter R, Dang Q, Cardenas ME. Topoisomerase II: its functions and phosphorylation. *Antonie Van Leeuwenhoek*. 1992;62: 15–24.
72. D'Incalci M. DNA-topoisomerase inhibitors. *Curr Opin Oncol*. 1993;5:1023–1028.
73. Ferguson LR, Baguley BC. Topoisomerase II enzymes and mutagenicity. *Environ Mol Mutagen*. 1994;24:245–261.
74. Larsen AK, Skladanowski A, Bojanowski K. The roles of DNA topoisomerase II during the cell cycle. *Prog Cell Cycle Res*. 1996;2:229–239.
75. Kato S, Kikuchi A. DNA topoisomerase: the key enzyme that regulates DNA super structure. *Nagoya J Med Sci*. 1998;61:11–26.
76. Wang JC. Cellular roles of DNA topoisomerases: a molecular perspective. *Nat Rev Mol Cell Biol*. 2002;3:430–440.
77. Gimenez-Abian JF, Clarke DJ. Replication-coupled topoisomerase II templates the mitotic chromosome scaffold? *Cell Cycle*. 2003;2:230–232.
78. Leppard JB, Champoux JJ. Human DNA topoisomerase I: relaxation, roles, and damage control. *Chromosoma*. 2005;114: 75–85.
79. Sharp SJ, Schaack J, Cooley L, et al. Structure and transcription of eukaryotic tRNA genes. *CRC Crit Rev Biochem*. 1985;19:107–144.
80. Persson BC. Modification of tRNA as a regulatory device. *Mol Microbiol*. 1993;8:1011–1016.
81. Green R, Noller HF. Ribosomes and translation. *Annu Rev Biochem*. 1997;66:679–716.
82. Sutherland GR, Richards RI. Simple tandem DNA repeats and human genetic disease. *Proc Natl Acad Sci USA*. 1995;92: 3636–3641.
83. Horii A, Han HJ, Shimada M, et al. Frequent replication errors at microsatellite loci in tumors of patients with multiple primary cancers. *Cancer Res*. 1994;54:3373–3375.
84. Loeb LA. Microsatellite instability: marker of a mutator phenotype in cancer. *Cancer Res*. 1994;54:5059–5063.
85. Mao L, Lee DJ, Tockman MS, et al. Microsatellite alterations as clonal markers for the detection of human cancer. *Proc Natl Acad Sci USA*. 1994;91:9871–9875.
86. Merlo A, Mabry M, Gabrielson E, et al. Frequent microsatellite instability in primary small cell lung cancer. *Cancer Res*. 1994;54: 2098–2101.
87. Wooster R, Cleton-Jansen AM, Collins N, et al. Instability of short tandem repeats (microsatellites) in human cancers. *Nat Genet*. 1994;6:152–156.
88. Fong KM, Zimmerman PV, Smith PJ. Microsatellite instability and other molecular abnormalities in non-small cell lung cancer. *Cancer Res*. 1995;55:28–30.
89. Miozzo M, Sozzi G, Musso K, et al. Microsatellite alterations in bronchial and sputum specimens of lung cancer patients. *Cancer Res*. 1996;56:2285–2288.
90. Bocker T, Diermann J, Friedl W, et al. Microsatellite instability analysis: a multicenter study for reliability and quality control. *Cancer Res*. 1997;57:4739–4743.
91. Dietmaier W, Wallinger S, Bocker T, et al. Diagnostic microsatellite instability: definition and correlation with mismatch repair protein expression. *Cancer Res*. 1997;57:4749–4756.
92. Lothe RA. Microsatellite instability in human solid tumors. *Mol Med Today*. 1997;3:61–68.
93. Arzimanoglou II, Gilbert F, Barber HR. Microsatellite instability in human solid tumors. *Cancer*. 1998;82:1808–1820.
94. Boland CR, Thibodeau SN, Hamilton SR, et al. A National Cancer Institute Workshop on Microsatellite instability for cancer detection and familial predisposition: development of international criteria for the determination of microsatellite instability in colorectal cancer. *Cancer Res*. 1998;58:5248–5257.
95. Boyer JC, Farber RA. Mutation rate of a microsatellite sequence in normal human fibroblasts. *Cancer Res*. 1998;58:3946–3949.
96. Hanford MG, Rushton BC, Gowen LC, Farber RA. Microsatellite mutation rates in cancer cell lines deficient or proficient in mismatch repair. *Oncogene*. 1998;16:2389–2393.
97. Jackson AL, Chen R, Loeb LA. Induction of microsatellite instability by oxidative DNA damage. *Proc Natl Acad Sci USA*. 1998;95:12468–12473.
98. Johannsdottir JT, Jonasson JG, Bergthorsson JT, et al. The effect of mismatch repair deficiency on tumourigenesis; microsatellite instability affecting genes containing short repeated sequences. *Int J Oncol*. 2000;16:133–139.
99. Kim WS, Park C, Hong SK, et al. Microsatellite instability (MSI) in non-small cell lung cancer (NSCLC) is highly associated with transforming growth factor-beta type II receptor (TGF-beta RII) frameshift mutation. *Anticancer Res*. 2000;20:1499–1502.
100. Biessmann H, Mason JM. Telomeric repeat sequences. *Chromosoma*. 1994;103:154–161.
101. Feng J, Funk WD, Wang SS, et al. The RNA component of human telomerase. *Science*. 1995;269:1236–1241.
102. Counter CM. The roles of telomeres and telomerase in cell life span. *Mutat Res*. 1996;366:45–63.
103. Wellinger RJ, Sen D. The DNA structures at the ends of eukaryotic chromosomes. *Eur J Cancer*. 1997;33:735–749.
104. Chakhparonian M, Wellinger RJ. Telomere maintenance and DNA replication: how closely are these two connected? *Trends Genet*. 2003;19:439–446.
105. Bayne S, Liu JP. Hormones and growth factors regulate telomerase activity in ageing and cancer. *Mol Cell Endocrinol*. 2005;240:11–22.
106. Blackburn EH. Telomeres and telomerase: their mechanisms of action and the effects of altering their functions. *FEBS Lett*. 2005;579:859–862.
107. Blasco MA. Telomeres and human disease: ageing, cancer and beyond. *Nat Rev Genet*. 2005;6:611–622.
108. Boukamp P, Popp S, Krunic D. Telomere-dependent chromosomal instability. *J Invest Dermatol Symp Proc*. 2005;10:89–94.
109. Brunori M, Luciano P, Gilson E, Geli V. The telomerase cycle: normal and pathological aspects. *J Mol Med*. 2005;83:244–257.

110. Dong CK, Masutomi K, Hahn WC. Telomerase: regulation, function and transformation. *Crit Rev Oncol Hematol.* 2005;54:85–93.
111. Jacobs JJ, de Lange T. p16INK4a as a second effector of the telomere damage pathway. *Cell Cycle.* 2005;4:1364–1368.
112. Opitz OG. Telomeres, telomerase and malignant transformation. *Curr Mol Med.* 2005;5:219–226.
113. Viscardi V, Clerici M, Cartagena-Lirola H, Longhese MP. Telomeres and DNA damage checkpoints. *Biochimie.* 2005;87:613–624.
114. Autexier C, Lue NF. The structure and function of telomerase reverse transcriptase. *Annu Rev Biochem.* 2006;75:493–517.
115. Bhattacharyya MK, Lustig AJ. Telomere dynamics in genome stability. *Trends Biochem Sci.* 2006;31:114–122.
116. Pallen CJ, Tan YH, Guy GR. Protein phosphatases in cell signaling. *Curr Opin Cell Biol.* 1992;4:1000–1007.
117. Boulikas T. Control of DNA replication by protein phosphorylation. *Anticancer Res.* 1994;14:2465–2472.
118. Berndt N. Protein dephosphorylation and the intracellular control of the cell number. *Front Biosci.* 1999;4:D22–D42.
119. Appella E, Anderson CW. Post-translational modifications and activation of p53 by genotoxic stresses. *Eur J Biochem.* 2001;268:2764–2772.
120. Obaya AJ, Sedivy JM. Regulation of cyclin-Cdk activity in mammalian cells. *Cell Mol Life Sci.* 2002;59:126–142.
121. Fu M, Wang C, Wang J, et al. Acetylation in hormone signaling and the cell cycle. *Cytokine Growth Factor Rev.* 2002;13:259–276.
122. Haglund K, Dikic I. Ubiquitylation and cell signaling. *EMBO J.* 2005;24:3353–3359.
123. Legube G, Trouche D. Regulating histone acetyltransferases and deacetylases. *EMBO Rep.* 2003;4:944–947.
124. Marmorstein R. Structural and chemical basis of histone acetylation. *Novartis Found Symp.* 2004;259:78–98.
125. Moore JD, Krebs JE. Histone modifications and DNA double-strand break repair. *Biochem Cell Biol.* 2004;82:446–452.
126. Peterson CL, Laniel MA. Histones and histone modifications. *Curr Biol.* 2004;14:R546–R551.
127. Quivy V, Calomme C, Dekoninck A, et al. Gene activation and gene silencing: a subtle equilibrium. *Cloning Stem Cells.* 2004;6:140–149.
128. Wang Y, Fischle W, Cheung W, et al. Beyond the double helix: writing and reading the histone code. *Novartis Found Symp.* 2004;259:3–17.
129. Fraga MF, Esteller M. Towards the human cancer epigenome: a first draft of histone modifications. *Cell Cycle.* 2005;4:1377–1381.
130. Khan AU, Krishnamurthy S. Histone modifications as key regulators of transcription. *Front Biosci.* 2005;10:866–872.
131. Verdone L, Caserta M, Di Mauro E. Role of histone acetylation in the control of gene expression. *Biochem Cell Biol.* 2005;83:344–353.
132. Yu Y, Waters R. Histone acetylation, chromatin remodelling and nucleotide excision repair: hint from the study on MFA2 in *Saccharomyces cerevisiae. Cell Cycle.* 2005;4:1043–1045.
133. Verdone L, Agricola E, Caserta M, Di Mauro E. Histone acetylation in gene regulation. *Brief Funct Genomic Proteomic.* 2006;5:209–221.
134. Haura EB, Turkson J, Jove R. Mechanisms of disease: insights into the emerging role of signal transducers and activators of transcription in cancer. *Nat Clin Pract Oncol.* 2005;2:315–324.
135. Wang JC. Finding primary targets of transcriptional regulators. *Cell Cycle.* 2005;4:356–358.
136. Wittenberg C, Reed SI. Cell cycle-dependent transcription in yeast: promoters, transcription factors, and transcriptomes. *Oncogene.* 2005;24:2746–2755.
137. Zaidi SK, Young DW, Choi JY, et al. The dynamic organization of gene-regulatory machinery in nuclear microenvironments. *EMBO Rep.* 2005;6:128–133.
138. Barrera LO, Ren B. The transcriptional regulatory code of eukaryotic cells – insights from genome-wide analysis of chromatin organization and transcription factor binding. *Curr Opin Cell Biol.* 2006;18:291–298.
139. Dillon N. Gene regulation and large-scale chromatin organization in the nucleus. *Chromosome Res.* 2006;14:117–126.
140. Maston GA, Evans SK, Green MR. Transcriptional regulatory elements in the human genome. *Annu Rev Genomics Hum Genet.* 2006;7:29–59.
141. Thomas MC, Chiang CM. The general transcription machinery and general cofactors. *Crit Rev Biochem Mol Biol.* 2006;41:105–178.
142. Engelkamp D, van Heyningen V. Transcription factors in disease. *Curr Opin Genet Dev.* 1996;6:334–342.
143. Tamura T, Konishi Y, Makino Y, Mikoshiba K. Mechanisms of transcriptional regulation and neural gene expression. *Neurochem Int.* 1996;29:573–581.
144. Bieker JJ, Ouyang L, Chen X. Transcriptional factors for specific globin genes. Ann NY Acad Sci. 1998;850:64–69.
145. Hertel KJ, Lynch KW, Maniatis T. Common themes in the function of transcription and splicing enhancers. *Curr Opin Cell Biol.* 1997;9:350–357.
146. Arnosti DN. Analysis and function of transcriptional regulatory elements: insights from *Drosophila. Annu Rev Entomol.* 2003;48:579–602.
147. Scannell DR, Wolfe K. Rewiring the transcriptional regulatory circuits of cells. *Genome Biol.* 2004;5:206.
148. Villard J. Transcription regulation and human diseases. *Swiss Med Wkly.* 2004;134:571–579.
149. Hampsey M. Molecular genetics of the RNA polymerase II general transcriptional machinery. *Microbiol Mol Biol Rev.* 1998;62:465–503.
150. Berk AJ. Activation of RNA polymerase II transcription. *Curr Opin Cell Biol.* 1999;11:330–335.
151. Berk AJ. TBP-like factors come into focus. *Cell.* 2000;103:5–8.
152. Green MR. TBP-associated factors (TAFIIs): multiple, selective transcriptional mediators in common complexes. *Trends Biochem Sci.* 2000;25:59–63.
153. Pugh BF. Control of gene expression through regulation of the TATA-binding protein. *Gene.* 2000;255:1–14.
154. Burley SK, Kamada K. Transcription factor complexes. *Curr Opin Struct Biol.* 2002;12:225–230.
155. Featherstone M. Coactivators in transcription initiation: here are your orders. *Curr Opin Genet Dev.* 2002;12:149–155.
156. Davidson I. The genetics of TBP and TBP-related factors. *Trends Biochem Sci.* 2003;28:391–398.
157. Hochheimer A, Tjian R. Diversified transcription initiation complexes expand promoter selectivity and tissue-specific gene expression. *Genes Dev.* 2003;17:1309–1320.
158. Asturias FJ. RNA polymerase II structure and organization of the preinitiation complex. *Curr Opin Struct Biol.* 2004;14:121–129.

159. Matangkasombut O, Auty R, Buratowski S. Structure and function of the TFIID complex. *Adv Protein Chem.* 2004;67:67–92.
160. Brady J, Kashanchi F. Tat gets the "green" light on transcription initiation. *Retrovirology.* 2005;2:69.
161. Thomas MC, Chiang CM. The general transcription machinery and general cofactors. *Crit Rev Biochem Mol Biol.* 2006;41:105–178.
162. Dang CV, Resar LM, Emison E, et al. Function of the c- Myc oncogenic transcription factor. *Exp Cell Res.* 1999;253:63–77.
163. Kuramoto N, Ogita K, Yoneda Y. Gene transcription through Myc family members in eukaryotic cells. *Jpn J Pharmacol.* 1999;80:103–109.
164. Grandori C, Cowley SM, James LP, Eisenman RN. The Myc/Max/Mad network and the transcriptional control of cell behavior. *Annu Rev Cell Dev Biol.* 2000;16:653–699.
165. Baudino TA, Cleveland JL. The Max network gone mad. *Mol Cell Biol.* 2001;21:691–702.
166. Eisenman RN. The Max network: coordinated transcriptional regulation of cell growth and proliferation. *Harvey Lect.* 2000–2001;96:1–32.
167. Luscher B. Function and regulation of the transcription factors of the Myc/Max/Mad network. *Gene.* 2001;277:1–14.
168. Zhou ZQ, Hurlin PJ. The interplay between Mad and Myc in proliferation and differentiation. *Trends Cell Biol.* 2001;11:S10–S14.
169. Lee LA, Dang CV. Myc target transcriptomes. *Curr Top Microbiol Immunol.* 2006;302:145–167.
170. Nair SK, Burley SK. Structural aspects of interactions within the Myc/Max/Mad network. *Curr Top Microbiol Immunol.* 2006;302:123–143.
171. Pirity M, Blanck JK, Schreiber-Agus N. Lessons learned from Myc/Max/Mad knockout mice. *Curr Top Microbiol Immunol.* 2006;302:205–234.
172. Rottmann S, Luscher B. The Mad side of the Max network: antagonizing the function of Myc and more. *Curr Top Microbiol Immunol.* 2006;302:63–122.

2

Receptors, Signaling Pathways, Cell Cycle, and DNA Damage Repair

Philip T. Cagle

Cell-Surface Receptors and Signal Transduction

Ligands are extracellular messenger molecules such as growth factors, inflammatory cytokines, and hormones that bind to specific receptors on the cell surface (i.e., growth factor receptors, cytokine receptors, and hormone receptors). Binding of the ligands to their receptors causes activation of second messengers in the cytosol and, eventually, activation of nuclear transcription factors. (Transcription factors are discussed in Chap. 1) The transcription factors then direct the transcription of a gene product as a result of the extracellular message (e.g., a growth factor may stimulate a growth factor receptor on the cell surface, causing activation of second messengers that eventually cause a transcription factor to cause transcription of a protein involved in cell growth). This cascade or activation and inactivation of protein messengers from the cell-surface receptors through proteins in the cytosol to the transcription factors in the nucleus is known as signal transduction. The series of steps that occurs during this process is called the signal transduction pathway or signaling pathway. Much of the activation and inactivation of proteins in signaling pathways occurs through reversible phosphorylation of tyrosine, serine, or threonine in the pathway proteins (see Chap. 1). Phosphorylation is accomplished by tyrosine kinases and serine/threonine kinases with phosphates donated from adenosine triphosphate (ATP) or guanosine triphosphate (GTP). Tyrosine kinases are much more common in signaling pathways than are serine/threonine kinases.[1-10] This discussion focuses on growth factor receptors and cytokine receptors and their associated signaling pathways.

Growth factor receptors are a common type of cell-surface receptor. Polypeptide growth factors such as epidermal growth factor (EGF) serve as ligands that bind to cell-surface receptor protein-tyrosine kinases, which causes activation of the receptor by dimerization, resulting in autophosphorylation. The activated receptor binds other proteins within the cell, leading to their phosphorylation and activation of their enzyme activity as part of the signaling pathway. The type I growth factor receptor tyrosine kinase family consists of epidermal growth factor receptor (EGFR), and ErbB1, ErbB2, ErbB3, and ErbB4 make up the type I growth factor receptor-tyrosine kinase family. In addition to EGF, EGFR has multiple ligands, including transforming growth factor-α (TGF-α).[11-16]

Several signaling pathways are well studied and important to human disease. Signaling pathways transmit the "message" from extracellular ligands such as growth factors, cytokines, and steroid hormones. Signaling pathways are involved in regulation of cell proliferation, cell differentiation, cell death or apoptosis, and cell survival. Some of the more noteworthy and established pathways are briefly reviewed. Most signaling pathways have multiple complex interactions and "cross-talk" with other pathways, so discussion of specific pathways is limited here to an abbreviated overview.

The mitogen-activated protein kinase (MAPK) family is involved in multiple signaling pathways influencing cell growth, differentiation, and apoptosis, including the Ras/Raf-1/MAPK pathway mentioned later. The MAPK family includes the extracellular signal-regulated kinases (ERK1 and ERK2); the c-Jun NH2-terminal kinases (JNK1, JNK2, and JNK3); and p38 (p38 MAP kinases α, β, and δ). The MAP kinase kinase kinases (MKKK) are activated by a wide range of agents, including growth factors, oxidative stress, inflammatory cytokines, and ultraviolet radiation. Activated MKKK activate the MAP kinase kinases (MKK), which subsequently activate the MAP kinases. Examples of MKKK include Raf-1, TGF-β-activated kinase (TAK), apoptosis signal-regulating kinase 1 (ASK1), MAP/ERK kinase kinases (MEKK), germinal center kinase (GCK), and p21-activated kinase (PAK). The ERKs have about 160 substrates and are antiapoptotic and involved in cell proliferation, cellular differentiation, and cell-cycle progression. On the other hand,

P.T. Cagle and T.C. Allen (eds.), *Basic Concepts of Molecular Pathology*, Molecular Pathology Library 2,
DOI: 10.1007/978-0-387-89626-7_2, © Springer Science+Business Media, LLC 2009

JNK and p38 are usually, but not exclusively, proapoptotic and have many complex effects on different cells and on the cell cycle.[17-25]

The Ras/Raf-1/MAPK pathway is significant in carcinogenesis. H-Ras, K-Ras, and N-Ras are members of the Ras family, a class of small GTP-binding proteins that are downstream targets of receptor tyrosine kinases. Ras is located at the plasma membrane inner surface. Activation of growth factor receptors converts Ras from its inactive guanosine diphosphate (GDP)-bound state to its active GTP-bound state. Activated Ras recruits Raf protein–serine/threonine kinase from the cytosol to the plasma membrane where kinases activate Raf. ERK, a member of the MAPK family, is activated by Raf through MAP/ERK kinase (MEK). In turn, the activated ERK phosphorylates and activates multiple other proteins, including other protein kinases. Activated ERK also translocates into the nucleus where it phosphorylates and activates transcription factors including Elk-1. Guanosine triphosphate hydrolysis by GTPase-activating proteins (GAPs) inactivates Ras. Ras is involved in many pathways, and epithelial cell proliferation is one of several possible results of Ras activation.[26-41]

The JAK/STAT pathway is linked with cytokine receptors. When ligands stimulate cytokine receptors, the signal transducers and activators of transcription (STAT) proteins associate with the activated cytokine receptors. The STATs are phosphorylated by the JAK nonreceptor protein tyrosine kinases that are members of the Janus kinase (JAK) family, undergo dimerization, and translocate into the nucleus where they function as transcription factors for their target genes. The STAT proteins can also be activated in growth factor receptor pathways.[42-49]

In the TGF-β/Smad pathway, cytokines in the transforming growth factor-β (TGF-β) superfamily inhibit the growth of many types of epithelial cells by formation of a complex of TGF-β type II and type I serine/threonine kinase receptors (TβRI and TβRII). The TGF-β1 ligand binds to TβRII, which subsequently phosphorylates and activates TβRI. Next, TRβI phosphorylates the receptor-regulated Smads (R-Smads) Smad2 and Smad3. Activated Smad2/3 complex with Smad4 (Co-Smad), translocate into the nucleus, and function as transcriptional modulators of TGF-1-regulated genes. On the other hand, the inhibitory Smads, Smad6 and Smad7, inhibit TGF-β1 signaling. The inhibitory Smads bind to Smad4 to prevent it from complexing with Smad2/3 or by binding to TRβI, blocking phosphorylation of Smad2 and Smad3.[50-59]

The Wnt/B/catenin pathway involves Wnt ("Wingless," derived from fruit fly studies), which binds to Frizzled cell-surface receptors. Signaling from Frizzled phosphorylates and activates Disheveled, which, in turn, inhibits the protein kinase glycogen synthase kinase-3 (GSK-3). β-Catenin phosphorylated by GSK-3 forms a complex with the adenomatous polyposis coli (APC) protein and the axin protein, restricting the quantity of free β-catenin in the cytosol. When activated Disheveled inhibits GSK-3, dephosphorylated β-catenin is freed from the APC–axin complex. This pathway is the classic Wnt signaling pathway, known as the canonical Wnt signaling pathway, and there are noncanonical Wnt signaling pathways that are not discussed here. β-Catenin associates with the TCF/LEF transcription factors, converting them to gene activators from gene repressors and, after translocation into the nucleus, binds to transcription factor TCF4, which induces Myc.[60-81] β-Catenin also has roles in cell adhesion, which is discussed in Chap. 3.

The PI3K/Akt/mTOR pathway regulates cell survival. Akt is a protein serine/threonine kinase. Cell membrane phosphatidylinositol 4,5-biphosphate in the cell is phosphorylated by phosphatidylinositol 3-kinase (PI3K), resulting in inositol 1,4,5-triphosphate (PIP3). The PIP3 binds to the protein serine/threonine kinase Akt and recruits it to the inner surface of the cell membrane. At the inner surface of the cell membrane, Akt is phosphorylated and activated. Activated Akt phosphorylates proteins directly involved in cell survival as well as transcription factors and other protein kinases.[82-90]

The nuclear factor-B (NF-B) transcription factor and NF-B signaling pathways regulate many proteins of the immune system, proteins that inhibit apoptosis and proteins which promote cell survival and proliferation. Nuclear factor-B consists of various dimers of the Rel protein family: Rel (c-Rel), RelA (p65), RelB, NF-B1 (p50 and its precursor p105), and NF-B2 (p52 and its precursor p100), of which the p50–p65 dimer is the most common. NF-B complexes bind to promoters to assist transcription in most situations, but homodimer complexes of p50 or p52 may inhibit transcription. Nuclear factor-B proteins are maintained in the cytoplasm in resting cells by associating with members of the inhibitory I.B family (I.B-a, I.B-β, and I.B-e).[91-100]

Inhibitory-B must be degraded for NF-B to be activated. Inhibitory-B kinases (IKKs) are activated by MAPKKK or by ligands for Toll-like receptors (TLRs), interleukin (IL)-1/IL-18 receptors, the TNF receptor superfamily, and B- and T-cell receptors. Activated IKKs phosphorylate I.B, which is subsequently bound by E3I.B ubiquitin ligase complex-TrCP-SCF, which ubiquinylates I.B. The 26S proteasome degrades the ubiquinylated I.B, releasing NF-B complex to translocate into the nucleus where it binds to specific B sites on DNA. The NF-B complex is a transcription factor that regulates expression of proinflammatory cytokines, chemokines, adhesion molecules, cyclooxygenase-2, inducible nitric oxide synthase, major histocompatibility complex, IL-2, IL-12, and interferon, in addition to antiapoptotic and apoptotic genes.[101-107] Apoptosis is discussed in Chap. 4.

In the Hedgehog-Patched-Smoothened signaling pathway, after attachment of a lipid, Sonic Hedgehog (Shh) polypeptide binds to Patched on the cell surface, which prevents inhibition of Smoothened (Smo), a G protein-coupled receptor, by Patched. Smoothened activates the serine/threonine kinase Fused and the zinc finger transcription factor Gli (first detected in gliomas as a mutation), which induces Wnt signaling.[108-116]

Notch is a receptor for direct cell-to-cell signaling. Delta binds to Notch, resulting in proteolytic cleavage of Notch. The intracellular domain of Notch is released and translocates into the nucleus, where it interacts with a transcription factor.[117-122]

The Cell Cycle

The cell cycle is the tightly regulated, sequential series of events or phases that govern cell proliferation, including preparation for DNA replication, DNA replication (see Chap. 1), preparation for cell division, cell division, and cell rest. The cell cycle provides orderly control of DNA replication and cell division in response to external and internal stimuli. The cell cycle is divided into several phases: G_0 (cell at rest), G_1 (preparation for DNA synthesis), S (DNA synthesis or replication), G_2, and M (mitosis with nuclear and cellular division). Progression through the series of steps in the cell cycle is tightly regulated by cyclin-dependent kinases (Cdks) after they form complexes with proteins called cyclins. These complexes activate and inactivate proteins by phosphorylation, including proteins that otherwise act as "brakes" on progression through the cell cycle and the proliferation process. There are many interacting pathways and positive and negative feedback loops that control passage through the cell cycle. The cell cycle may be stimulated appropriately or inappropriately in various inflammatory diseases, and loss of cell-cycle regulation is a very important step in uncontrolled cell proliferation during carcinogenesis.

Checkpoints in the cell cycle ordinarily prevent the passage of damaged DNA to daughter cells. Checkpoints temporarily arrest the cell cycle at specific steps in the cell cycle to allow repair of damaged DNA or programmed cell death or apoptosis if the damage is too severe to be repaired (discussed in Chap. 4). The major checkpoint in the cell cycle is the restriction point where "commitment" to the cell cycle occurs in G_1, as is further discussed below. In addition to the G_1–S checkpoint, there are an S-phase checkpoint and a G_2–M checkpoint.[123-142]

The Rad9–Rad1–Hus1 heterotrimer complex (9-1-1 complex) and the Rad17–RFC complex are damage sensor proteins that detect DNA damage at the checkpoints. The 9-1-1 complex is loaded around DNA by the Rad17–RFC complex. The ATR [ataxia-telangiectasiamutated (ATM) and Rad3-related protein kinase]-mediated and ATM-mediated phosphorylation and activation of Chk1 and Chk2 follow, and ATM and Chk2 phosphorylate and stabilize p53. Cyclin-dependent kinases are inactivated by the regulation of Cdc25, Wee1, and p53, which causes cell-cycle arrest. DNA damage repair is discussed later.[143-146] After DNA damage repair, the DNA damage checkpoint is silenced, and the cell cycle restarts in a process called recovery, involving polo-like kinase (Plk1).[147,148]

Growth factor signaling initiates the cell cycle and maintains the transition through the G_1 phase. When the cell passes through the restriction point of the cell cycle, the cell no longer requires growth factor signaling to complete the cell cycle, and the cell is "committed" to the cell cycle. Passage through the restriction point depends on phosphorylation of the retinoblastoma (Rb) gene product, pRb. The Rb product governs progression past the restriction point of the cell cycle and governs the expression of genes involved in DNA synthesis.[149-152] Activation of cyclin D is necessary for progression of the cell cycle.[153-159]

In response to stimuli for mitosis, such as growth factor signaling, complexes of cyclin D with Cdk4 and Cdk6 phosphorylate pRB during G_1 in response to stimuli for mitosis. In addition, cyclin E–Cdk2 complexes phosphorylate pRb just before the S phase. Families of Cdk inhibitors (the INK4 family, including p16INK4, and the p21WAF1/Cip1/p27Kip1/p57Kip2 family) control these cyclin–Cdk complexes.[160-171] Phosphorylation inactivates pRb in G_1, releasing E2F transcription factors[172-182] that activate transcription of numerous genes involved in DNA replication, such as c-Myc,[183-187] initiating the S phase. As complexes with Cdks, cyclin A functions in both G_1–S phase transition and in mitosis,[188] and cyclin B is involved in entry into mitosis.[189]

One of the primary roles of p53, the product of the TP53 gene, is to "protect" the DNA through the arrest of the cell cycle at checkpoints in response to DNA damage or to help induce apoptosis when damage is beyond repair. Because of these roles, p53 has been referred to as the guardian of the genome. Part of the p53 arrest of the cell cycle is by activation of p21WAF1, which blocks cyclin–Cdk complexes necessary for cell-cycle progression.[190-196] A member of the TGF-β superfamily of cytokines, TGF-β1, is involved in inhibition of cell-cycle progression.[197,198]

It is apparent that Rb and p53 play very crucial roles in the management of the cell cycle. Abnormalities of Rb and p53 are the most common abnormalities associated with the cell-cycle dysregulation of malignancy. However, because there are so many redundancies, interacting pathways, and positive and negative feedback loops, there are many other abnormalities that can produce effects similar to the direct loss of Rb or p53.

The CDKN2A gene encodes for two completely unrelated protein products, p16INK4A, the Cdk inhibitor[199-203] mentioned earlier,[204-209] and p14ARF. Both these proteins are transcribed from different exons of the CDKN2A gene. Abnormalities of either of these genes or their products can produce effects similar to abnormalities of the Rb or p53 genes themselves. By inhibiting Cdk4/6 kinase, p16 blocks phosphorylation and inactivation of pRb. Loss of p16 function results in loss of Rb function because Cdk4/6 kinase inactivation of pRb is not blocked. The other CDKN2A product, p14, destabilizes the MDM2 protein, which binds and degrades p53. Excessive levels of MDM2 resulting from loss of p14 function cause excess degradation of p53 and loss of p53 function. Abnormalities of CDKN2A, p16, p14, and MDM2 and other genes and their products upstream or downstream of Rb and p53 can produce loss of control of the cell cycle similar to the direct loss of Rb and p53.

DNA Damage Repair

DNA is regularly damaged by endogenous factors (such as oxygen radicals), extracellular factors (such as chemicals, radiation, ultraviolet light), and errors in replication (such as stalled replication forks). As a result of exposure to these harmful agents, DNA undergoes depurination, deamination, hydrolysis, and nonenzymatic methylation (alkylation), which attach chemical groups called adducts to the DNA. DNA damage repair during cell-cycle checkpoints typically involves excision of the damaged DNA and filling of the resultant gap by newly synthesized DNA using the undamaged complementary DNA strand as template. Depending on the type of damage, there are several DNA repair pathways. The DNA damage repair pathways are important in individual susceptibility to cancer and in response to therapy.

The base excision repair (BER) pathway repairs small lesions such as oxidized or reduced single bases and fragmented or nonbulky adducts. In the BER pathway, a single damaged base is excised by base-specific DNAglycosylases (e.g., oxidized 8-oxoguanine is excised by 8-oxoguanine DNA glycosylase or OGG1). Some glycosylates are bifunctional and have an apurinic/apyrimidinic lyase activity to incise the phosphodiester bond of the intact apurinic/apyrimidinic site. An apurinic/apyrimidinic endonuclease (APE1/APEX1) is required by monofunctional glycosylates to incise the apurinic/apyrimidinic site. DNA polymerase fills in the single nucleotide gap[210-216], and a DNA ligase III/X-ray repair cross-complementing group 1 (XRCC1) complex seals the nick.[217-222]

The nucleotide excision repair (NER) repairs lesions large enough to deform the DNA helical structure, such as pyrimidine dimers, bulky chemical adducts, and cross-links, by excising damaged bases as part of an oligonucleotide. Xeroderma pigmentosum (XP) proteins are an important part of the NER pathway. A protein complex including xeroderma pigmentosum group C protein (XPC) and hHR23B recognizes helical distortion by bulky chemical adducts. TFIIH is composed of nine protein subunits, including p62, p52, p44, p34, Cdk7, cyclin H, MAT1, and the two DNA helicases XPD (xeroderma pigmentosum group D protein), also known as ERCC2 (excision repair cross-complementation group 2), and XPB (xeroderma pigmentosum group B protein), also known as ERCC3 (excision repair cross-complementation group 3). TFIIH, xeroderma pigmentosum group A protein (XPA), and replication protein A (RPA) accumulate at the damage site, and the XPD and XPB helicases of TFIIH unwind the DNA double helix. This mechanism permits excision of the damaged single-stranded DNA fragment (usually about 27–30 bp) by a complex that includes ERCC1 and xeroderma pigmentosum group F protein (XPF). DNA polymerases synthesize a new strand of DNA using the undamaged complementary DNA strand as template to complete the repair process.[223-226]

Mismatch repair genes (MMR) participate in an excision repair pathway by scanning newly replicated DNA for mismatched base pairs such as deamination of a nucleotide into a different nucleotide. Heterodimers from MMR (including MLH1, MSH2, PMS1, and PMS2) cause cell-cycle arrest, permitting DNA repair.[227-239]

The enzyme O6-meG–DNA methyltransferase (MGMT/AGT) repairs O6-meG and other alkylated bases in the direct damage reversal (DR) pathway.[240-249] O6-Methylguanine (O6-meG) formed by alkylating compounds in tobacco smoke may mismatch with thymine during DNA replication if not repaired.

The DNA damage response or DSB repair (double-strand break repair) pathway (or during the S-phase checkpoint, the DNA replication stress response pathway) occurs in response to DSB DNA damage. The DSB repair pathway includes a cascade of events: sensing of the DNA damage and transduction of the damage signal to multiple pathways (cell-cycle checkpoints, DNA repair, responses to telomere maintenance, and apoptosis) (see Chap. 4). The DSB repair process involves many genes and their products, including the MRE11–Rad50–NBS1 complex (MRN); X-ray repair cross-complementing (XRCC); the PI3K-like protein kinases (PIKKs) DNA-PKcs, ATM (mutated in ataxia telangiectasia), and ATR (ATM–Rad3related); and ATM substrates NBS1 (Nijmegen breakage syndrome protein 1), SMC1 (structural maintenance of chromosomes 1), Chk1, Chk2, MRE11, p53, MDM2, BRCA1 (breast cancer protein 1), BRCA2/FANCD1 (breast cancer protein 2/Fanconi anemia protein D1), and FANCD2 (Fanconi anemia protein D2).[250-269]

The ATM pathway reacts to DSBs in all phases of the cell cycle. The ATR pathway reacts to DSBs more slowly than ATM and reacts to factors that impede the function of replication forks. The ATM pathway activates many downstream proteins of the ATR pathway. In response to replication stress, ATM and ATR activate members of the Chk kinase family. The ATM phosphorylates Chk2, which, in turn, phosphorylates p53 and Cdc25A, blocking Cdk2.[250-269]

References

1. Williams LT, Escobedo JA, Fantl WJ, et al. Interactions of growth factor receptors with cytoplasmic signaling molecules. *Cold Spring Harbor Symp Quant Biol*. 1991;56:243–250.
2. Fantl WJ, Escobedo JA, Martin GA, et al. Distinct phosphotyrosines on a growth factor receptor bind to specific molecules that mediate different signaling pathways. *Cell*. 1992;69:413–423.
3. Hunter T, Lindberg RA, Middlemas DS, et al. Receptor protein tyrosine kinases and phosphatases. *Cold Spring Harbor Symp Quant Biol*. 1992;57:25–41.
4. Fantl WJ, Johnson DE, Williams LT. Signalling by receptor tyrosine kinases. *Annu Rev Biochem*. 1993;62:453–481.

5. Johnson GL, Vaillancourt RR. Sequential protein kinase reactions controlling cell growth and differentiation. *Curr Opin Cell Biol.* 1994;6:230–238.
6. van der Geer P, Hunter T, Lindberg RA. Receptor protein-tyrosine kinases and their signal transduction pathways. *Annu Rev Cell Biol.* 1994;10:251–337.
7. Schlessinger J. Cell signaling by receptor tyrosine kinases. *Cell.* 2000;103:211–225.
8. Gavi S, Shumay E, Wang HY, Malbon CC. G-protein-coupled receptors and tyrosine kinases: crossroads in cell signaling and regulation. *Trends Endocrinol Metab.* 2006;17:48–54.
9. Li E, Hristova K. Role of receptor tyrosine kinase transmembrane domains in cell signaling and human pathologies. *Biochemistry.* 2006;45:6241–6251.
10. Perona R. Cell signalling: growth factors and tyrosine kinase receptors. *Clin Transl Oncol.* 2006;8:77–82.
11. Tiganis T. Protein tyrosine phosphatases: dephosphorylating the epidermal growth factor receptor. *IUBMB Life.* 2002;53:3–14.
12. Jorissen RN, Walker F, Pouliot N, et al. Epidermal growth factor receptor: mechanisms of activation and signalling. *Exp Cell Res.* 2003;284:31–53.
13. Bazley LA, Gullick WJ. The epidermal growth factor receptor family. *Endocr Relat Cancer.* 2005;12(suppl 1):S17–S27.
14. Normanno N, Bianco C, Strizzi L, et al. The ErbB receptors and their ligands in cancer: an overview. *Curr Drug Targets.* 2005;6:243–257.
15. Zaczek A, Brandt B, Bielawski KP. The diverse signaling network of EGFR, HER2, HER3 and HER4 tyrosine kinase receptors and the consequences for therapeutic approaches. *Histol Histopathol.* 2005;20:1005–1015.
16. Warren CM, Landgraf R. Signaling through ERBB receptors: multiple layers of diversity and control. *Cell Signal.* 2006;18: 923–933.
17. Bagrodia S, Derijard B, Davis RJ, Cerione RA. Cdc42 and PAK mediated signaling leads to Jun kinase and p38 mitogen-activated protein kinase activation. *J Biol Chem.* 1995;270:27995–27998.
18. Xia Z, Dickens M, Raingeaud J, et al. Opposing effects of ERK and JNKp38 MAP kinases on apoptosis. *Science.* 1995;270: 1326–1331.
19. Ichijo H, Nishida E, Irie K, et al. Induction of apoptosis by ASK1, a mammalian MAPKKK that activates SAPK/JNK and p38 signaling pathways. *Science.* 1997;275:90–94.
20. Wilkinson MG, Millar JB. SAPKs and transcription factors do the nucleocytoplasmic tango. *Genes Dev.* 1998;12:1391–1397.
21. Davis RJ. Signal transduction by the JNK group of MAP kinases. *Cell.* 2000;103:239–252.
22. Wada T, Penninger JM. Mitogen-activated protein kinases in apoptosis regulation. *Oncogene.* 2004;23:2838–2849.
23. Bradham C, McClay DR. p38 MAPK in development and cancer. *Cell Cycle.* 2006;5:824–828.
24. MacCorkle RA, Tan TH. Mitogen-activated protein kinases in cell-cycle control. *Cell Biochem Biophys.* 2005;43:451–461.
25. Yoon S, Seger R. The extracellular signal-regulated kinase: multiple substrates regulate diverse cellular functions. *Growth Factors.* 2006;24:21–44.
26. Magnuson NS, Beck T, Vahidi H, et al. The Raf-1 serine/threonine protein kinase. *Semin Cancer Biol.* 1994;5:247–253.
27. Williams NG, Roberts TM. Signal transduction pathways involving the Raf proto-oncogene. *Cancer Metastasis Rev.* 1994;13:105–116.
28. Burgering BM, Bos JL. Regulation of Ras-mediated signalling: more than one way to skin a cat. *Trends Biochem Sci.* 1995;20: 18–22.
29. Morrison DK. Mechanisms regulating Raf-1 activity in signal transduction pathways. *Mol Reprod Dev.* 1995;42:507–514.
30. Morrison DK, Cutler RE. The complexity of Raf-1 regulation. *Curr Opin Cell Biol.* 1997;9:174–179.
31. Dhillon AS, Kolch W. Untying the regulation of the Raf-1 kinase. *Arch Biochem Biophys.* 2002;404:3–9.
32. Bernards A, Settleman J. GAP control: regulating the regulators of small GTPases. *Trends Cell Biol.* 2004;14:377–385.
33. Bernards A, Settleman J. GAPs in growth factor signalling. *Growth Factors.* 2005;23:143–149.
34. Chan A. Teaching resources. Ras-MAPK pathways. *Sci STKE.* 2005;2005(271):tr5.
35. Hancock JF, Parton RG. Ras plasma membrane signalling platforms. *Biochem J.* 2005;389(pt 1):1–11.
36. Kranenburg O. The KRAS oncogene: past, present, and future. *Biochim Biophys Acta.* 2005;1756:81–82.
37. McCudden CR, Hains MD, Kimple RJ, et al. G-protein signaling: back to the future. *Cell Mol Life Sci.* 2005;62:551–577.
38. Mitin N, Rossman KL, Der CJ. Signaling interplay in Ras superfamily function. *Curr Biol.* 2005;15:R563–R574.
39. Philips MR. Compartmentalized signalling of Ras. *Biochem Soc Trans.* 2005;33(pt 4):657–661.
40. Wennerberg K, Rossman KL, Der CJ. The Ras superfamily at a glance. *J Cell Sci.* 2005;118(pt 5):843–846.
41. Mor A, Philips MR. Compartmentalized Ras/MAPK signaling. *Annu Rev Immunol.* 2006;24:771–800.
42. Pellegrini S, Dusanter-Fourt I. The structure, regulation and function of the Janus kinases (JAKs) and the signal transducers and activators of transcription (STATs). *Eur J Biochem.* 1997;248: 615–633.
43. Liu KD, Gaffen SL, Goldsmith MA. JAK/STAT signaling by cytokine receptors. *Curr Opin Immunol.* 1998;10:271–278.
44. Shuai K. The STAT family of proteins in cytokine signaling. *Prog Biophys Mol Biol.* 1999;71:405–422.
45. Boudny V, Kovarik J. JAK/STAT signaling pathways and cancer. Janus kinases/signal transducers and activators of transcription. *Neoplasma.* 2002;49:349–355.
46. Kisseleva T, Bhattacharya S, Braunstein J, Schindler CW. Signaling through the JAK/STAT pathway, recent advances and future challenges. *Gene.* 2002;285:1–24.
47. O'Shea JJ, Gadina M, Schreiber RD. Cytokine signaling in 2002: new surprises in the Jak/Stat pathway. *Cell.* 2002;109(suppl): S121–S131.
48. Rawlings JS, Rosler KM, Harrison DA. The JAK/STAT signaling pathway. *J Cell Sci.* 2004;117(pt 8):1281–1283.
49. Hebenstreit D, Horejs-Hoeck J, Duschl A. JAK/STATdependent gene regulation by cytokines. *Drug News Perspect.* 2005;18: 243–249.
50. Lutz M, Knaus P. Integration of the TGF-beta pathway into the cellular signalling network. *Cell Signal.* 2002;14:977–988.
51. Mehra A, Wrana JL. TGF-beta and the Smad signal transduction pathway. *Biochem Cell Biol.* 2002;80:605–622.
52. Cohen MM Jr. TGF beta/Smad signaling system and its pathologic correlates. *Am J Med Genet A.* 2003;116:1–10.
53. Derynck R, Zhang YE. Smad-dependent and Smad-independent pathways in TGF-beta family signalling. *Nature.* 2003;425: 577–584.

54. Chin D, Boyle GM, Parsons PG, Coman WB. What is transforming growth factor-beta (TGF-beta)? *Br J Plast Surg*. 2004;57: 215–221.
55. ten Dijke P, Hill CS. New insights into TGF-beta-Smad signalling. *Trends Biochem Sci*. 2004;29:265–273.
56. Feng XH, Derynck R. Specificity and versatility in TGF-beta signaling through Smads. *Annu Rev Cell Dev Biol*. 2005;21: 659–693.
57. Park SH. Fine tuning and cross-talking of TGF-beta signal by inhibitory Smads. *J Biochem Mol Biol*. 2005;38:9–16.
58. Massague J, Seoane J, Wotton D. Smad transcription factors. *Genes Dev*. 2005;19:2783–2810.
59. Massague J, Gomis RR. The logic of TGF-beta signaling. *FEBS Lett*. 2006;580:2811–2820.
60. Gumbiner BM. Signal transduction of beta-catenin. *Curr Opin Cell Biol*. 1995;7:634–640.
61. Shimizu H, Julius MA, Giarre M, et al. Transformation by Wnt family proteins correlates with regulation of betacatenin. *Cell Growth Differ*. 1997;8:1349–1358.
62. Boutros M, Mlodzik M. Dishevelled: at the crossroads of divergent intracellular signaling pathways. *Mech Dev*. 1999;83: 27–37.
63. Miller JR, Hocking AM, Brown JD, Moon RT. Mechanism and function of signal transduction by the Wnt/betacatenin and Wnt/Ca^{2+} pathways. *Oncogene*. 1999;18:7860–7872.
64. Hinoi T, Yamamoto H, Kishida M, et al. Complex formation of adenomatous polyposis coli gene product and axin facilitates glycogen synthase kinase-3 beta-dependent phosphorylation of beta-catenin and downregulates beta-catenin. *J Biol Chem*. 2000;275:34399–34406.
65. Polakis P. Wnt signaling and cancer. *Genes Dev*. 2000;14: 1837–1851.
66. Doble BW, Woodgett JR. GSK-3: tricks of the trade for a multitasking kinase. *J Cell Sci*. 2003;116:1175–1186.
67. Lee E, Salic A, Kruger R, et al. The roles of APC and Axin derived from experimental and theoretical analysis of the Wnt pathway. *PLoS Biol*. 2003;1:E10.
68. van Es JH, Barker N, Clevers H. You Wnt some, you lose some: oncogenes in the Wnt signaling pathway. *Curr Opin Genet Dev*. 2003;13:28–33.
69. Veeman MT, Axelrod JD, Moon RT. A second canon. Functions and mechanisms of beta-catenin-independent Wnt signaling. *Dev Cell*. 2003;5:367–377.
70. Logan CY, Nusse R. The Wnt signaling pathway in development and disease. *Annu Rev Cell Dev Biol*. 2004;20:781–810.
71. Malbon CC. Frizzleds: new members of the superfamily of G-protein-coupled receptors. *Front Biosci*. 2004;9:1048–1058.
72. Nelson WJ, Nusse R. Convergence of Wnt, beta-catenin, and cadherin pathways. *Science*. 2004;303:1483–1487.
73. Tolwinski NS, Wieschaus E. Rethinking WNT signaling. *Trends Genet*. 2004;20:177–181.
74. Bejsovec A. Wnt pathway activation: new relations and locations. *Cell*. 2005;120:11–14.
75. Senda T, Shimomura A, Iizuka-Kogo A. Adenomatous polyposis coli (Apc) tumor suppressor gene as a multifunctional gene. *Anat Sci Int*. 2005;80:121–131.
76. Takada R, Hijikata H, Kondoh H, Takada S. Analysis of combinatorial effects of Wnts and Frizzleds on betacatenin/armadillo stabilization and Dishevelled 77 phosphorylation. *Genes Cells*. 2005;10:919–928.
77. Cadigan KM, Liu YI. Wnt signaling: complexity at the surface. *J Cell Sci*. 2006;119:395–402.
78. Kikuchi A, Kishida S, Yamamoto H. Regulation of Wnt signaling by protein–protein interaction and posttranslational modifications. *Exp Mol Med*. 2006;38:1–10.
79. Malbon CC, Wang HY. Dishevelled: a mobile scaffold catalyzing development. *Curr Top Dev Biol*. 2006;72:153–166.
80. Pongracz JE, Stockley RA. Wnt signalling in lung development and diseases. *Respir Res*. 2006;7:15.
81. Tian Q. Proteomic exploration of the Wnt/beta-catenin pathway. *Curr Opin Mol Ther*. 2006;8:191–197
82. Franke TF, Kaplan DR, Cantley LC. PI3K: downstream AKTion blocks apoptosis. *Cell*. 1997;88:435–437.
83. Wymann MP, Pirola L. Structure and function of phosphoinositide 3-kinases. *Biochim Biophys Acta*. 1998;1436: 127–150.
84. Krasilnikov MA. Phosphatidylinositol-3 kinase dependent pathways: the role in control of cell growth, survival, and malignant transformation. *Biochemistry (Mosc)*. 2000;65:59–67.
85. Cantley LC. The phosphoinositide 3-kinase pathway. *Science*. 2002;296:1655–1657.
86. Franke TF, Hornik CP, Segev L, et al. PI3K/Akt and apoptosis: size matters. *Oncogene*. 2003;22:8983–8998.
87. Liang J, Slingerland JM. Multiple roles of the PI3K/ PKB (Akt) pathway in cell cycle progression. *Cell Cycle*. 2003;2:339–345.
88. Asnaghi L, Bruno P, Priulla M, Nicolin A. mTOR: a protein kinase switching between life and death. *Pharmacol Res*. 2004;50:545–549.
89. Osaki M, Oshimura M, Ito H. PI3K-Akt pathway: its functions and alterations in human cancer. *Apoptosis*. 2004;9:667–676.
90. Henson ES, Gibson SB. Surviving cell death through epidermal growth factor (EGF) signal transduction pathways: implications for cancer therapy. *Cell Signal*. 2006;18:2089–2097.
91. Liang Y, Zhou Y, Shen P. NF-kappaB and its regulation on the immune system. *Cell Mol Immunol*. 2004;1:343–350.
92. Xiao W. Advances in NF-kappaB signaling transduction and transcription. *Cell Mol Immunol*. 2004;1:425–435.
93. Courtois G. The NF-kappaB signaling pathway in human genetic diseases. *Cell Mol Life Sci*. 2005;62:1682–1691.
94. Moynagh PN. The NF-kappaB pathway. *J Cell Sci*. 2005;118(pt 20):4589–4592.
95. Zingarelli B. Nuclear factor-kappaB. *Crit Care Med*. 2005;33(12 suppl):S414–S416.
96. Bubici C, Papa S, Pham CG, et al. The NF-kappaB-mediated control of ROS and JNK signaling. *Histol Histopathol*. 2006;21:69–80.
97. Campbell KJ, Perkins ND. Regulation of NF-kappaB function. *Biochem Soc Symp*. 2006;73:165–180.
98. Hoffmann A, Baltimore D. Circuitry of nuclear factor kappaB signaling. *Immunol Rev*. 2006;210:171–186.
99. Piva R, Belardo G, Santoro MG. NF-kappaB: a stress-regulated switch for cell survival. *Antioxid Redox Signal*. 2006;8: 478–486.
100. Vermeulen L, Vanden Berghe W, Haegeman G. Regulation of NF-kappaB transcriptional activity. *Cancer Treat Res*. 2006;130:89–102.
101. Karin M. The beginning of the end: IkappaB kinase (IKK) and NF-kappaB activation. *J Biol Chem*. 1999;274:27339–27342.
102. Karin M. How NF-kappaB is activated: the role of the IkappaB kinase (IKK) complex. *Oncogene*. 1999;18:6867–6874.

103. Rothwarf DM, Karin M. The NF-kappa B activation pathway: a paradigm in information transfer from membrane to nucleus. *Sci STKE.* 1999;1999:RE1.
104. Senftleben U, Karin M. The IKK/NF-kappa B pathway. *Crit Care Med.* 2002;30(1 suppl):S18–S26.
105. Hayden MS, Ghosh S. Signaling to NF-kappaB. *Genes Dev.* 2004;18:2195–2224.
106. Viatour P, Merville MP, Bours V, Chariot A. Phosphorylation of NF-kappaB and IkappaB proteins: implications in cancer and inflammation. *Trends Biochem Sci* 2005;30:43–52.
107. Gloire G, Dejardin E, Piette J. Extending the nuclear roles of IkappaB kinase subunits. *Biochem Pharmacol.* 2006;72(9): 1081–1089.
108. Kalderon D. Similarities between the Hedgehog and Wnt signaling pathways. *Trends Cell Biol.* 2002;12:523–531.
109. Mullor JL, Sanchez P, Altaba AR. Pathways and consequences: Hedgehog signaling in human disease. *Trends Cell Biol.* 2002;12:562–569.
110. Cohen MM Jr. The hedgehog signaling network. *Am J Med Genet A.* 2003;123:5–28.
111. McMahon AP, Ingham PW, Tabin CJ. Developmental roles and clinical significance of hedgehog signaling. *Curr Top Dev Biol.* 2003;53:1–114.
112. Wetmore C. Sonic hedgehog in normal and neoplastic proliferation: insight gained from human tumors and animal models. *Curr Opin Genet Dev.* 2003;13:34–42.
113. Lum L, Beachy PA. The Hedgehog response network: sensors, switches, and routers. *Science.* 2004;304:1755–1759.
114. Ogden SK, Ascano M Jr, Stegman MA, Robbins DJ. Regulation of Hedgehog signaling: a complex story. *Biochem Pharmacol.* 2004;67:805–814.
115. Yu TC, Miller SJ. The hedgehog pathway: revisited. *Dermatol Surg.* 2004;30:583–584.
116. Neumann CJ. Hedgehogs as negative regulators of the cell cycle. *Cell Cycle.* 2005;4:1139–1140.
117. Baron M, Aslam H, Flasza M, et al. Multiple levels of Notch signal regulation (review). *Mol Membr Biol.* 2002;19:27–38.
118. Baron M. An overview of the Notch signalling pathway. *Semin Cell Dev Biol.* 2003;14:113–119.
119. Collins BJ, Kleeberger W, Ball DW. Notch in lung development and lung cancer. *Semin Cancer Biol.* 2004;14:357–364.
120. Hansson EM, Lendahl U, Chapman G. Notch signaling in development and disease. *Semin Cancer Biol.* 2004;14:320–328.
121. Bianchi S, Dotti MT, Federico A. Physiology and pathology of notch signalling system. *J Cell Physiol.* 2006;207:300–308.
122. Wilson A, Radtke F. Multiple functions of Notch signaling in self-renewing organs and cancer. *FEBS Lett.* 2006;580: 2860–2868.
123. Hartwell LH, Weinert TA. Checkpoints: controls that ensure the order of cell cycle events. *Science.* 1989;246:629–634.
124. Pardee AB. G1 events and regulation of cell proliferation. *Science.* 1989;246:603–608.
125. Kastan MB, Kuerbitz SJ. Control of G1 arrest after DNA damage. *Environ Health Perspect.* 1993;101(suppl 5):55–58.
126. Sherr CJ. G1 phase progression: cycling on cue. *Cell.* 1994;79: 551–555.
127. Elledge SJ. Cell cycle checkpoints: preventing an identity crisis. *Science.* 1996;274:1664–1672.
128. Sanchez I, Dynlacht BD. Transcriptional control of the cell cycle. *Curr Opin Cell Biol.* 1996;8:318–324.
129. O'Connor PM. Mammalian G1 and G2 phase checkpoints. *Cancer Surv.* 1997;29:151–182.
130. Mercer WE. Checking on the cell cycle. *J Cell Biochem Suppl.* 1998;30-31:50–54.
131. Weinert T. DNA damage checkpoints update: getting molecular. *Curr Opin Genet Dev.* 1998;8:185–193.
132. Johnson DG, Walker CL. Cyclins and cell cycle checkpoints. *Annu Rev Pharmacol Toxicol.* 1999;39:295–312.
133. Clarke DJ, Gimenez-Abian JF. Checkpoints controlling mitosis. *Bioessays.* 2000;22:351–363.
134. Nyberg KA, Michelson RJ, Putnam CW, Weinert TA. Toward maintaining the genome: DNA damage and replication checkpoints. *Annu Rev Genet.* 2002;36:617–656.
135. Shreeram S, Blow JJ. The role of the replication licensing system in cell proliferation and cancer. *Prog Cell Cycle Res.* 2003;5:287–293.
136. Lisby M, Rothstein R. DNA damage checkpoint and repair centers. *Curr Opin Cell Biol.* 2004;16:328–334.
137. Lukas J, Lukas C, Bartek J. Mammalian cell cycle checkpoints: signalling pathways and their organization in space and time. *DNA Repair (Amst).* 2004;3:997–1007.
138. Stark GR, Taylor WR. Analyzing the G2/M checkpoint. *Methods Mol Biol.* 2004;280:51–82.
139. Branzei D, Foiani M. The DNA damage response during DNA replication. *Curr Opin Cell Biol.* 2005;17:568–575.
140. Macaluso M, Montanari M, Cinti C, Giordano A. Modulation of cell cycle components by epigenetic and genetic events. *Semin Oncol.* 2005;32:452–457.
141. Musgrove EA. Cyclins: roles in mitogenic signaling and oncogenic transformation. *Growth Factors.* 2006;24:13–19.
142. Niida H, Nakanishi M. DNA damage checkpoints in mammals. *Mutagenesis.* 2006;21:3–9.
143. Burtelow MA, Roos-Mattjus PM, Rauen M, et al. Reconstitution and molecular analysis of the hRad9-hHus1hRad1 (9-1-1) DNA damage responsive checkpoint complex. *J Biol Chem.* 2001;276:25903–25909.
144. Bao S, Lu T, Wang X, et al. Disruption of the Rad9/Rad1/Hus1 (9-1-1) complex leads to checkpoint signaling and replication defects. *Oncogene.* 2004;23:5586–5593.
145. Parrilla-Castellar ER, Arlander SJ, Karnitz L. Dial 9–1-1 for DNA damage: the Rad9-Hus1-Rad1 (9–1-1) clamp complex. *DNA Repair (Amst).* 2004;3:1009–1014.
146. Majka J, Burgers PM. Function of Rad17/Mec3/Ddc1 and its partial complexes in the DNA damage checkpoint. *DNA Repair.* 2005;4:1189–1194.
147. van Vugt MA, Medema RH. Checkpoint adaptation and recovery: back with Polo after the break. *Cell Cycle.* 2004;3:1383–1386.
148. van Vugt MA, Bras A, Medema RH. Restarting the cell cycle when the checkpoint comes to a halt. *Cancer Res.* 2005;65:7037–7040.
149. Pardee AB. A restriction point for control of normal animal cell proliferation. *Proc Natl Acad Sci USA.* 1974;71:1286–1290.
150. Campisi J, Medrano EE, Morro G, Pardee AB. Restriction point control of cell growth by a labile protein: evidence for increased stability in transformed cells. *Proc Natl Acad Sci USA.* 1982;79:436–440.
151. Blagosklonny MV, Pardee AB. The restriction point of the cell cycle. *Cell Cycle.* 2002;1:103–110.
152. Boonstra J. Progression through the G1-phase of the ongoing cell cycle. *J Cell Biochem.* 2003;90:244–252.

153. Baldin V, Lukas J, Marcote MJ, et al. Cyclin D1 is a nuclear protein required for cell cycle progression in G1. *Genes Dev.* 1993;7:812–821.
154. Dowdy SF, Hinds PW, Louie K, et al. Physical interaction of the retinoblastoma protein with human D cyclins. *Cell.* 1993;73:499–511.
155. Kato J, Matsushime H, Hiebert SW, et al. Direct binding of cyclin D to the retinoblastoma gene product (pRb) and pRb phosphorylation by the cyclin D-dependent kinase CDK4. *Genes Dev.* 1993;7:331–342.
156. Sewing A, Burger C, Brusselbach S, et al. Human cyclin D1 encodes a labile nuclear protein whose synthesis is directly induced by growth factors and suppressed by cyclic AMP. *J Cell Sci.* 1993;104:545–555.
157. Lukas J, Muller H, Bartkova J, et al. DNA tumor virus oncoproteins and retinoblastoma gene mutations share the ability to relieve the cell's requirement for cyclin D1 function in G1. *J Cell Biol.* 1994;125:625–638.
158. Xiao ZX, Ginsberg D, Ewen M, Livingston DM. Regulation of the retinoblastoma protein-related protein p107 by G1 cyclin-associated kinases. *Proc Natl Acad Sci USA.* 1996;93: 4633–4637.
159. Ortega S, Malumbres M, Barbacid M. Cyclin D-dependent kinases, INK4 inhibitors and cancer. *Biochim Biophys Acta.* 2002;1602:73–87.
160. El-Deiry WS, Tokino T, Velculescu VE, et al. WAF1, a potential mediator of p53 tumor suppression. *Cell.* 1993;75:817–825.
161. Polyak K, Kato JY, Solomon MJ, et al. p27Kip1, a cyclin-Cdk inhibitor, links transforming growth factor-beta and contact inhibition to cell cycle arrest. *Genes Dev.* 1994;8: 9–22.
162. Biggs JR, Kraft AS. Inhibitors of cyclin-dependent kinase and cancer. *J Mol Med.* 1995;73:509–514.
163. Datto MB, Li Y, Panus JF, et al. Transforming growth factor beta induces the cyclin-dependent kinase inhibitor p21 through a p53-independent mechanism. *Proc Natl Acad Sci USA.* 1995;92: 5545–5549.
164. Datto MB, Yu Y, Wang XF. Functional analysis of the transforming growth factor beta responsive elements in the WAF1/Cip1/p21 promoter. *J Biol Chem.* 1995;270:28623–28628.
165. Quelle DE, Zindy F, Ashmun RA, Sherr CJ. Alternative reading frames of the INK4a tumor suppressor gene encode two unrelated proteins capable of inducing cell cycle arrest. *Cell.* 1995;83:993–1000.
166. Yeudall WA, Jakus J. Cyclin kinase inhibitors add a new dimension to cell cycle control. *Eur J Cancer B Oral Oncol.* 1995;31B:291–298.
167. Serrano M, Lee H, Chin L, et al. Role of the INK4a locus in tumor suppression and cell mortality. *Cell.* 1996;85:27–37.
168. Craig C, Kim M, Ohri E, et al. Effects of adenovirus-mediated p16INK4A expression on cell cycle arrest are determined by endogenous p16 and Rb status in human cancer cells. *Oncogene.* 1998;16:265–272.
169. Niculescu AB III, Chen X, Smeets M, et al. Effects of p21(Cip1/Waf1) at both the G1/S and the G2/M cell cycle transitions: pRb is a critical determinant in blocking DNA replication and in preventing endoreduplication. *Mol Cell Biol.* 1998;18:629–643.
170. Sherr CJ, Roberts JM. CDK inhibitors: positive and negative regulators of G1-phase progression. *Genes Dev.* 1999;13: 1501–1512.
171. Ohtani N, Yamakoshi K, Takahashi A, Hara E. The p16INK4a-RB pathway: molecular link between cellular senescence and tumor suppression. *J Med Invest.* 2004;51:146–153.
172. Chellapan SP. The E2F transcription factor: role in cell cycle regulation and differentiation. *Mol Cell Differ.* 1994;2: 201–220.
173. Schwarz JK, Bassing CH, Kovesdi I, et al. Expression of the E2F1 transcription factor overcomes type beta transforming growth factor-mediated growth suppression. *Proc Natl Acad Sci USA.* 1995;92:483–487.
174. Hurford RK Jr, Cobrinik D, Lee MH, Dyson N. pRB and p107/p130 are required for the regulated expression of different sets of E2F responsive genes. *Genes Dev.* 1997;11:1447–1463.
175. Ohtani K. Implication of transcription factor E2F in regulation of DNA replication. *Front Biosci.* 1999;4:D793–D804.
176. Humbert PO, Verona R, Trimarchi JM, et al. E2f3 is critical for normal cellular proliferation. *Genes Dev.* 2000;14:690–703.
177. Ren B, Cam H, Takahashi Y, et al. E2F integrates cell cycle progression with DNA repair, replication, and G(2)/M checkpoints. *Genes Dev.* 2002;16:245–256.
178. Schlisio S, Halperin T, Vidal M, Nevins JR. Interaction of YY1 with E2Fs, mediated by RYBP, provides a mechanism for specificity of E2F function. *EMBO J.* 2002;21:5775–5786.
179. Stevaux O, Dyson NJ. A revised picture of the E2F transcriptional network and RB function. *Curr Opin Cell Biol.* 2002;14:684–691.
180. Mundle SD, Saberwal G. Evolving intricacies and implications of E2F-1 regulation. *EMBO J.* 2003;17:569–574.
181. Rogoff HA, Kowalik TF. Life, death and E2F: linking proliferation control and DNA damage signaling via E2F1. *Cell Cycle.* 2004;3:845–846.
182. Korenjak M, Brehm A. E2F–Rb complexes regulating transcription of genes important for differentiation and development. *Curr Opin Genet Dev.* 2005;15:520–527.
183. Henriksson M, Luscher B. Proteins of the Myc network: essential regulators of cell growth and differentiation. *Adv Cancer Res.* 1996;68:109–182.
184. Schmidt EV. MYC family ties. *Nat Genet.* 1996;14:8–10.
185. Amati B, Alevizopoulos K, Vlach J. Myc and the cell cycle. *Front Biosci.* 1998;3:D250–D268.
186. Burgin A, Bouchard C, Eilers M. Control of cell proliferation by Myc proteins. *Results Probl Cell Differ.* 1998;22:181–197.
187. Matsumura I, Tanaka H, Kanakura Y. E2F1 and c-Myc in cell growth and death. *Cell Cycle.* 2003;2:333–338.
188. Yam CH, Fung TK, Poon RY. Cyclin A in cell cycle control and cancer. *Cell Mol Life Sci.* 2002;59:1317–1326.
189. Porter LA, Donoghue DJ. Cyclin B1 and CDK1: nuclear localization and upstream regulators. *Prog Cell Cycle Res.* 2003;5:335–347.
190. Livingstone LR, White A, Sprouse J, et al. Altered cell cycle arrest and gene amplification potential accompany loss of wild-type p53. *Cell.* 1992;70:923–935.
191. Harper JW, Adami GR, Wei N, et al. The p21 cdk-interacting protein Cip1 is a potent inhibitor of G1 cyclin-dependent kinases. *Cell.* 1993;75:805–816.
192. Xiong Y, Hannon GJ, Zhang H, et al. p21 is a universal inhibitor of cyclin kinases. *Nature (Lond).* 1993;366:701–704.

193. Chen CY, Oliner JD, Zhan Q, et al. Interactions between p53 and MDM2 in a mammalian cell cycle checkpoint pathway. *Proc Natl Acad Sci USA*. 1994;91:2684–2688.
194. El-Deiry WS, Harper JW, O'Connor PM, et al. WAF1/CIP1 is induced in p53-mediated G1 arrest and apoptosis. *Cancer Res*. 1994;54:1169–1174.
195. Chen X, Bargonetti J, Prives C. p53, through p21 (WAF1/CIP1), induces cyclin D1 synthesis. *Cancer Res*. 1995;55:4257–4263.
196. Del Sal G, Murphy M, Ruaro E, et al. Cyclin D1 and p21/waf1 are both involved in p53 growth suppression. *Oncogene*. 1996;12:177–185.
197. Laiho M, DeCaprio JA, Ludlow JW, et al. Growth inhibition by TGF-beta linked to suppression of retinoblastoma protein phosphorylation. *Cell*. 1990;62:175–185.
198. Ewen ME, Sluss HK, Whitehouse LL, Livingston DM. TGF beta inhibition of Cdk4 synthesis is linked to cell cycle arrest. *Cell*. 1993;74:1009–1020.
199. Chin L, Pomerantz J, DePinho RA. The INK4a/ARF tumor suppressor: one gene – two products – two pathways. *Trends Biochem Sci*. 1998;23:291–296.
200. Stott FJ, Bates S, James MC, et al. The alternative product from the human CDKN2A locus, p14(ARF), participates in a regulatory feedback loop with p53 and MDM2. *EMBO J*. 1998;17:5001–5014.
201. James MC, Peters G. Alternative product of the p16/CKDN2A locus connects the Rb and p53 tumor suppressors. *Prog Cell Cycle Res*. 2000;4:71–81.
202. Weber HO, Samuel T, Rauch P, Funk JO. Human p14(ARF) mediated cell cycle arrest strictly depends on intact p53 signaling pathways. *Oncogene*. 2002;21:3207–3212.
203. Satyanarayana A, Rudolph KL. p16 and ARF: activation of teenage proteins in old age. *J Clin Invest*. 2004;114:1237–1240.
204. Foulkes WD, Flanders TY, Pollock PM, Hayward NK. The CDKN2A (p16) gene and human cancer. *Mol Med*. 1997;3:5–20.
205. Serrano M. The tumor suppressor protein p16INK4a. *Exp Cell Res*. 1997;237:7–13.
206. Carnero A, Hannon GJ. The INK4 family of CDK inhibitors. *Curr Top Microbiol Immunol*. 1998;227:43–55.
207. Huschtscha LI, Reddel RR. p16(INK4a) and the control of cellular proliferative life span. *Carcinogenesis*. 1999;20:921–926.
208. Roussel MF. The INK4 family of cell cycle inhibitors in cancer. *Oncogene*. 1999;18:5311–5317.
209. Shapiro GI, Edwards CD, Rollins BJ. The physiology of p16(INK4A)-mediated G1 proliferative arrest. *Cell Biochem Biophys*. 2000;33:189–197.
210. Laval J, Jurado J, Saparbaev M, Sidorkina O. Antimutagenic role of base-excision repair enzymes upon free radical-induced DNA damage. *Mutat Res*. 1998;402:93–102.
211. Boiteux S, Radicella JP. Base excision repair of 8-hydroxyguanine protects DNA from endogenous oxidative stress. *Biochimie*. 1999;81:59–67.
212. Boiteux S, Radicella JP. The human OGG1 gene: structure, functions, and its implication in the process of carcinogenesis. *Arch Biochem Biophys*. 2000;377:1–8.
213. Nishimura S. Mammalian Oggl/Mmh gene plays a major role in repair of the 8-hydroxyguanine lesion in DNA. *Prog Nucleic Acid Res Mol Biol*. 2001;68:107–123.
214. Nishimura S. Involvement of mammalian OGG1(MMH) in excision of the 8-hydroxyguanine residue in DNA. *Free Radic Biol Med*. 2002;32:813–821.
215. Fortini P, Pascucci B, Parlanti E, et al. 8-Oxoguanine DNA damage: at the crossroad of alternative repair pathways. *Mutat Res*. 2003;531:127–139.
216. Nakabeppu Y, Tsuchimoto D, Furuichi M, Sakumi K. The defense mechanisms in mammalian cells against oxidative damage in nucleic acids and their involvement in the suppression of mutagenesis and cell death. *Free Radic Res*. 2004;38:423–429.
217. Thompson LH, West MG. XRCC1 keeps DNA from getting stranded. *Mutat Res*. 2000;459:1–18.
218. Tomkinson AE, Chen L, Dong Z, et al. Completion of base excision repair by mammalian DNA ligases. *Prog Nucleic Acid Res Mol Biol*. 2001;68:151–164.
219. Caldecott KW. XRCC1 and DNA strand break repair. *DNA Repair*. 2003;2:955–969.
220. Dianov GL, Sleeth KM, Dianova II, Allinson SL. Repair of abasic sites in DNA. *Mutat Res*. 2003;531:157–163.
221. Malanga M, Althaus FR. The role of poly(ADP-ribose) in the DNA damage signaling network. *Biochem Cell Biol*. 2005;83:354–364.
222. Williams RS, Bernstein N, Lee MS, et al. Structural basis for phosphorylation-dependent signaling in the DNA-damage response. *Biochem Cell Biol*. 2005;83:721–727.
223. Johnson RT, Squires S. The XPD complementation group. Insights into xeroderma pigmentosum, Cockayne's syndrome and trichothiodystrophy. *Mutat Res*. 1992;273:97–118.
224. Wood RD. DNA damage recognition during nucleotide excision repair in mammalian cells. *Biochimie*. 1999;81:39–44.
225. Bernstein C, Bernstein H, Payne CM, Garewal H. DNA repair/pro-apoptotic dual-role proteins in five major DNA repair pathways: fail-safe protection against carcinogenesis. *Mutat Res*. 2002;511:145–178.
226. Chen J, Suter B. Xpd, a structural bridge and a functional link. *Cell Cycle*. 2003;2:503–506.
227. MacPhee DG. Mismatch repair as a source of mutations in non-dividing cells. *Genetica* 1996;97:183–195.
228. Peltomaki P. DNA mismatch repair gene mutations in human cancer. *Environ Health Perspect*. 1997;105(suppl 4):775–780.
229. Kirkpatrick DT. Roles of the DNA mismatch repair and nucleotide excision repair proteins during meiosis. *Cell Mol Life Sci*. 1999;55:437–449.
230. Kolodner RD, Marsischky GT. Eukaryotic DNA mismatch repair. *Curr Opin Genet Dev*. 1999;9:89–96.
231. Harfe BD, Jinks-Robertson S. Mismatch repair proteins and mitotic genome stability. *Mutat Res*. 2000;451:151–167.
232. Harfe BD, Jinks-Robertson S. DNA mismatch repair and genetic instability. *Annu Rev Genet*. 2000;34:359–399.
233. Aquilina G, Bignami M. Mismatch repair in correction of replication errors and processing of DNA damage. *J Cell Physiol*. 2001;187:145–154.
234. Hsieh P. Molecular mechanisms of DNA mismatch repair. *Mutat Res*. 2001;486:71–87.
235. Schofield MJ, Hsieh P. DNA mismatch repair: molecular mechanisms and biological function. *Annu Rev Microbiol*. 2003;57:579–608.

236. Isaacs RJ, Spielmann HP. A model for initial DNA lesion recognition by NER and MMR based on local conformational flexibility. *DNA Repair.* 2004;3:455–464.
237. Stojic L, Brun R, Jiricny J. Mismatch repair and DNA damage signalling. *DNA Repair.* 2004;3:1091–1101.
238. Kunkel TA, Erie DA. DNA mismatch repair. *Annu Rev Biochem.* 2005;74:681–710.
239. Jun SH, Kim TG, Ban C. DNA mismatch repair system. Classical and fresh roles. *FEBS J.* 2006;273:1609–1619.
240. Montesano R, Becker R, Hall J, et al. Repair of DNA alkylation adducts in mammalian cells. *Biochimie.* 1985;67:919–928.
241. D'Incalci M, Citti L, Taverna P, Catapano CV. Importance of the DNA repair enzyme O6-alkyl guanine alkyltransferase (AT) in cancer chemotherapy. *Cancer Treat Rev.* 1988;15:279–292.
242. Pegg AE, Byers TL. Repair of DNA containing O6-alkylguanine. *FASEB J.* 1992;6:2302–2310.
243. Sekiguchi M, Nakabeppu Y, Sakumi K, Tuzuki T. DNA-repair methyltransferase as a molecular device for preventing mutation and cancer. *J Cancer Res Clin Oncol.* 1996;122:199–206.
244. Pieper RO. Understanding and manipulating O6-methylguanine-DNA methyltransferase expression. *Pharmacol Ther.* 1997;74: 285–297.
245. Sekiguchi M, Sakumi K. Roles of DNA repair methyltransferase in mutagenesis and carcinogenesis. *Jpn J Hum Genet.* 1997;42:389–399.
246. Yu Z, Chen J, Ford BN, et al. Human DNA repair systems: an overview. *Environ Mol Mutagen.* 1999;33:3–20.
247. Kaina B, Ochs K, Grosch S, et al. BER, MGMT, and MMR in defense against alkylation-induced genotoxicity and apoptosis. *Prog Nucleic Acid Res Mol Biol.* 2001;68:41–54.
248. Drablos F, Feyzi E, Aas PA, et al. Alkylation damage in DNA and RNA-repair mechanisms and medical significance. *DNA Repair.* 2004;3:1389–1407.
249. Gerson SL. MGMT: its role in cancer aetiology and cancer therapeutics. *Nat Rev Cancer.* 2004;4:296–307.
250. Varon R, Vissinga C, Platzer M, et al. Nibrin, a novel DNA double-strand break repair protein, is mutated in Nijmegen breakage syndrome. *Cell.* 1998;93:467–476.
251. Buscemi G, Savio C, Zannini L, et al. CHK2 activation dependence on NBS1 after DNA damage. *Mol Cell Biol.* 2001;21:5214–5222.
252. Xu B, Kim S, Kastan MB. Involvement of BRCA1 in S-phase and G(2)-phase checkpoints after ionizing irradiation. *Mol Cell Biol.* 2001;21:3445–3450.
253. D'Amours D, Jackson SP. The MRE11 complex: at the crossroads of DNA repair and checkpoint signalling. *Nat Rev Mol Cell Biol.* 2002;3:317–327.
254. Girard PM, Riballo E, Begg AC, et al. NBS1 promotes ATM dependent phosphorylation events including those required for G1/S arrest. *Oncogene.* 2002;21:4191–4199.
255. Huang J, Dynan WS. Reconstitution of the mammalian DNA double-strand break end-joining reaction reveals a requirement for an MRE11/RAD50/NBS1-containing fraction. *Nucleic Acids Res.* 2002;30:667–674.
256. Nakanishi K, Taniguchi T, Ranganathan V, et al. Interaction of FANCD2 and NBS1 in the DNA damage response. *Nat Cell Biol.* 2002;4:913–920.
257. Osborn AJ, Elledge SJ, Zou L. Checking on the fork: the DNA-replication stress-response pathway. *Trends Cell Biol.* 2002;12:509–516.
258. Yazdi PT, Wang Y, Zhao S, et al. SMC1 is a downstream effector in the ATM/NBS1 branch of the human S-phase checkpoint. *Genes Dev.* 2002;16:571–582.
259. Carson CT, Schwartz RA, Stracker TH, et al. The MRE11 complex is required for ATM activation and the G2/M checkpoint. *EMBO J.* 2003;22:6610–6620.
260. Goodarzi AA, Block WD, Lees-Miller SP. The role of ATM and ATR in DNA damage-induced cell cycle control. *Prog Cell Cycle Res.* 2003;5:393–411.
261. Shiloh Y. ATM and related protein kinases: safeguarding genome integrity. *Nat Rev Cancer.* 2003;3:155–168.
262. Uziel T, Lerenthal Y, Moyal L, et al. Requirement of the MRN complex for ATM activation by DNA damage. *EMBO J.* 2003;22:5612–5621.
263. Abraham RT. PI 3-kinase related kinases: "big" players in stress-induced signaling pathways. *DNA Repair.* 2004;3:883–887.
264. Lee JH, Paull TT. Direct activation of the ATM protein kinase by the MRE11/RAD50/NBS1 complex. *Science.* 2004;304:93–96.
265. Matsuura S, Kobayashi J, Tauchi H, Komatsu K. Nijmegen breakage syndrome and DNA double strand break repair by NBS1 complex. *Adv Biophys.* 2004;38:65–80.
266. Lavin MF, Birrell G, Chen P, et al. ATM signaling and genomic stability in response to DNA damage. *Mutat Res.* 2005;569: 123–132.
267. Lee JH, Paull TT. ATM activation by DNA double-strand breaks through the MRE11-RAD50-NBS1 complex. *Science.* 2005;308:551–554.
268. O'Driscoll M, Jeggo PA. The role of double-strand break repair—insights from human genetics. *Nat Rev Genet.* 2006;7: 45–54.
269. Zhang Y, Zhou J, Lim CU. The role of NBS1 in DNA double strand break repair, telomere stability, and cell cycle checkpoint control. *Cell Res.* 2006;16:45–54.

3 Cell Adhesion Molecules

Timothy Craig Allen and Philip T. Cagle

Cell adhesion molecules, also termed cell adhesion receptors, are one of three classes of macromolecules – along with extracellular matrix molecules and adhesion plaque proteins – that mediate cell adhesion, an activity which is critical for the commencement and maintenance of the three-dimensional structure and normal function of tissues.[1,2] Cell adhesion molecules are predominantly transmembrane glycoproteins that mediate binding to extracellular matrix molecules or to associated receptors on other cells, in a manner that determines the specificity of cell–cell or cell–extracellular matrix interactions.[1] There are five families of adhesion receptors – integrins, cadherins, immunoglobulin cell adhesion molecules (Ig CAMs), selectins, and CD44.[1,3–9] Complexes formed by cell adhesion receptors are not static, but are dynamic units capable of obtai-ning and incorporating extracellular environmental signals, and are indeed the foundation of two-way signaling between the cell and its external environment.[1,10] These cell adhesion molecule families are also involved with signaling between the interior and exterior of the cell, and as such are important in cell growth, proliferation, spatial organization, motility, migration, signaling, differentiation, apoptosis, and gene transcription in normal physiological growth and development as well as in pathological conditions such as inflammation and wound healing.[5,9]

The intricate processes involved in cancer invasion and metastasis require tumor cell detachment from the primary tumor, then tumor cell entry and exit from the lymphatic or vascular systems in turn, culminating in tumor cell growth at distant tissue sites.[5,11] These steps can only take place with the dysregulation of normal cell–cell adhesion and cell–matrix interactions that are mediated by the cell adhesion receptor families.[5] An increased understanding of cell adhesion molecules and their intercellular and extracellular interactions may provide opportunities for the development of diagnostic and therapeutic modalities for cancer.[8]

Integrins

The integrin superfamily, the largest known family of cell adhesion molecules, are widely expressed receptors via which cells attach to extracellular matrices or to other cells.[8,9] They are expressed in various combinations in all cells.[9,12] Integrins are transmembrane glycoproteins that form heterodimers consisting of one alpha (α) and one beta (β) subunit.[3,8,9] Both subunits are required for interaction of integrins with the cytoskeleton and extracellular matrix, and they collaborate to bind extracellular matrix molecules, the specificity of which is determined by the subunit combination and cell type.[8,9] The specific combination of the subunits determines the ligand-binding specificity, and various specific combinations of α- and β-subunits exhibit binding specificity for collagen receptors, fibronectin receptors, laminin receptors, vitronectin receptors, and other integrin receptors.[3] Most integrins are substrate adhesion molecules that mediate interactions between cells and extracellular matrix components; however, some integrins are cell–cell adhesion molecules (CAMs).[3] Lymphocytes express integrins that mediate heterotypic cell–cell adhesion, binding some Ig CAMs.[3] The integrin receptor family plays a critical role in complex cellular events such as cellular differentiation, proliferation, and migration and is involved in biological processes related to organogenesis, wound healing, and the altered adhesive and invasive properties of tumor cells.[13–15] Integrins transmit signals bidirectionally across plasma membranes.[16] Alterations in integrin secretion or functional activity may regulate the development and progression of cancers, and changes in integrins in vivo leading to more aggressive tumor behavior have been identified in cancers of the lung, breast, colon, prostate, stomach, pancreas, liver, kidney, ovary, skin, and endometrium.[3,17–28] β Integrins have been associated with tumor cell migration in cell lines including fibrosarcoma, bladder cancer, and colon cancer.[3,29] Integrin subunits have

P.T. Cagle and T.C. Allen (eds.), *Basic Concepts of Molecular Pathology*, Molecular Pathology Library 2,
DOI: 10.1007/978-0-387-89626-7_3, © Springer Science+Business Media, LLC 2009

been identified immunohistochemically in bronchial epithelium, endothelium, and smooth muscle.[3] In melanomas, increased expression of αvβ3 has been shown to correlate with a more malignant phenotype.[9,30] Integrins may be important in the development of drug resistance in cancer cells.[9] The inhibition of αvβ3 integrin has been shown to enhance the antitumor effect of radiotherapy.[9] Integrin alterations may be necessary for the development of cisplatin resistance in cancer cells.[9]

β1 Integrins

β1 Integrins make up the largest number of integrins, and there are at least seven receptors in this subfamily, each with different ligand specificity.[8] α4β1, α5β1, α6β1, and αnβ1 are among the most studied of these receptors.[8] α4β1 Integrin is a cell adhesion receptor, primarily found on lymphocytes, monocytes, and eosinophils, that may be a target for treatment of chronic inflammatory diseases.[8,31] The regulation of β1 integrins by cytokines and other inflammatory mediators is important for eosinophil adhesion to endothelium and transendothelial migration.[8,32,33] α4 Integrins are heterodimeric cell surface molecules important for leukocyte–cell and leukocyte–matrix adhesion interactions.[8]

CXCR4 Chemokine Receptor

Cancer cells must pass through blood or lymphatic circulation and through their vessel walls to metastasize. Metastatic tumor cells are thought to co-opt signals normally controlling leukocyte movement, such as chemokine-mediated cell migration.[34] Emigration from vessels is regulated by a sequence of distinct molecular signals, one of which involves chemokines that activate integrins and direct migration of leukocytes.[34] The chemokine stromal cell-derived factor-1 (SDF-1/CXCL12) is a CXC chemokine expressed in bone marrow stromal cells.[34] CXCR4, the receptor for CXCL12, has a critical role for the homing of hematopoietic stem cells in the bone marrow microenvironment.[34] The CXCR4/CXCL12 axis may regulate migration and metastasis of several cancers, and the neutralization of the CXCR4/CXCL12 axis has been shown in vivo to inhibit or attenuate metastases of breast cancer.[34]

Integrin-Linked Kinase

Integrin-linked kinase (ILK) is an ubiquitously expressed serine/threonine kinase that binds to the cytoplasmic domain of the β1 integrin subunit.[28] It is involved in the regulation of cell survival and plays a role in cancer, with its upregulation identified in cancers of the colon, breast, prostate, and lung, and in melanoma.[28,35–38] ILK is upregulated, often in a stage- and grade-dependent manner, and is associated with metastases.[28] ILK is a central structural and signaling protein that serves as a link between the extracellular matrix and the cytoskeleton via integrins, paxillin, and parvins, and with adaptor proteins particularly interesting new cysteine-histidine-rich protein (PINCH) and Nck2, and between integrins and receptor tyrosine kinases.[28,39,40] It also activates various signaling pathways through its kinase activity.[28]

Focal Adhesion Kinase (FAK)

As integrin receptors are not catalytically active, they must recruit and activate other signaling molecules. FAK, a nonreceptor tyrosine kinase enriched in focal adhesions and ubiquitously expressed during development, plays a key role in cell migration, proliferation, and survival.[16,28] FAK may be a critical mediator of integrin signaling and has been linked to the integrin- and growth factor receptor-signaling pathways that regulate several biological processes concerning neoplastic transformation, invasion, and metastasis, including cell adhesion, migration, and apoptosis.[16,41–43] FAK phosphorylation and kinase activity are regulated by integrin-mediated and matrix-dependent cell adhesion in tumor cells of many cancers.[16,44,45] FAK overexpression and phosphorylation are related to incre-ased cell tumor cell motility, invasion, and cytoskeleton alteration.[16]

FAK is important for tumor progression. Elevated FAK mRNA levels have been found in human carcinomas and in acute lymphoblastic leukemias.[28,46,47] The gene encoding FAK, termed protein-tyrosine kinase-2, located on chromosome 8q24, is increased in number in tumor cells.[28,48] FAK overexpression in cancer contributes to an invasive phenotype resulting from increased formation of invadopodia.[28,49,50] Further examination of the role of FAK role in cancer progression and invasion may allow for the development of targeted cancer therapy.[28]

Integrins and adhesion of cells to extracellular membranes confer higher resistance to ionizing radiation and cytotoxic drugs.[28] The survival-promoting effects of integrins has been identified in tumor response to conventional chemotherapeutic agents such as paclitaxel.[28,51,52] Integrins, as well as integrin-associated proteins, modulate the radiation response of cells. ILK has been found in high levels in radiosensitive tumor cells.[28,53] Because integrin-associated proteins are involved in all major signal transduction pathways regarding cell proliferation and survival, they are strong potential candidates for targeted therapy.[28]

Cadherins

The cadherin family consists of calcium-dependent cell–cell adhesion molecules that are highly conserved transmembrane glycoproteins with similar domains for homophilic binding, calcium binding, and interaction with intracellular proteins.[6,54,55] The cytoplasmic domain of cadherins interacts with catenins (α-catenin, β-catenin, γ-catenin), and the resulting complexes associate with cortical actin filaments.[6,56]

The cadherin–catenin interaction is necessary for cadherin-mediated adhesion and association of the complexes with the cytoskeleton.[6,57] The cadherin family contains 16 members, of which the most significant is E-cadherin (epithelial cadherin), which is found in epithelial tissues and is involved in formation and maintenance of cell histoarchitecture.[6] Other important cadherins include N-cadherin (neural cadherin), found in neural and muscle tissues, P-cadherin (placental cadherin), R-cadherin (retinal cadherin), and VE-cadherin (vascular endothelial cadherin).[6,58] Loss of function or secretion of the E-cadherin–catenin complex, or any of its components, eliminates a cell's capacity to adhere, with resulting loss of normal tissue architecture.[6] Altered or absent E-cadherin expression has been identified in various cancers, including stomach, head and neck, bladder, prostate, breast, and colon.[6,59–63] α-Catenin absence or alteration has been identified in breast, gastric, and esophageal cancers,[6,64,65] and alteration of β-catenin expression or phosphorylation has been found in cancers of the esophagus, stomach, and colon.[6,66] Reduced E-cadherin expression has also been identified in lung cancer.[6] Assembly and disassembly of the cadherin–catenin complex, composed of E-cadherin, β-catenin, and α-catenin, is highly regulated and is essential for proper cell–cell adhesion and development, the establishment of cellular polarity, and the maintenance of tissue organization.[67] Loss of the complex is important in cancer development.[67]

β-Catenin

β-Catenin is a cytoplasmic plaque protein involved in linking cadherin family receptors to the actin cytoskeleton.68 It also is a cotranscriptional activator of genes in the nucleus together with lymphoid enhancer factor 1/T-cell factor.[68] It is important not only in the cadherin–catenin complex but also as a critical component of the Wnt signaling pathway.[67,68] In the presence of Wnt, stabilized β-catenin is translocated to the nucleus where it binds T-cell factor/lymphoid enhancer-binding factor-1 transcription factors, inducing target gene expression, including the cancer-related genes MYC, cyclin D1, and matrix metalloproteinase 7.[67,69] Both loss of β-catenin expression, and gain-of-function mutations, are common in human cancers, resulting in loss of cell–cell adhesion and increased gene transcription.[67,70] Its functions are probably controlled not by the regulation of the transcriptional level of its gene but rather by controlling the level of its cytoplasmic pool.[68]

α-Catenin

α-Catenin was first examined in human cancer cell lines such as the lung cancer cell line PC-9, which has a homozygous deletion of part of the α-catenin gene.[67,71] α-Catenin is involved in cancer progression via both canonical and noncanonical roles.[67] Loss of α-catenin expression causes increased cell proliferation, and it is involved in apoptosis and growth factor signaling.[67,72–74] Dissociation of α-catenin from the cadherin–catenin complex may lead to actin bundling and suppression of Arp2/3-mediated actin assembly, possibly causing decreased cell migration and stabilization of cell–cell adhesion.[67] As such, α-catenin in cancers may cause rampant Arp2/3-mediated actin polymerization and increased lamellipodial membrane activity, with resulting decreased cell–cell adhesion, increased cell migration, and tumor cell invasion.[67]

E-Cadherin

Cadherins are essential for tight junctions between cells, and E-cadherin is the cadherin most strongly expressed in epithelial cells.[54,75,76] Cadherins form a complex with cytoplasmic proteins known as catenins, and the resulting complex, along with other cytoskeletal components such as actin, constitutes the intercellular adherence junction.[54,75–77] Transmembrane single-chain glycoproteins, cadherins have an extracellular domain, a transmembrane domain, and an intracellular domain, and play a primary role in maintaining physical cell–cell adhesion by mediating calcium-dependent homotypic intercellular adhesion in epithelial issues.[8,78,79] Cadherin-mediated cell adhesion suppresses invasion of cancer cells in vitro, and dysfunction of the E-cadherin system correlates with cancer invasion in human cancers.[75,80–84] The human lung cancer cell line PC9 expresses an aberrant α-catenin mRNA, and the cells have very loose cell–cell associations.[75,85,86] α-Cadherin is considered by some authors to be indispensable for cadherin-mediated cell–cell adhesion.[75] Researchers have identified reduced or heterogeneous E-cadherin expression and/or α-catenin expression in undifferentiated invasive cancers, and impaired expression of E-cadherin and α-catenin has been associated with increased lymph node metastases in breast, esophageal, and head and neck cancers.[60,63,64,75] The prognostic value of reduced E-cadherin expression in cancer patients has not been widely studied.[59,61,75,87–89]

The relationship between E-cadherin and α-catenin is mediated by β-catenin, and β-catenin in turn mediates the interactions of the cadherin–catenin complex with the c-erbB-2 gene product and epidermal growth factor receptor (EGFR).[75,90–93] APC protein, a tumor suppressor gene product, interacts with β-catenin and plakoglobin and is important in the E-cadherin-mediated cell adhesion system, influencing tumor invasion and metastasis.[75]

Cadherins and Cyclooxygenase 2

Cyclooxygenase 2 (COX-2) and its metabolite prostaglandin E_2 (PGE_2) are critical for regulating diverse cellular functions under both physiological and pathological conditions.[94–97] COX-2 is overexpressed in human non-small cell lung cancers, and its inhibition causes tumor reduction in vivo in murine lung cancer models.[94,98] COX-2 activity is identified throughout the progression of a premalignant lesion to the metastatic phenotype.[94,99] Higher COX-2 expression has been identified in lung adenocarcinoma lymph node metastases.[94,99] COX-2 overexpression has been associated with angiogenesis, decreased host immunity, and enhanced invasion and metastasis, and therefore COX-2 has been considered to have an important role in multiple pathways in lung cancer carcinogenesis, suggesting it has a multifa-ceted role in conferring malignant and metastatic phenotypes.[94,100–104] COX-2 may be a central element in orchestrating the multiple genetic alterations required for lung cancer invasion and metastasis.[94,104] COX-2-dependent invasive capacity in lung cancer is caused by PGE_2-mediated regulation of CD44 and matrix metalloproteinase-2.[94] Normal cell–cell adhesion disruption leads to enhanced tumor cell migration and proliferation, with resulting invasion and metastasis.[94,105,106] Downregulating the cadherin family or catenin family members, or activation of signaling pathways that prevent cell–cell cadherin junction assembly, can cause this disruption.[94,105] As such, extracellular matrix and cell–cell adhesion are significant barriers to tumor metastasis.[94,105] The E-cadherin–catenin complex is required for intercellular adhesiveness and normal tissue architecture maintenance.[94,105] E-cadherin reduction has been linked to tumor invasion, metastasis, and poor prognosis.[94,107] E-cadherin loss, along with increased COX-2 expression, has been identified in familial adenomatous polyposis.[94,103]

Cadherins and EGFR

Epidermal growth factor receptor (EGFR) is overexpressed in many non-small cell lung cancers (NSCLC), and treatment with the EGFR tyrosine kinase inhibitors gefitinib and erlotinib has shown improved survival in some chemotherapy-resistant NSCLC patients.[108–111] Nonetheless, about half these NSCLC patients have tumor progression within 8 months and show no treatment benefit.[108] Activating mutations in the EGFR tyrosine kinase domain have been shown to increase EGFR copy number and/or expression of EGFR protein correlating with response and survival after EGFR tyrosine kinase inhibitor therapy.[108,112,113] EGFR activation and signaling through its downstream targets are modulated by E-cadherin; specifically, E-cadherin inhibits EGFR ligand activation and enhances AKT activation in neighboring cells.[108,110–113] High phosphorylated AKT levels may predict tumor response to EGFR tyrosine kinase inhibitors.[110,114] In lung cancer cell lines, E-cadherin expression is regulated by β-catenin signaling and by zinc finger proteins, including the Slug/Snail family, SIP1 and ZEB1.[108,115] These transcription factors regulate gene expression by interaction with two 5′-CACCTG (E-box) promoter sequences.[108,116] This regulation is facilitated by interaction with the transcriptional corepressor CtBP, which recruits histone deacetylases (HDAC), causing chromatin condensation and gene silencing.[108,117]

H-Cadherin

The H-cadherin gene, a cadherin superfamily member that has been isolated and mapped to 16q24, lacks the cytoplasmic domain, in contrast to E-cadherin, P-cadherin, and N-cadherin.[118] Genetic abnormalities in the H-cadherin gene have been identified in human cancer cell lines and in lung, stomach, and ovarian cancers.[118]

Epithelial–Mesenchymal Transition

Epithelial–mesenchymal transition, a primary cellular process involved in remodeling tissues during development, plays an important role in the progression of cancers toward a more aggressive phenotype that is associated with increased motility and invasiveness.[119,120] Changes in cadherin expression accompany this process, with loss of E-cadherin and upregulation of N-cadherin leading to increased invasive progression, possibly because of a functional cooperation between N-cadherin and the GFGR, which may potentiate ERK signaling to upregulate matrix metalloproteases and downregulate E-cadherin, yielding localized epithelial–mesenchymal transformation.[119]

Selectins

The selectin family of adhesion molecules, consisting of L(eukocyte) selectin, E(ndothelial) selectin, and P(latelet) selectin, are calcium-dependent type I transmembrane glycoproteins with extracellular lectin-like domains that interact, for example, with sialylated carbohydrate-determinant and mucin-like glycoproteins.[121] They have been studied predominantly by examination of the recruitment of leukocytes from the circulation.[4,121] Adhesion with selectins is calcium dependent, and the ligands are cell-surface glycans possessing a specific sialyl-Lewis X-type structure also found in blood group antigens.[1,4] P selectin binds to P-selectin glycoprotein ligand (PSGL-1), L selectin interacts with GlyCAM-1 and CD34, and E selectin possibly reacts with ESL-1.[4] Selectins, similar to Ig CAMs, are expressed on cell surfaces in low levels.[4] P selectin is recruited from storage in Weibel–Palade bodies to the cell surface after inflammatory stimuli activate cells, and E selectin is synthesized and transported to the cell surface upon exposure to inflammatory mediators.[4,122] L selectin, present on leukocyte surfaces, is shed from the cell surface into circulation.[4,123] Enhanced cell-surface selectin expression is related to slowing and rolling of leukocytes at the endothelial cell wall.[4,124] Targeted disruption of P selectin in mice confirms the selectin–ligand interaction is vital for leukocyte recruitment from the circulation, and this step has been examined for possible therapeutic intervention for inflammatory diseases, with antibodies against selectins and against carbohydrate moieties used to block inflammatory cell recruitment in animal models.[4,125,126]

The selectin family of cell adhesion molecules interacts with their cognate glycoprotein ligands to mediate tethering, rolling, and weak adhesion.[127] The integrin family of cell adhesion molecules interacts with their ligands of the Ig CAM superfamily to mediate firm adhesion and signal transduction, eventually triggering shape changes of the adherent leukocytes and transendothelial migration.[127] P selectin rapidly translocates to the cell surface by exocytosis and mediates leukocyte rolling to the activated endothelial cells and heterotypic aggregation of activated platelets to leukocytes, upon thrombogenic and inflammatory challenges.[127,128] P selectin interacts with a leukocyte surface sialomucin, PSGL-1, expressed on a variety of human leukocytes.[129] P selectin has been found to bind to several human cancers and cancer cell lines, including NSCLC and squamous cell carcinoma (SCC), colon cancer, breast cancer, malignant melanoma, gastric cancer, and neuroblastoma.[127,130] NKI-4 cells, a human malignant melanoma cell line, bind P selectin and express glycoprotein ligands for P selectin that are functionally and structurally distinct from leukocyte PSGL-1.[127,130]

Sialyl Lewis x (sLe-x) and sailyl Lewis a (sLe-a) are cancer-associated carbohydrate antigens involved in metastasis.[131] Both serve as ligands for P selectin, I selectin, and E selectin that are found on the surfaces of platelets, leukocytes, and endothelial cells, and both mediate adhesion of tumor cells to endothelial cells.[131,132] Their importance in tumor metastasis is supported by the findings that (1) their antibodies block tumor cell adhesion on endothelial cells in vitro, (2) their expression is associated with increased metastatic potential of tumor cells, and (3) their antibodies have inhibitory effects on angiogenesis.[131,133] Increased sLe-x and sLe-a antigen expression is often found in cancers.[131,134] A correlation between their expression and poor prognosis has been shown in breast and colon cancers, but its correlation in lung cancer is controversial.[131,134]

E selectin is the endothelial ligand for sialyl carbohydrate antigens expressed on the surface of tumor cells.[135] Its secretion is reportedly stimulated by tumor cells, and serum titers have correlated with prognosis in breast cancer patients, and, with a limited number of studies, in NSCLC patients.[135–137] D'Amico et al. studied preoperative and postoperative serum levels of E selectin, CD44, basic epidermal growth factor (BEGF), hepatocyte growth factor, basic fibroblast growth factor, urokinase plasminogen activator, and urokinase plasminogen activator receptor (uPAR), by enzyme-linked immunosorbent assay.[135] The authors found that decreasing serum E selectin levels and increasing CD44 and uPAR levels were significantly associated with increased recurrence risk.[135] CD44 is discussed in more detail below.

P selectin, present on activated platelets and endothelial cells after thrombogenic and inflammatory challenges, interacts with cancer cells in a manner that promotes tumor metastasis.[138,139] Although P selectin has several roles in tumor metastasis, its role in the initial step of metastasis, compared to other adhesion molecules, is singular and necessary for the metastatic process.[138,140] As such interference with the P selectin–tumor cell interaction may attenuate the ability for long-term organ colonization with tumor cells.[138] Heparin, aside from its anticoagulant effects, interferes with tumor cell–platelet association via antiplatelet agents targeted to P selectin. P selectin-mediated interaction has been shown to potentially inhibit both spontaneous and experimental metastasis in vivo; however, the bleeding risk from heparin administration is significant.[138] Modified heparins, with diminished anticoagulant activity but with retained antimetastatic properties, have recently been developed.[138]

Ig-Like CAMS

Ig CAMs are a superfamily of cell adhesion molecules that have diverse structures and functions, but which all contain one or more of a common Ig-like repeat characterized by two cysteines separated by 55–75 amino acids.[4,124] The Ig-like domains are expressed on the extracellular domain of the protein, and typically these molecules span the cell membrane and contain only a short cytoplasmic tail.[4,124] The molecules are important in nervous system development, embryonic development, and immune and inflammatory responses.[4,124,141] NCAM is a family member that is critical for the preservation and integrity of the nervous system.[4,141] Ig CAMS called ICAMS are important for cellular adhesion in the immune system. ICAMs, including VCAM-1 and MAdCAM-1, serve as ligands for integrins.[4,123,125] PECAM, another Ig CAM, is vital for leukocyte adhesion, and is important in cell–cell contact to promote adhesion between endothelial cells and leukocytes.[4] Modulation in ICAM expression may help regulate the type of leukocyte recruited to an area of inflammation and the temporal pattern of leukocyte recruitment.[4,123] ICAM-3, found on leukocytes and endothelial cells, interacts with lymphocyte function-associated antigen-1 and is involved in leukocyte intercellular adhesion.[129] CAM-3 may also be important for angiogenesis.[167,168] ICAM-3 may play a role in tumor progression, and studies have shown a relationship between ICAM-3 and some diseases.[129,142] Radiation-resistant cervical cancer cases have shown increased ICAM-3 expression in tumor stromal endothelial cells and lymphocytes.[129,143]

Some adhesion molecules serve not only as adhesion substances but also as regulators of other cell functions by influencing signaling, a process termed "outside-to-in signaling." [144] β1 Integrins, β2 integrins, and CD28 induce costimulatory signals in the binding of T cells to antigen-presenting cells via multiple cellular signaling molecules, including FAKs, causing cell activation and cytokine production.[144,145] Adhesion molecules are regulated by intracellular signaling induced by various cellular stimuli in the process termed "inside-to-out signaling."[144] ICAM-1 expression is highly regulated by locally produced inflammatory cytokines such as interleukin-1β, tumor necrosis factor-α, interleukin-6, and interferon-γ.[144,146] The ICAM-1/leukocyte function-associated antigen-1 pathway regulates important cell–cell interactions such as leukocyte adhesion and migration, including tumor cell killing by natural killer cells and cytotoxic T lymphocytes.[144,147] While several tumor cells highly express ICAM-1, a potent ligand for leukocyte function-associated antigen-1 on cytotoxic T lymphocytes in vitro, many tumor cells remain viable against killing by cytotoxic T lymphocytes in vivo.[144]

Cd44

Adhesion molecule CD44, a cell-surface transmembrane glycoprotein, is distributed extensively and can be detected on lymphocytes and fibroblasts.[148–150] CD44 is a principal cell-surface receptor for hyaluronan, a major extracellular matrix component, and it communicates cell–matrix interactions into cells via "outside-in signaling."[151] CD44 assists in cell-to-cell interactions and cell-to-matrix interactions; it has been identified in several isoforms on tumor cells, and serum levels have been shown to be prognostically relevant in stomach cancer patients.[135,152] It is involved in lymphocyte activation, recirculation, and homing, adhesion of extracellular matrix, angiogenesis, cell proliferation, and cell migration.[149,153] Tumor expression of CD44 has been found to be an independent predictor of cancer recurrence with both colorectal cancer and NSCLC.[135,154] Several isoforms are expressed by tumor cells, and serum levels have been reported to have prognostic significance in gastric cancer.[135,150] CD44v expression, related to tumor progression, metastasis, and prognosis, has been detected in cancers of the lung, colon, esophagus, liver, cervix, and kidney, as well as reticulo-sarcoma.[148]

The secretion of E-selectin, an endothelial ligand for sailyl carbohydrate antigens expressed on the surface of tumor cells, is reportedly stimulated by tumor cells, and serum E-selectin titers have reportedly correlated with prognosis in breast cancer, and, in a few small studies, in NSCLC.[135] D'Amico et al., examining 196 stage I NSCLC patients, examined serial serum samples and found that decreasing serum E-selectin levels ($P = 0.002$), increasing serum C44 levels ($P = 0.001$), and increasing serum urokinase plasminogen activator receptor ($P = 003$) predicted NSCLC recurrence before clinical or radiographic determination.[135] Future research regarding such markers might one day allow for the prediction of NSCLC recurrence in patients who have undergone resection therapy, so that earlier and improved systemic therapy might be provided.

CD44, expressed in a large number of normal and malignant tissues, interacts with its ligands, hyaluronate and osteopontin, and is involved in lymphocyte homing, as well as T-lymphocyte activation and tumor metastasis.[155] Several CD44 isoforms have been identified, arising from mRNA alternative spicing, with CD44s, also termed CD44h, being the isoform expressed with hematopoietic cells, fibroblasts, glial cells, and melanoma cells.[155] The isoform generally identified on epithelial cells and epithelial tumor cells is high molecular weight CD44, containing one or more alternatively spliced exons.[155] CD44 isoform alterations are associated with the transformation of normal cells into cancer cells.[155] CD44 isoforms, including high molecular weight CD44 isoforms such as CD44v6, have been shown to be overexpressed in non-small cell lung cancers.[155] CD44 isoform function in tumors varies with the organ involved.[155] CD44S overexpression in melanoma and lymphoma enhances tumorigenicity, whereas upregulation of CD44S in colon cancer reduces tumorigenicity.[155] CD44 expression reduction in non-small cell lung cancer has been shown to be associated with lymph node metastases and shortened disease-free survival.[155] The interplay between CD44 and extracellular matrix proteins such as osteopontin may regulate osteoclastic motility, responsible for osteolytic metastases.[156]

Conclusion

Further research regarding the various mechanisms of action of each of the cell adhesion molecules, and their interactions with other molecular pathways, in the development and spread of cancer is ongoing. For example, the effect of glycosylation of cell adhesion molecules on their cell–cell and cell–extracellular matrix interactions are being studied.[157] Ultimately, more effective cancer prognostic markers and therapeutic regimens may arise from these studies.

References

1. Aplin AE, Howe A, Alahari SK, Juliano RL. Signal transduction and signal modulation by cell adhesion receptors: the role of integrins, cadherins, immunoglobulin-cell adhesion molecules, and selectins. *Pharmacol Rev*. 1998;50:197–263.
2. Gumbiner BM. Cell adhesion: the molecular basis of tissue architecture and morphogenesis. *Cell*. 1996;84:345–357.
3. Gogali A, Charalabopoulos K, Constantopoulos S. Integrin receptors in primary lung cancer. *Exp Oncol*. 2004;26:106–110.
4. Petruzzelli L, Takami M, Humes HD. Structure and function of cell adhesion molecules. *Am J Med*. 1999;106:467–476.
5. Nair KS, Naidoo R, Chetty R. Expression of cell adhesion molecules in oesophageal carcinoma and its prognostic value. *J Clin Pathol*. 2005;58:343–351.
6. Charalabopoulos K, Gogali A, Kostoula OK, Constantopoulos SH. Cadherin superfamily of adhesion molecules in primary lung cancer. *Exp Oncol*. 2004;26:256–260.
7. Meyer T, Hart IR. Mechanisms of tumour metastasis. *Eur J Cancer*. 1998;34:214–221.
8. Mousa SA. Cell adhesion molecules: potential therapeutic & diagnostic implications. *Mol Biotechnol*. 2008;38:33–40.
9. Lyons AJ, Jones J. Cell adhesion molecules, the extracellular matrix and oral squamous carcinoma. *Int J Oral Maxillofac Surg*. 2007;36:671–679.
10. Rosales C, O'Brian V, Kornberg L, Juliano RL. Signal transduction by cell adhesion receptors. *Biochim Biophys Acta*. 1995;1242: 77–98.
11. Nicolson GL. Cancer metastasis: tumor cell and host organ properties important in metastasis to specific secondary sites. *Biochim Biophys Acta*. 1988;948:175–224.
12. Hynes RO. Integrins: versatility, modulation, and signalling in cell adhesion. *Cell*. 1992;69:11–25.

13. Han SW, Roman J. COX-2 inhibitors suppress integrin α5 expression in human lung carcinoma cells through activation of Erk: involvement of Sp1 and AP-1 sites. *Int J Cancer*. 2005;116: 536–546.
14. Watt FM. Role of integrins in regulating epidermal adhesion, growth and differentiation. *EMBO J*. 2002;21:3919–3926.
15. Okegawa T, Li Y, Pong RC, Hsieh JT. Cell adhesion proteins as tumor suppressors. *J Urol*. 2002;167:1836–1843.
16. Mukhopadhyay NK, Gordon GJ, Chen CJ, et al. Activation of focal adhesion kinase in human lung cancer cells involves multiple and potentially parallel signaling events. *J Cell Mol Med*. 2005;9:387–397.
17. Damjanovich L, Albelda SM, Mette SA. Distribution of integrin cell adhesion receptors in normal and malignant lung tissue. *Am J Respir Cell Mol Biol*. 1992;6:197–206.
18. Hanby AN, Gilet CE, Pignatelli M, Stamp GW. Beta1 and beta4 integrin expression in metacarn and formalin fixed material from in situ ductal carcinoma of the breast. *J Pathol*. 1993;171:257–262.
19. Kitayama J, Nayawa H, Nakayama H, et al. Functional expression of beta1 and beta2 integrins on tumor infiltrating lymphocytes (TILs) in colorectal cancer. *J Gastroenterol*. 1999;34:327–333.
20. Bankhof H, Stein V, Remberger K. Differential expression of a6 and a2 very late-antigen integrins in the normal, hyperplastic and neoplastic prostate. *Hum Pathol*. 1993;24:243–248.
21. Damkisson YP, Wilding JC, Filipe M, Hall PA, Pignatelli M. Cell–matrix interactions in gastric carcinoma. *J Pathol*. 1993; 169:120.
22. Elenrieder V, Alder G, Gress TM. Invasion and metastasis in pancreatic cancer. *Ann Oncol*. 1999;4:46–50.
23. Volpes R, Van der Oord J, Pesmet VJ. Distribution of the VLA family of integrins in normal and pathological human liver tissue. *Gastroenterology*. 1991;101:200–206.
24. Bichler KH, Wechsel HW. The problematic nature of metastasized cell carcinoma. *Anticancer Res*. 1999;19:1463–1466.
25. Stamb GW, Pignatelli M. Distribution of β1, α1, α2 and α3 integrin chains in basal cell carcinomas. *J Pathol*. 1991;103:307–313.
26. Strobel T, Cannisha SA. Beta-1 integrins partly mediate binding of ovarian cancer cells to perimetral mesothelium in vitro. *Gynecol Oncol*. 1999;73:362–367.
27. Vessey BA, Albelda S, Buck CA, et al. Distribution of integrin cell adhesion molecules in endometrial cancer. *Am J Pathol*. 1995;146:717–726.
28. Hehlgans S, Haase M, Cordes N. Signalling via integrins: implications for cell survival and anticancer strategies. *Biochim Biophys Acta*. 2007;1775:163–180.
29. Yamada KM, Kennedy DW, Yamada SS, et al. Monoclonal antibody and synthetic peptide inhibitors of human tumor cell migration. *Cancer Res*. 1990;50:4485–4496.
30. Albelda SM, Mette SA, Elder DE, et al. Integrin distribution in malignant melanoma: association of the beta 3 subunit with tumor progression. *Cancer Res*. 1990;50:6757–6764.
31. Hamann A, Andrew DP, Pablonski-Westrich D, Holzmann B, Butcher EC. Role of alpha-4-integrins in lymphocyte homing to mucosal tissues in vivo. *J Immunol*. 1994;152:3282–3293.
32. Yednock TA, Cannon C, Fritz LC, et al. Prevention of experimental autoimmune encephalomyelitis by antibodies against alpha-4-beta-1 integrin. *Nature*. 1992;356:63–66.
33. Springer TA. Traffic signals on endothelium for lymphocyte recirculation and leukocyte emigration. *Annu Rev Physiol*. 1995;57:827–872.
34. Hartmann TJ, Burger JA, Gloded A, Fujii N, Burger M. CXCR4 chemokine receptor and integrin signaling co-operate in mediating adhesion and chemoresistance in small cell lung cancer (SCLC) cells. *Oncogene*. 2005;24:4462–4471.
35. Dai DL, Makretsov N, Campos EI, et al. Increased expression of integrin-linked kinase is correlated with melanoma progression and poor patient survival. *Clin Cancer Res*. 2003;9:4409–4414.
36. Graff JR, Deddens JA, Konicek BW, et al. Integrin-linked kinase expression increases with prostate tumor grade. *Clin Cancer Res*. 2001;7:1987–1991.
37. Marotta A, Tan C, Gray V, et al. Dysregulation of integrin-linked kinase (ILK) signaling in colonic polyposis. *Oncogene*. 2001;20:6250–6257.
38. Takanami I. Increased expression of integrin-linked kinase is associated with shorter survival in non-small cell lung cancer. *BMC Cancer*. 2005;5:1.
39. Dedhar S. Cell-substrate interactions and signaling through ILK. *Curr Opin Cell Biol*. 2000;12:250–256.
40. Wu C, Dedhar S. Integrin-linked kinase (ILK) and its interactors: a new paradigm for the coupling of extracellular matrix to actin cytoskeleton and signaling complexes. *J Cell Biol*. 2001;155: 505–510.
41. Liotta LA, Stetler-Stevenson WG. Tumor invasion and metastasis: an imbalance of positive and negative regulation. *Cancer Res*. 1991;51:5054–5059.
42. Furuta SM, Ilic D, Kanazawa S, et al. Mesodermal defect in late phase of gastrulation by a targeted mutation of focal adhesion kinase FAK. *Oncogene*. 1995;11:1989–1995.
43. Carelli S, Zadra G, Vaira V, et al. Up-regulation of focal adhesion kinase in non-small cell lung cancer. *Lung Cancer*. 2006;53: 263–271.
44. Sanders MA, Basson MD. Collagen IV-dependent ERK activation in human Caco-2 intestinal epithelial cells requires focal adhesion kinase. *J Biol Chem*. 2000;275:38040–38047.
45. Guan JL, Shalloway D. Regulation of pp1256FAK both by cellular adhesion and by oncogenic transformation. *Nature*. 1992;358: 690–692.
46. Bhattacharjee A, Richards WG, Staunton J, et al. Classification of human lung carcinomas by mRNA expression profiling reveals distinct adenocarcinoma subclasses. *Proc Natl Acad Sci USA*. 2001;98:13790–13795.
47 Yeoh EJ, Ross ME, Shurtleff SA, et al. Classification, subtype discovery, and prediction of outcome in pediatric acute lymphoblastic leukemia by gene expression profiling. Cancer Cell. 2002; 133-143.
48. Agochiya M, Brunton VG, Owens DW, et al. Increased dosage and amplification of the focal adhesion kinase gene in human cancer cells. *Oncogene*. 1999;18:5646–5653.
49. Hauck CR, Hsia DA, Ilic D, Schlaepfer DD. V-Src SH3-enhanced interaction with focal adhesion kinase at beta 1 integrin-containing invadopodia promotes cell invasion. *J Biol Chem*. 2002;277:12487–12490.
50. Hsia DA, Mitra SK, Hauck CR, et al. Differential regulation of cell motility and invasion by FAK. *J Cell Biol*. 2003;160:753–767.
51. Sethi T, Rintoul RC, Moore SM, et al. Extracellular matrix proteins protect small cell lung cancer cells against apoptosis: a

mechanism for small cell lung cancer growth and drug resistance in vivo. *Nat Med.* 1999;5:662–668.
52. Aoudjit F, Vuori K. Integrin signaling inhibits paclitaxel-induced apoptosis in breast cancer cells. *Oncogene* 2001;20: 4995–5004.
53. Cordes N. Overexpression of hyperactive integrin-linked kinase leads to increased cellular radiosensitivity. *Cancer Res.* 2004;64: 5683–5692.
54. Takeichi M. Cadherin cell adhesion receptors as morphogenetic regulator. *Science.* 1991;251:1451–1459.
55. Kemler R. From cadherins to catenins: cytoplasmic protein interactions and regulation of cell adhesion. *Trends Genet.* 1993;9: 317–321.
56. Ozawa M, Baribault H, Kemler R. The cytoplasmic domain of the cell adhesion molecule uvamorulin associates with three independent proteins structurally related in different species. *EMBO J.* 1989;8:1711–1717.
57. Hirano S, Kinoto N, Shimoyama Y, Hirohashi S, Takeichi M. Identification of a neural α-catenin as a key regulator of cadherin function and multicellular organization. *Cell.* 1992;70:293–301.
58. Breier G, Breviario F, Caveda L, et al. Molecular cloning and expression of murine vascular endothelial cadherin in early stage development of cardiovascular system. *Blood.* 1996;87:630–641.
59. Mayer B, Johnson JP, Leitl F, et al. E-cadherin expression in primary and metastatic gastric cancer: down-regulation correlates with cellular dedifferentiation and glandular disintegration. *Cancer Res.* 1993;53:1690–1695.
60. Schipper JH, Frixen UH, Behrens J, et al. E-cadherin expression in squamous cell carcinomas of head and neck: inverse correlation with tumor dedifferentiation and lymph node metastasis. *Cancer Res.* 1991;51:6328–6337.
61. Bringuier PP, Umbas R, Schaafsma HE, et al. Decreased E-cadherin immunoreactivity correlates with poor survival in patients with bladder tumors. *Cancer Res.* 1993;53:2341–2345.
62. Dorudi S, Sheffield JP, Poulsom R, Northover JM, Hart IR. E-cadherin expression in colorectal cancer. An immunocytochemical and in situ hybridization study. *Am J Pathol.* 1993;142:981–986.
63. Oka H, Shiozaki H, Kobayashi K, et al. Expression of E-cadherin cell adhesion molecules in human breast cancer tissues and its relationship to metastasis. *Cancer Res.* 1993;53:1696–1701.
64. Ochiai A, Akimoto S, Shimoyama Y, et al. Frequent loss of α-catenin expression in scirrhous carcinomas with scattered cell growth. *Jpn J Cancer Res.* 1994;85:266–273.
65. Kadowaki T, Shiozaki H, Inoue M, et al. E-cadherin and α-catenin expression in human esophageal cancer. *Cancer Res.* 1994;54: 291–296.
66. Takayama T, Shinozaki H, Shibamoto S, et al. β-Catenin expression in human cancer. *Am J Pathol.* 1996;148:39–46.
67. Benjamin JM, Nelson WJ. Bench to bedside and back again: molecular mechanisms of alpha-catenin function and roles in tumorigenesis. *Semin Cancer Biol.* 2008;18:53–64.
68. Gavert N, Ben-Ze'ev A. Beta-catenin signaling in biological control and cancer. *J Cell Biochem.* 2007;1102:820–828.
69. Avizienyte E, Frame MC. Src and FAK signalling controls adhesion fate and the epithelial-to-mesenchymal transition. *Curr Opin Cell Biol.* 2005;17:542–547.
70. Hajra KM, Fearon ER. Cadherin and catenin alterations in human cancer. *Genes Chromosomes Cancer.* 2002;34:255–268.
71. Kanai Y, Oda T, Shimoyama Y, et al. Alterations of the cadherin–catenin cell adhesion system in cancers. *Princess Takamatsu Symp.* 1994;24:51–62.
72. Watabe M, Nagafuchi A, Tsukita S, Takeichi M. Induction of polarized cell-cell association and retardation of growth by activation of the E-cadherin-catenin adhesion system in a dispersed carcinoma line. *J Cell Biol.* 1994;127:247–256.
73. Lien WH, Klezovich O, Fernandez TE, Delrow J, Vasioukhin V. Alpha E-catenin controls cerebral cortical size by regulating the hedgehog signaling pathway. *Science.* 2006;311:1609–1612.
74. Vasioukhin V, Bauer C, Degenstein L, Wise B, Fuchs E. Hyperproliferation and defects in epithelial polarity upon conditional ablation of alpha-catenin in skin. *Cell.* 2001;104:605–617.
75. Kato Y, Hirano T, Yoshida K, et al. Frequent loss of E-cadherin and/or catenins in intrabronchial lesions during carcinogenesis of the bronchial epithelium. *Lung Cancer.* 2005;48:323–330.
76. Takeichi M. Functional correlation between cell adhesive properties and some cell surface proteins. *J Cell Biol.* 1997;75:464–474.
77. Hirano S, Nose A, Hatta K, et al. Calcium dependent cell-cell adhesion molecules (cadherins). *J Cell Biol.* 1987;105:2501–2510.
78. Scher RL, Koch WM, Richtsmeier WJ. Induction of the intercellular adhesion molecule (ICAM-1) on squamous carcinoma by interferon gamma. *Arch Otolaryngol Head Neck Surg.* 1993;119: 432–438.
79. Takeichi M. Cadherins: a molecular family important in selective cell–cell adhesion. *Annu Rev Biochem.* 1990;59:237–252.
80. Behrens J, Mareel MM, Van Roy FM, Birchmeier W. Dissecting tumor cell invasion: epithelial cells acquire invasive properties after the loss of uvomorulin-mediated cell–cell adhesion. *J Cell Biol.* 1989;108:2435–2447.
81. Frixen UH, Behrens J, Sachs M, et al. E-cadherin-mediated cell–cell adhesion prevents invasiveness of human carcinoma cells. *J Cell Biol.* 1991;113:173–185.
82. Vleminckx K, Vakaet L Jr, Mareel M, Fiers W, Van Roy F. Genetic manipulation of E-cadherin expression by epithelial tumor cells reveals an invasion suppressor role. *Cell.* 1991;66:107–119.
83. Hirohashi S. Inactivation of the E-cadherin-mediated cell adhesion system in human cancers. *Am J Pathol.* 1998;153:333–339.
84. Akimoto S, Ochiai A, Inomata M, Hirohashi S. Expression of cadherin–catenin cell adhesion molecules, phosphorylated tyrosine residues and growth factor receptor-tyrosine kinases in gastric cancer. *Jpn J Cancer Res.* 1998;89:829–836.
85. Shimoyama Y, Nagafuchi A, Fujita S, et al. Cadherin dysfunction in a human cancer line: possible involvement of loss of α-catenin expression in reduced cell-cell adhesiveness. *Cancer Res.* 1992;52:5770–5774.
86. Oda T, Kanai Y, Shimoyama Y, et al. Cloning of the human α-catenin cDNA and its aberrant mRNA in a human cancer cell line. *Biochem Biophys Res Commun.* 1993;193: 897–904.
87. Mattijssen V, Peters HM, Schalwijk L, et al. E-cadherin expression in head and neck squamous-cell carcinoma is associated with clinical outcome. *Int J Cancer.* 1993;55:580–585.
88. Umbas R, Isaacs WB, Bringuier PP, et al. Decreased E-cadherin expression is associated with poor prognosis in prostate cancer. *Cancer Res.* 1994;15:3929–3933.

89. Nakanishi Y, Ochiai A, Akimoto S, et al. Expression of E-cadherin, α-catenin, β-catenin, and plakoglobinin esophageal carcinomas and its prognostic significance: immunohistochemical analysis of 96 lesions. *Oncology*. 1997;54:158–165.
90. Oyama T, Kanai Y, Ochiai A, et al. A truncated β-catenin disrupts the interaction between E-cadherin and α-catenin: a cause of loss of intercellular adhesiveness in human cancer cell lines. *Cancer Res*. 1994;54:6282–6287.
91. Hoschuetzky H, Aberle H, Kemler R. β-Catenin mediates the interaction of the cadherin-catenin complex with epidermal growth factor receptor. *J Cell Biol*. 1994;127:1375–1380.
92. Ochiai A, Akimoto S, Kanai Y, et al. c-erbB-2 gene product associates with catenins in human cancer cells. *Biochem Biophys Res Commun*. 1994;205:73–78.
93. Kanai Y, Ochiai A, Shibata T, et al. c-erbB-2 gene product directly associates with β-catenin and plakoglobin. *Biochem Biophys Res Commun*. 1995;208:1067–1072.
94. Dohadwala M, Yan SC, Luo J, et al. Cyclooxygenase-2-dependent regulation of E-cadherin: prostaglandin E2 induces transcriptional repressors ZEB1 and Snail in non-small cell lung cancer. *Cancer Res*. 2006;66:5338–5345.
95. Brown JR, DuBois RN. Cyclooxygenase as a target in lung cancer. *Clin Cancer Res*. 2004;10:4266s–4269s.
96. Dubinett SM, Sharma S, Huang M, et al. Cyclooxygenase-2 in lung cancer. *Prog Exp Tumor Res*. 2003;37:138–162.
97. Dannenberg AJ, Zakim D. Chemoprevention of colorectal cancer through inhibition of cyclooxygenase-2. *Semin Oncol*. 1999;26:499–504.
98. Pold M, Zhu L, Sharma S, et al. Cyclooxygenase-2-dependent expression of angiogenic CXC chemokines ENA-78/CXC ligand (CXCL) 5 and interleukin 8/CXCL8 in human non-small cell lung cancer. *Cancer Res*. 2004;64:1853–1860.
99. Hilda T, Yatabe Y, Achiwa H, et al. Increased expression of cyclooxygenase 2 occurs frequently in human lung cancers, specifically in adenocarcinomas. *Cancer Res*. 1998;58:3761–3764.
100. Huang M, Stolina M, Sharma S, et al. Non-small cell lung cancer cyclooxygenase-2 dependent regulation of cytokine balance in lymphocytes and macrophages: up-regulation of interleukin 10 and down-regulation of interleukin 12 production. *Cancer Res*. 1998;58:1208–1216.
101. Sharma S, Yang SC, Zhu L, et al. Tumor cyclooxygenase-2/prostaglanding E2-dependent promotion of FOXP3expression and $CD4^+$ $CD25^+$ T regulatory cell activities in lung cancer. *Cancer Res*. 2005;65:5211–5220.
102. Sharma S, Zhu L, Yang SC, et al. Cyclooxygenase 2 inhibition promotes IFN-γ-dependent enhancement of antitumor responses. *J Immunol*. 2005;175:813–819.
103. Jungck M, Grunhage F, Spengler U, et al. E-cadherin expression is homogeneously reduced in adenoma from patients with familial adenomatous polyposis: an immunohistochemical study of E-cadherin, β-catenin and cyclooxygenase-2 expression. *Int J Colorectal Dis*. 2004;19:438–445.
104. Riedl K, Krysan K, Pold M, et al. Multifaceted roles of cyclooxygenase-2 in lung cancer. *Drug Resist Updat*. 2004;7:169–184.
105. Cavallaro U, Christofori G. Cell adhesion and signaling by cadherins and Ig-CAMs in cancer. *Natl Rev Cancer*. 2004;4:118–132.
106. Choi YS, Shim YM, Kim SH, et al. Prognostic significance of E-cadherin and β-catenin in resected stage I non-small cell lung cancer. *Eur J Cardiothorac Surg*. 2003;24:441–449.
107. Liu D, Huang C, Kameyama K, et al. E-cadherin expression associated with differentiation and prognosis in patients with non-small cell lung cancer. *Ann Thorac Surg*. 2001;71:949–951.discussion 954-955.
108. Witta SE, Gemmill RM, Hirsch FR, et al. Restoring E-cadherin expression increases sensitivity to epidermal growth factor receptor inhibitors in lung cancer cell lines. *Cancer Res*. 2006;66:944–950.
109. Fukuoka M, Yano S, Giaccone G, et al. Multi-institutional randomized phase II trial of gefitinib for previously treated patients with advanced non-small cell lung cancer (the IDEAL 1 Trial). *J Clin Oncol*. 2003;21:2237–2246.
110. Kris MG, Natale RB, Herbst RS, et al. Efficacy of gefitinib, an inhibitor of the epidermal growth factor receptor tyrosine kinase, in symptomatic patients with non-small cell lung cancer: a randomized trial. *JAMA*. 2003;290:2149–2158.
111. Shepherd FA, Rodrigues P, Jose C, et al. The National Cancer Institute of Canada Clinical Trials Group. Erlotinib in previously treated non-small-cell lung cancer. *N Engl J Med*. 2005;353:123–132.
112. Paez JG, Janne PA, Lee JC, et al. EGFR mutations in lung cancer: correlation with clinical response to gefitinib therapy. *Science*. 2004;304:1497–1500.
113. Tsao MS, Sakurada A, Cutz JC, et al. Erlotinib in lung cancer – molecular and clinical predictors of outcome. *N Engl J Med*. 2005;353:133–144.
114. Cappuzzo F, Magrini E, Ceresoli GL, et al. Akt phosphorylation and gefitinib efficacy in patients with advanced non-small-cell lung cancer. *J Natl Cancer Inst*. 2004;96:1133–1141.
115. Ohira T, Gemmill RM, Ferguson K, et al. WNT7a induces E-cadherin in lung cancer cells. *Proc Natl Acad Sci USA*. 2003;100:10429–10434.
116. Verschuren K, Remacle JE, Collart C, et al. SIP1, a novel zinc finger/homeodomain repressor, interacts with Smad proteins and binds to 5'-CACCT sequences in candidate target genes. *J Biol Chem*. 1999;27:20489–20498.
117. Chinnadurai G. CtBP, an unconventional transcriptional corepressor in development and oncogenesis. *Mol Cell*. 2002;9:213–224.
118. Zhong Y, Delgado Y, Gomez J, Lee SW, Perez-Soler R. Loss of H-cadherin protein expression in human non-small cell lung cancer is associated with tumorigenicity. *Clin Cancer Res*. 2001;7:1683–1687.
119. Agiostratidou G, Julit J, Phillips GR, Hazan RB. Differential cadherin expression: potential markers for epithelial to mesenchymal transformation during tumor progression. *J Mammary Gland Biol Neoplasia*. 2007;12:127–133.
120. Thiery JP. Epithelial–mesenchymal transitions in development and pathologies. *Curr Opin Cell Biol*. 2003;15:740–746.
121. Kneuer C, Ehrhardt C, Radomski MW, Bakowsky U. Selectins-potential pharmacological targets. *Drug Discov Today* 2006;11:1034–1040.
122. McEver RP. Properties of GMP-140, an inducible granule membrane protein of platelets and endothelium. *Blood Cells*. 1990;16:73–80.

123. Carlos TM, Harlan JM. Leukocyte-endothelial adhesion molecules. *Blood.* 1994;84:2068–2101.
124. Springer TA. Adhesion receptors of the immune system. *Nature.* 1990;346:425–434.
125. Springer TA. Traffic signals for lymphocyte recirculation and leukocyte emigration: the multistep paradigm. *Cell.* 1994;76: 301–314.
126. Lowe JB, Ward PA. Therapeutic inhibition of carbohydrate–protein interactions in vivo. *J Clin Invest.* 1997;100:S47–S51.
127. Li L, Short HJ, Qian KX, et al. Characterization of glycoprotein ligands for P-selectin on a human small cell lung cancer cell line NCI-H345. *Biochem Biophys Res Commun.* 2001;288:637–644.
128. Tedder TF, Steeber DA, Chen A, Engle P. The selectins: vascular adhesion molecules. *FASEB J.* 1995;9:866–873.
129. Kim YG, Kim MJ, Lim JS, et al. ICAM-3-induced cancer cell proliferation through the PI3K/Akt pathway. *Cancer Lett.* 2006;239:103–110.
130. Stone JP, Wagner DD. P-selectin mediates adhesion of platelets to neuroblastoma and small cell lung cancer. *J Clin Invest.* 1993;92:804–813.
131. Yu CJ, Shih JY, Lee YC, et al. Sialyl Lewis antigens: association with MUC5AC protein and correlation with post-operative recurrence of non-small cell lung cancer. *Lung Cancer.* 2005; 47:59–67.
132. Rosen SD, Bertozzi CR. The selectins and their ligands. *Curr Opin Cell Biol.* 1994;6:663–673.
133. Izumi Y, Taniuchi Y, Tsuji T, et al. Characterization of human colon carcinoma variant cells selected for sialyl Le^x carbohydrate antigen: liver colonization and adhesion to vascular endothelial cells. *Exp Cell Res.* 1995;216:215–221.
134. Laack E, Nikbakht H, Peters A, et al. Expression of CEACAM1 in adenocarcinoma of the lung: a factor of independent prognostic significance. *J Clin Oncol.* 2002;20:4279–4284.
135. D'Amico TA, Brooks KR, Joshi MBM, et al. Serum protein expression predicts recurrence in patients with early-stage lung cancer after resection. *Ann Thorac Surg.* 2006;81: 1982–1987.
136. Rosselli M, Mineo TC, Martini F, et al. Soluble selectin levels in patients with lung cancer. *Int J Biol Marker.* 2002;17:56–62.
137. Tsumatori G, Ozeki Y, Takagi K, Ogata T, Tanaka S. Relation between the serum E-selectin level and the survival rate of patients with resected non-small cell lung cancers. *Jpn J Cancer Res.* 1999;90:301–307.
138. Gao Y, Wei M, Zheng S, Ba X. Chemically modified heparin inhibits the in vitro adhesion of nonsmall cell lung cancer cells to P-selectin. *J Cancer Res Clin Oncol.* 2006;132:257–264.
139. Borsig L, Wong R, Hynes RQ, Varki NM, Varki A. synergistic effects of L- and P-selectin in facilitating tumor metastasis can involve non-mucin ligands and implicate leukocytes as enhancers of metastasis. *Proc Natl Acad Sci. USA.* 2002;99:2193–2198.
140. Borsig L, Wong R, Feramisco J, et al. Heparin and cancer revisited: mechanistic connections involving platelets, P-selectin, carcinoma mucins, and tumor metastasis. *Proc Natl Acad Sci USA.* 2001;98:3352–3357.
141. Tessier-Lavigne M, Goodman CS. The molecular biology of axon guidance. *Science.* 1996;274:1123–1133.
142. van Buul JD, Mul FP, van der Schoot CE, Hordijk PL. ICAM-3 activation modulates cell–cell contacts of human bone marrow endothelial cells. *J Vasc Res.* 2004;41:28–37.
143. Chung YM, Kim GB, Park CS, et al. Increased expression of ICAM3 is associated with radiation resistance in cervical cancer. *Int J Cancer.* 2005;117:194–201.
144. Yasuda M, Tanaka Y, Tamura M, et al. Stimulation of β1 integrin down-regulates ICAM-1 expression and ICAM-1-dependent adhesion of lung cancer cells through focal adhesion kinase. *Cancer Res.* 2001;61:2022–2030.
145. Schwartz RH. Models of T cell anergy: is there a common molecular mechanism? *J Exp Med.* 1996;184:1–8.
146. Shen J, Devery JM, King NJ. Adherence status regulates the primary cellular activation responses to the *Flavivirus*, West Nile. *Immunology.* 1995;84:254–264.
147. Mukai S, Kagamu H, Shu S, Plautz GE. Critical role of CD11a (LFA-1) in therapeutic efficacy of systemically transferred antitumor effector T cells. *Cell Immunol.* 1999;192:122–132.
148. Liu YJ, Yan PS, Li J, Jia JF. Expression and significance of CD44s, CD44v6, and nm23 mRNA in human cancer. *World J Gastroenterol.* 2005;11:6601–6606.
149. Liu J, Jiang G. CD44 and hematologic malignancies. *Cell Mol Immunol.* 2006;3:359–365.
150. Georgolios A, Batistatou A, Charalabopoulos A, Manolopoulos L, Charalabopoulos K. The role of CD44 adhesion molecule in oral cavity cancer. *Exp Oncol.* 2006;28:94–98.
151. Yasuda M, Nakano K, Yasumoto K, Tanaka Y. CD44: functional relevance to inflammation and malignancy. *Histol Histopathol.* 2002;17:945–950.
152. Guo, YJ, Liu G, Wang X, et al. Potential use of soluble CD44 in serum as indicator of tumor burden and metastasis in patients with gastric or colon cancer. *Cancer Res.* 1994;54:422–426.
153. Adamia S, Maxwell CA, Pilarski LM. Hyaluronan and hyaluronan synthases: potential therapeutic targets in cancer. *Curr Drug Targets Cardiovasc Haematol Disord.* 2005;5:3–14.
154. Wielenga VJM, Heider KH, Offerhaus JA, et al. Expression of cD44 variant proteins in human colorectal cancer is related to tumor progression. *Cancer Res.* 1993;53:4754–4756.
155. Takahashi K, Takahashi F, Hirama M, Tanabe KK, Fukuchi Y. Restoration of CD44S in non-small cell lung cancer cells enhanced their susceptibility to the macrophage cytotoxicity. *Lung Cancer.* 2003;41:145–153.
156. Li M, Amizuka N, Takeuchi K, et al. Histochemical evidence of osteoclastic degradation of extracellular matrix in osteolytic metastasis originating from human lung small carcinoma SBC-50 cells. *Micro Res Tech.* 2006;69:73–83.
157. Zhao Y, Sato Y, Isaji T, et al. Branched N-glycans regulate the biological functions of integrins and cadherins. *FEBS J.* 2008;275:1939–1948.

4
Apoptosis and Cell Death

Pothana Saikumar and Manjeri A. Venkatachalam

Introduction

Cell death plays a central role in multicellular organisms during their early development in sculpting the body parts and in their adult life by controlling cell numbers (homeostasis), and cell death further protects the organism overall by removing all cells damaged by disease, aging, infection, genetic mutation, and exposure to toxic agents. Kerr et al[1] coined the term "apoptosis," referring to a particular morphology of physiological cell death involving cell shrinkage, nuclear condensation, membrane blebbing, and cellular and nuclear fragmentation into membrane-bound apoptotic bodies (Fig. 4.1). The lipid changes that occur in the membrane eventually lead to phagocytosis of the apoptotic cellular fragments. Often apoptosis is used synonymously with programmed cell death, implying that death results from the regulated activation of a preexisting death program that is encoded in the genome. The first evidence of a genetic program that orchestrates physiological cell death came from the developmental studies of the nematode *Caenorhabditis elegans*.[2] Our knowledge of cell death and the mechanisms of its regulation increased dramatically in the past two decades with the discovery of several death genes in *C. elegans*[3] and their counterparts in mammals. Now it is clear that apoptotic cell death has important biological roles not only in development and homeostasis but also in the pathogenesis of several disease processes. Dysregulation of apoptosis is found in a wide spectrum of human diseases, including cancer, autoimmune diseases, neurodegenerative diseases, ischemic diseases, and viral infections.[4] Interestingly, death programs other than apoptosis with different morphological features were also considered in controlling the cell numbers. These death programs could be directed by the condemned cell itself or by neighboring cells, with or without the help of humoral factors.

Alternative Forms of Cell Death

Previously, cell death was broadly categorized into only two distinct types: apoptosis and necrosis. However, it has become increasingly evident that such a categorization is an oversimplification. There are 12 different types of cell death that have been described in the literature, which can be grouped into only five major types: apoptosis, necrosis, autophagy, paraptosis, and autoschizis. Other forms of death could be classified under one of these headings. For example, anoikis and oncosis are forms of apoptosis (triggered by cell detachment) and necrosis, respectively. Because of overlapping and shared signaling pathways among different death programs, it is difficult to provide exclusive definitions for each one of these cell death programs. However, there is evidence that inhibition of one form of cell death may lead to another. Although the term "programmed cell death" was used specifically to describe apoptosis, other forms of cell death may also fall into this category, if gene activation is required for their execution.

Necrosis

Necrosis results from a variety of accidental and lethal actions by toxins or physical stimuli or in association with pathological conditions, such as ischemia. Necrosis is characterized by cellular edema, dissolution of nuclear chromatin, disruption of the plasma membrane, and release of intracellular contents into the extracellular space, resulting in inflammation (see Fig. 4.1). In contrast, membrane damage occurs very late in the apoptotic process, where dead cells are engulfed by neighboring cells or phagocytes, leading to little or no inflammation (see Fig. 4.1). Although necrosis has mostly been regarded as an accidental form of cell death, more recent

P.T. Cagle and T.C. Allen (eds.), *Basic Concepts of Molecular Pathology*, Molecular Pathology Library 2,
DOI: 10.1007/978-0-387-89626-7_4, © Springer Science+Business Media, LLC 2009

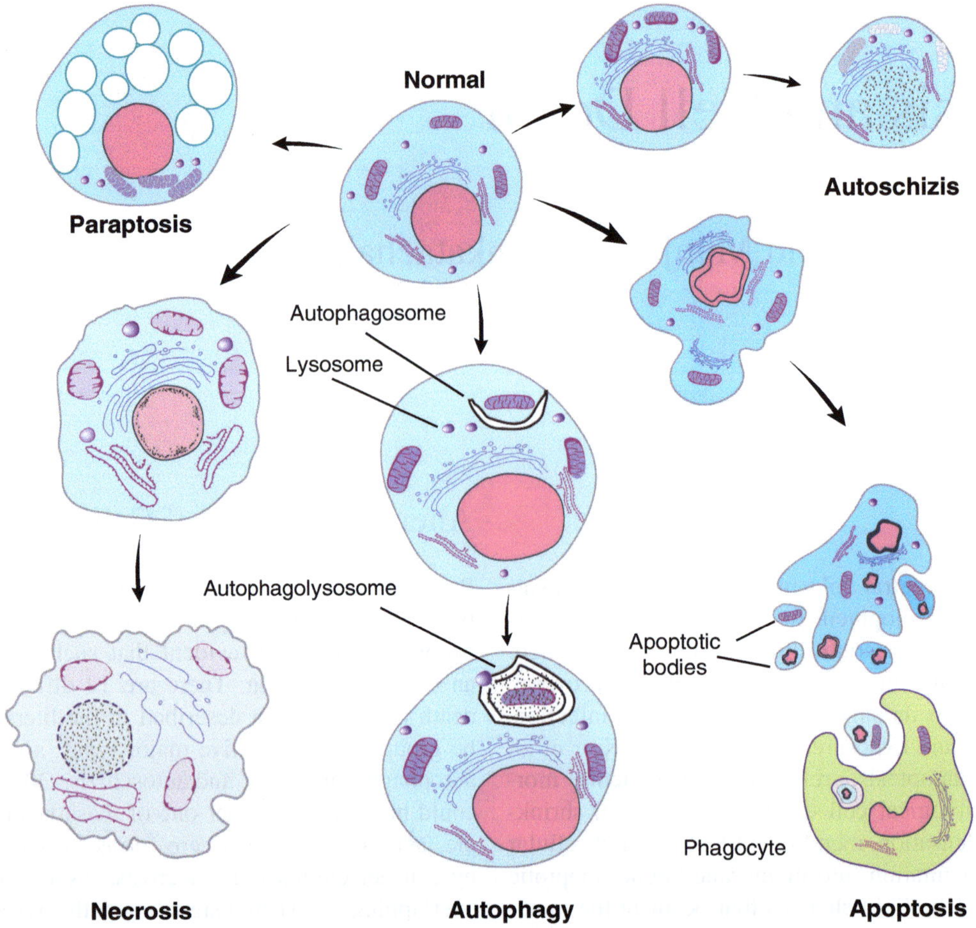

FIG. 4.1. Morphological features of cell death. Necrosis: Cells die by necrosis, and their organelles are characteristically swollen. There is early membrane damage with eventual loss of plasma membrane integrity and leakage of cytosol into extracellular space. Despite early clumping, the nuclear chromatin undergoes lysis (karyolysis). Apoptosis: Cells die by type I programmed cell death (also called apoptosis); they are shrunken and develop blebs containing dense cytoplasm. Membrane integrity is not lost until after cell death. Nuclear chromatin undergoes striking condensation and fragmentation. The cytoplasm becomes divided to form apoptotic bodies containing organelles and/or nuclear debris. Terminally, apoptotic cells and fragments are engulfed by phagocytes or surrounding cells. Autophagy: Cells die by type II programmed cell death, which is characterized by the accumulation of autophagic vesicles (autophagosomes and autophagolysosomes). One feature that distinguishes apoptosis from autophagic cell death is the source of the lysosomal enzymes used for most of the dying cell degradation. Apoptotic cells use phagocytic cell lysosomes for this process, whereas cells with autophagic morphology use the endogenous lysosomal machinery of dying cells. Paraptosis: Cells die by type III programmed cell death, which is characterized by extensive cytoplasmic vacuolization and swelling and clumping of mitochondria, along with absence of nuclear fragmentation, membrane blebbing, or apoptotic body formation. Autoschizis: In this form of cell death, the cell membrane forms cuts or schisms that allow the cytoplasm to leak out. The cell shrinks to about one-third of its original size, and the nucleus and organelles remain surrounded by a tiny ribbon of cytoplasm. After further excisions of cytoplasm, the nuclei exhibit nucleolar segregation, and chromatin decondensation followed by nuclear karyorrhexis and karyolysis.

data have suggested that necrosis can also occur as a programmed form of cell death. There is growing evidence that necrotic and apoptotic forms of cell death may share some similarities.[5] Nevertheless, necrosis has been shown to occur in cells having defective apoptotic machinery or upon inhibition of apoptosis,[6] and this form of cell death is emerging as an important therapeutic tool for cancer treatment.[7] Some forms of cell death may have a mixture of morphological features of both necrosis and apoptosis and are referred to as "aponecrosis."[8]

Autophagy

Autophagy, which is referred to as "*macroautophagy*" or *type II programmed cell death*,[9] is characterized by sequestration of cytoplasm and organelles by double or multimembrane structures called *autophagic vacuoles*, followed by degradation of the contents of these vacuoles by fusing to the cell's own lysosomes (see Fig. 4.1). Although the precise role of autophagy in cell death is not clear, yet it has long been regarded as a cell survival mechanism whereby starving cells produce energy-generating basic molecules from complex polymers such as proteins, lipids, polysaccharides, and nucleic acids through sequestration and degradation of some of the cytoplasm and organelles. In this regard, it is argued that autophagy may help cancer cells to survive under nutrient-limiting and low-oxygen conditions and against ionizing radiation.[10,11] However, recent observations that there is decreased autophagy during experimental carcinogenesis and heterologous disruption of *Beclin 1* (Atg6),[12–14] an autophagy gene, in breast and ovarian cancers suggest that breakdown of autophagic machinery may contribute to development of cancer.[15] In addition, Atg6 heterozygous mice develop tumors spontaneously, suggesting tumor suppressor activity of Atg6.[12] Similarly, mice deficient in another autophagy gene, Atg4C, a cysteine protease, are susceptible to chemical carcinogenesis.[16] However, the mechanism of how autophagy suppresses tumorigenesis is still unclear. Other studies have shed some light on the relationship between autophagy and apoptosis. For example, reduction of *Atg7* and *beclin 1* inhibited zVAD-induced death in human U937 cells[17] and knockdown of Atg5 and *beclin 1* protected Bax/Bak double-knockout cells from staurosporine- or etoposide-induced nonapoptotic death.[18,19] However, these studies were done in cells whose apoptotic pathways had been compromised. Thus, it remains to be seen whether cells with intact apoptotic machinery can also die by autophagy and whether apoptosis-competent cells lacking autophagy genes will be resistant to other death stimuli. In the in vivo setting, the complicated relationship between autophagy and apoptosis needs to be clarified.

Because some autophagic cells would undergo caspase-independent gene-activated cell death but not display any of the characteristic ultrastructural features of apoptosis including DNA laddering, autophagy is still considered as programmed cell death. Similar to apoptosis, cells do require de novo gene expression with an increase in the expression of the ubiquitin-like gene.[20,21] The ubiquitin-like protein conjugation system and formation of protein complexes that direct membrane docking and fusion of the lysosomes and vacuoles are main components of autophagy. Although the molecular details are still being elucidated, this process appears to be regulated by various kinases, phosphatases, and guanosine triphosphatases (GTPases).[22,23]

Paraptosis

Paraptosis was recently described as a form of cell death characterized by extensive cytoplasmic vacuolation (see Fig. 4.1) involving swelling of mitochondria and endoplasmic reticulum. This form of cell death was not inhibited by the caspase inhibitors but is inhibited by translation and transcription inhibitors, cycloheximide and actinomycin D, suggesting a requirement for new protein synthesis.[24] Paraptosis has been shown to be triggered by the tumor necrosis factor (TNF) receptor family TAJ/TROY and insulin-like growth factor I receptor. TAJ/TROY-induced paraptotic cell death was enhanced by overexpression of programmed cell death 5 (PDCD5).[25] This form of cell death was shown to be mediated by mitogen-activated protein (MAP) kinases and inhibited by AIP1/ALIX, a protein interacting with the calcium-binding death-related protein ALG-2.[26] Inhibition by Na^+/H^+ exchanger also led to cell death resembling paraptosis.[27] Recently, glucocorticoids were shown to induce retinal toxicity by mechanisms associated with paraptosis.[28] Although this form of cell death has been shown to be evoked by a variety of reagents, the exact molecular mechanism is far from clear.

Cytoplasmic vacuolation, as seen in paraptosis, has been shown to occur in a wide range of cell lines either spontaneously or as induced by a variety of stimuli. The extent to which a cell becomes vacuolated depends on the cell type. The process of vacuolation seems to follow a definite pattern, with the vacuole number and size increasing gradually. Cells can recover from vacuolation up to a certain threshold, beyond which they succumb to death.[29] The presence of a definite underlying program that decides whether a cell undergoing cytoplasmic vacuolation should be destined to death is currently unknown. However, there are instances in which a decidedly vacuolated cell undergoes cell death, which suggests the existence of some underlying programs. During the salivary gland development in metamorphosis of blowfly larva (*Calliphora vomitoria*), cells are eliminated by a process of intensive vacuolation in a particular order.[30,31] Furthermore, in *Dictyostelium discoideum*, stalk cells undergo extensive cytoplasmic vacuolation while proceeding toward programmed cell death.[32] In addition, cytoplasmic vacuolation was also observed in aspirates of lobular and ductal breast carcinoma.[33] These studies further emphasize that cytoplasmic vacuolation-induced cell death has a physiological role and warrants further investigation. The most studied cytoplasmic vacuolation-induced cell death is autophagy.

Autoschizis

A new form of cell death, which has been shown to be triggered by oxidative stress,[34] differs from apoptosis and necrosis and is characterized by reduction of cytoplasm to a narrow rim around the nucleus with chromatin marginating the entire nucleus from inside.[35] Mitochondria and other organelles aggregate around the nucleus as a consequence of cytoskeletal damage and loss of cytoplasm. Interestingly, the rough endoplasmic reticulum is preserved until the late stages of autoschizis, in which cells fragment and the nucleolus becomes condensed and breaks into smaller fragments.[35] The nuclear envelope dissipates eventually with the remaining organelles after cell death. In this type of death, cells lose cytoplasm by self-morsellation or self-excision (see Fig. 4.1). Autoschizis usually affects contiguous groups of cells both in vitro and in vivo but can also occasionally affect scattered individual cells trapped in subcapsular sinuses of lymph nodes.[36]

Apoptosis

Genetic studies in the nematode worm *C. elegans* led to the initial characterization of apoptosis. Activation of specific death genes during the development of this worm results in death of exactly 131 cells, leaving 959 cells intact.[3] Further studies revealed that apoptosis can be divided into three successive stages: (1) commitment phase, in which death is initiated by specific extracellular or intracellular signals; (2) execution phase; and (3) cleanup phase, in which dead cells are removed by other cells with eventual degradation of the dead cells in the lysosomes of phagocytic cells.[37] The apoptotic machinery is conserved through evolution from worm to human.[38] In *C. elegans*, execution of apoptosis is mediated by CED-3 and CED-4 proteins. Commitment to a death signal results in the activation of CED-3 by CED-4 binding. The CED-9 protein prevents activation of CED-3 by binding to CED-4.[39]

Mechanisms of Apoptosis

The mechanisms of apoptosis are very complex and involve a cascade of energy-requiring molecular events. Past research indicates that there are two main apoptotic pathways: the extrinsic or death receptor pathway, and the intrinsic or mitochondrial pathway. An additional pathway also operates during T-cell-mediated cytotoxicity involving perforin-granzyme-mediated cell killing. The perforin/granzyme pathway can induce apoptosis via either granzyme B or granzyme A. All three pathways then converge on the same terminal execution pathway, which is initiated by the cleavage and activation of effector caspase-3 and -7, resulting in the degradation of cytoplasmic, cytoskeletal, and nuclear proteins, fragmentation of nuclear DNA, formation of apoptotic bodies (membrane blebbing), expression of cell-surface ligands for phagocytosis, and finally uptake by phagocytic cells.

Caspases

The executioners in both intrinsic and extrinsic pathways of cell death are the caspases,[40] which are cysteine proteases with specificity to cleave their substrates after aspartic acid residues. The first member of the caspase family, caspase-1, was initially known as interleukin-1β-converting enzyme (ICE), an enzyme required for the maturation of IL-1β. Later, the C. *elegans* cell death gene ced-3 was found to be similar to ICE and a developmentally regulated protein Nedd2 (now referred to as caspase-2). Subsequently, many caspase genes have been cloned from mammalian and nonmammalian sources. The central role of caspases in apoptosis is underscored by the observation that apoptosis and all classic changes associated with apoptosis can be blocked by inhibition of caspase activity. To date, 12 mammalian caspases (caspase-1 to -10, caspase-14, and mouse caspase-12) have been identified. Caspase-13 was later found to represent a bovine homologue, and caspase-11 appears to be a murine homologue of human caspase-4 and -5, respectively. These caspases have been broadly categorized into initiators (caspase-2, -8, -9, -10), effectors or executioners (caspase-3, -6, -7), and inflammatory caspases (caspase-1, -4, -5). While murine caspase-11 is reported to regulate apoptosis and cytokine maturation during septic shock, caspase-12 appears to mediate endoplasmic reticulum-specific apoptosis and cytotoxicity by amyloid-β. Caspase-14 is highly expressed in embryonic tissues but not in adult tissues.[41]

Caspases are normally produced as inactive zymogens containing an N-terminal prodomain followed by a large and a small subunit that constitute the catalytic core of the protease. They have been categorized into two distinct classes: initiator and effector caspases. The upstream initiator caspases contain long N-terminal prodomains and one of the two characteristic protein–protein interaction motifs: the death effector domain (DED; caspase-8 and -10) and the caspase activation and recruitment domain (CARD; caspase-1, -2, -4, -5, -9, and -12). The downstream effector caspases (caspase-3, -6, and -7) are characterized by the presence of a short prodomain. Apart from the structural differences, a prominent difference between initiator and effector caspases is their basal state. Both the zymogen and the activated forms of effector caspases exist as constitutive homodimers, whereas initiator caspase-9 exists predominantly as a monomer both before and after proteolytic processing.[42] Initiator caspase-8 has been reported to exist in equilibrium between monomers and homodimers.[43] Although the initiator caspases are capable of autocatalytic activation, the activation of effector caspases requires formation of oligomeric complexes with their adapter proteins and often intrachain cleavage within the initiator caspase.

Caspases have also been divided into three categories based on substrate specificity. Group I members (caspase-1, -4, and -5) have a substrate specificity for the WEHD sequence with high promiscuity; group II members (caspase-2, -3, and -7 and CED-3) prefer the DEXD sequence and have an absolute requirement for aspartate (D) at P4; and members of group III (caspase-6, -8, and -9 and the "aspase" granzyme B) have a preference for the (I/L/V)EXD sequences. Several reports have suggested a role for group I members in inflammation and that of group II and III members in apoptotic signaling events.

The roles of various caspases in apoptotic pathways and their relative importance for animal development have been examined in genetic studies involving knockout of different caspase genes (Table 4.1). A caspase-1 (interleukin-1β-converting enzyme, ICE) knockout study suggested that ICE plays an important role in inflammation

TABLE 4.1. Summary of effects of loss of function of caspases in vivo.

Genotype	Phenotype
Caspase-1$^{-/-}$	Viable mice. No defects in cell death. Defects in pro-IL-1b and pro-IL-18 processing and resistant to lipopolysaccharide (LPS)-induced septic shock
Caspase-2$^{-/-}$	Viable and fertile mice. No defects in cell death except during development. Excess germ cells in the female mutant ovaries. Oocytes and neurons show resistance to apoptosis in vitro. Mutant mouse embryonic fibroblasts (MEFs) show some resistance to killing by heat shock and some drugs
Caspase-3$^{-/-}$	Postnatal death in mixed background mutants (129/SvJ and C57BL/6). Mice are viable in C57BL/6 background. Decreased apoptosis in brain
Caspase-7$^{-/-}$	Viable mice. Mild resistance to apoptosis in mutant MEFs
Caspase-3$^{-/-}$ and caspase-7$^{-/-}$	Double-knockout (DKO) mice die perinatally. Exencephaly in about 10% of DKO embryos. Resistant to apoptosis. Defects in cardiac development. MEFs show resistance to apoptosis
Caspase-8$^{-/-}$	Embryonic lethal. Defects in cardiac and T-cell development. Impaired heart muscle development and congested accumulation of erythrocytes. Degeneration of yolk sac vasculature and circulatory failure. Defective hematopoietic progenitor function and macrophage differentiation. Mutant MEFs are resistant to death receptor-mediated apoptosis
Caspase-9$^{-/-}$	Perinatal lethal. Defective brain development associated with decreased apoptosis. Mutant MEFs are resistant to apoptosis
Caspase-11$^{-/-}$	Viable mice. Defects in IL-1 production. Defects in LPS-induced apoptosis
Caspase-12$^{-/-}$	Viable mice. Reduced endoplasmic reticulum (ER) stress-induced apoptosis. Improved bacterial clearance and resistance to sepsis

by activating cytokines such as IL-1β and IL-18. However, caspase-1 was not required to mediate apoptosis under normal circumstances and did not have a major role during development.[44] Surprisingly, ischemic brain injury was significantly reduced in caspase-1 knockout mice compared with wild-type mice,[45] suggesting that inflammation may contribute to ischemic injury. Caspase-3 deficiency leads to impaired brain development and premature death. Also, functional caspase-3 is required for some typical hallmarks of apoptosis such as formation of apoptotic bodies, chromatin condensation, and DNA fragmentation in many cell types.[46] Lack of caspase-8 results in the death of embryos at day 11 with abnormal formation of the heart,[47] suggesting that caspase-8 is required for cell death during mammalian development. In support of this finding, knockout of FADD, which is required for caspase-8 activation, resulted in fetal death with signs of abdominal hemorrhage and cardiac failure.[48] Moreover, caspase-8-deficient cells did not die in response to signals from members of the TNF receptor family.[47] However, cells lacking either FADD or caspase-8, which are resistant to TNF-α-mediated or CD95-mediated death, are susceptible to chemotherapeutic drugs, serum deprivation, ceramide, γ-irradiation, and dexamethasone-induced killing.[48] In contrast, caspase-9 has a key role in apoptosis induced by intracellular activators, particularly those that cause DNA damage. Deletion of caspase-9 resulted in perinatal lethality, apoptotic failure in developing neurons, enlarged brains, and craniofacial abnormalities.[49] In caspase-9-deficient cells, caspase-3 was not activated, suggesting that caspase-9 is upstream of caspase-3 in the apoptotic cascade. As a consequence, caspase-9-deficient cells are resistant to dexamethasone or irradiation, whereas they retain their sensitivity to TNF-α-induced or CD95-induced death[49] because of the presence of caspase-8, the initiator caspase involved in death receptor signaling that can also activate caspase-3. Overall, these observations support the idea that different death signaling pathways converge on downstream effector caspases (Fig. 4.2). Indeed, caspase-3 is regarded as one of the key executioner molecules activated by apoptotic stimuli originating either at receptors for exogenous molecules or within cells through the action of drugs, toxins, or radiation.

A cascade of caspases plays the central executioner role by cleaving various mammalian cytosolic and nuclear proteins that play roles in cell division, maintenance of cytoskeletal structure, DNA replication and repair, RNA splicing, and other cellular processes. This proteolytic carnage produces the characteristic morphological changes of apoptosis. Once the caspase cascade is initiated, the process of cell death has crossed the point of no return.

Extrinsic Death Pathway

The extrinsic pathway involves binding of death ligands such as tumor necrosis factor-α (TNF-α), CD95 ligand (Fas ligand), and TNF-related apoptosis-inducing ligand (TRAIL) to their cognate cell-surface receptors TNFR1, CD95/Fas, TRAIL-R1, TRAIL-R2, and the DR series of receptors,[50] resulting in the activation of initiator caspase-8 (also known as FADD-homologous ICE/CED-3-like protease, or FLICE) and subsequent activation of effector caspase-3 (see Fig. 4.2).[51] The cytoplasmic domains of death receptors contain the "death domain," which plays a crucial role in transmitting the signal from the cell's surface to intracellular signaling molecules. Binding of the ligands to their cognate receptors results in receptor trimerization and recruitment of adapter proteins to the cell membrane, which involves homophilic interactions between death domains of the receptors and the adapter proteins. The adapter protein for the receptors TNFR1 and DR3 is TNFR-associated death domain protein

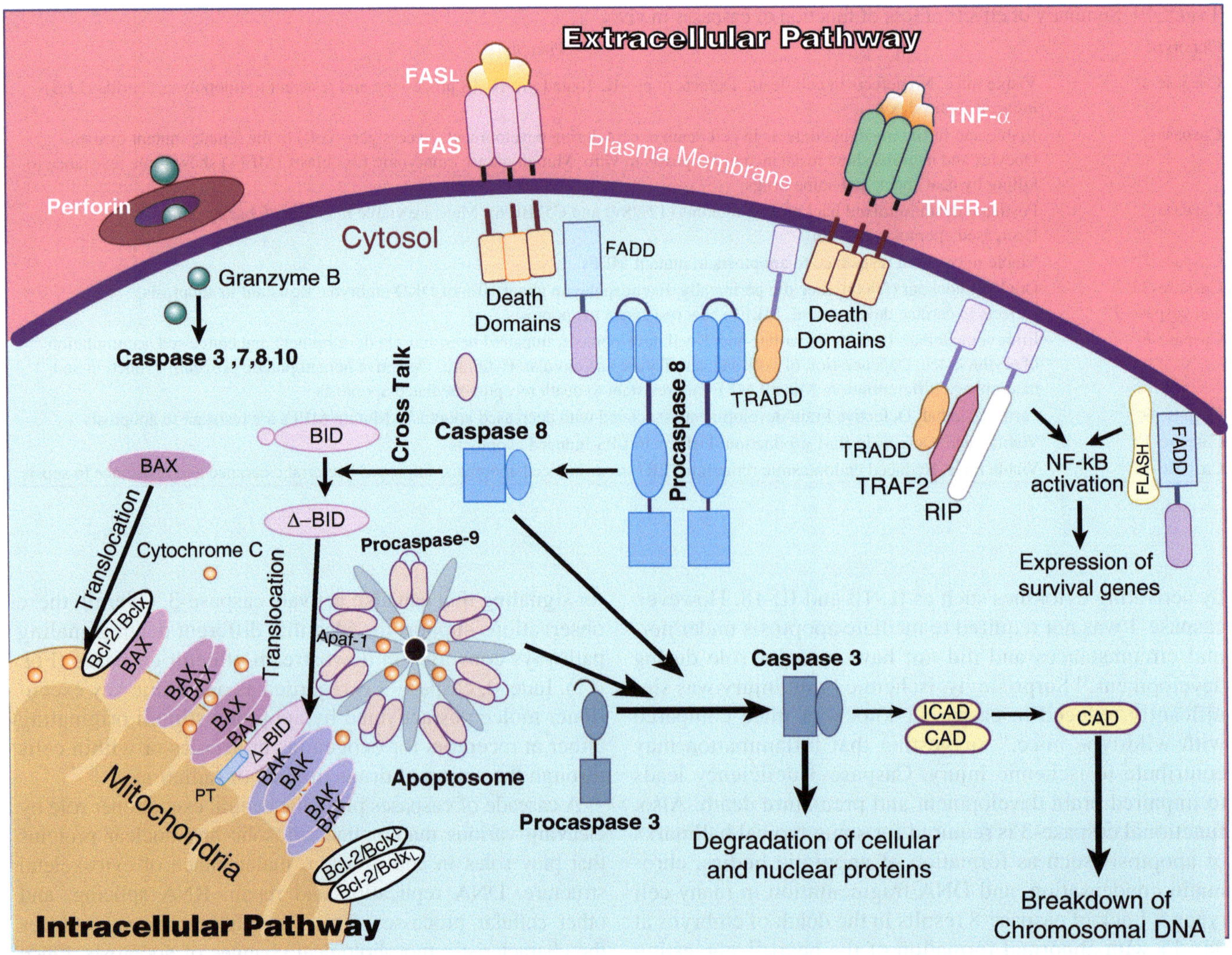

FIG. 4.2. Schematic representation of apoptotic signaling pathways. *Extracellular Pathway:* Following the binding of peptides such as tumor necrosis factor (TNF)-• or Fas ligand (FASL), the receptors oligomerize and recruit adapter proteins [Fas-associated death domain (FADD), tumor necrosis factor receptor (TNFR)-associated death domain (TRADD)] to form death-inducing signaling complexes, causing the activation of the initiator caspase-8, which sequentially activates effector caspases (e.g., caspase-3). Other adapter proteins (FLASH), inhibitory proteins (FLIP), or proteins involved in survival pathways as well as death mechanisms (receptor interaction protein, RIP) may participate in complex mechanisms that determine life or death. The TNF-•-TNFR1 complex can also elicit an antiapoptotic response by recruiting TRAF2, which results in NF-κB-mediated upregulation of antiapoptotic genes. In cytotoxic T-lymphocyte-induced death, granzyme B, which enters the cell through membrane channels formed by the protein perforin, activates caspases by cleaving them directly or indirectly. *Intracellular Pathways:* Lack of survival stimuli (withdrawal of growth factor, hypoxia, genotoxic substances, etc.) is thought to generate apoptotic signals through poorly defined mechanisms, which lead to translocation of proapoptotic proteins such as Bax to the outer mitochondrial membrane. In some cases, transcription mediated by p53 may be required to induce proteins such as Bax. Translocated Bax undergoes conformational changes in the outer membrane to form oligomeric structures (pores) that leak cytochrome *c* from mitochondria into the cytosol. Formation of a ternary complex of cytochrome *c*, the adapter protein Apaf-1, and the initiator caspase-9 called the "apoptosome" results in the activation of caspase-9, followed by sequential activation of effector caspase(s) such as caspase-3 and others. The actions of caspases, endonucleases, and possibly other enzymes lead to cellular disintegration. For example, the endonuclease CAD (caspase-activated DNAse) becomes activated when it is released from its inhibitor ICAD upon cleavage of ICAD by an effector caspase. Antiapoptotic proteins such as Bcl-2 and Bcl-xL inhibit the membrane-permeabilizing effects of Bax and other proapoptotic proteins. Cross-talk between extra- and intracellular pathways occurs through caspase-8-mediated Bid cleavage, which yields a 15-kDa protein that migrates to mitochondria and releases cytochrome *c*, thereby setting in motion events that lead to apoptosis via caspase-9.

(TRADD)[52] and that for Fas, TRAIL-R1, TRAIL-R2, and DR4 is Fas-associated death domain protein (FADD).[53]

The receptor–ligand and FADD complex in turn recruit caspase-8 to the activated receptor, resulting in the formation of death-inducing signaling complex (DISC) and subsequent activation of caspase-8 through oligomerization and self-cleavage. Depending on the cell type and/or apoptotic stimulus, caspase-8 can also be activated by caspase-6.[54] Activated caspase-8 then activates effector caspase-3. In some cell types, cleavage of caspase-3 by caspase-8 also requires a mitochondrial amplification loop involving cleavage of proapoptotic protein Bid by caspase-8 and its translocation to the mitochondrial membrane, triggering the release of apoptogenic proteins from mitochondria into cytosol (see Fig. 4.2). In these cell types, overexpression of *Bcl-2* and *Bcl-xL* can block CD95-induced apoptosis.[55]

Tumor necrosis factor-α is produced by T cells and activated macrophages in response to infection. Although TNF-α-mediated signaling can be propagated through either TNFR1 or TNFR2 receptors, the majority of biological functions are initiated by TNFR1.[56] Binding of TNF-α to TNFR1 causes release of inhibitory protein silencer of death domain protein (SODD) from TNFR1, which enables recruitment of adapter protein TRADD. Signaling induced by activation of TNFR1 or DR3 diverges at the level of TRADD. In one pathway, nuclear translocation of the transcription factor nuclear factor-κB (NF-κB) and activation of c-Jun N-terminal kinase (JNK) are initiated, which results in the induction of a number of proinflammatory and immunomodulatory genes.[57] In another pathway, TNF-α signaling is coupled to Fas signaling events through interaction of TRADD with FADD.[58] The TNFR1–TRADD complex can alternatively engage TRAF2 protein, resulting in activation of transcription factor c-Jun, which is involved in survival signaling. Furthermore, binding of receptor interaction protein (RIP) to TNFR1 through TRADD results in the activation of transcription factor NF-κB, which suppresses apoptosis through transcriptional upregulation of antiapoptotic molecules such as TRAF1, TRAF2, cIAP1, cIAP2, and FLIP. The FLICE-associated huge protein (FLASH) was initially identified to be a CED-4 homologue interacting with the DED of caspase-8 and was shown to modulate Fas-mediated activation of caspase-8.[59] *FLASH* is also required for the antiapoptotic effects of TNF that occur in some cells, where it participates in the activation of the NF-κB transcription factor.[60] Another class of protein, FLIP (FLICE inhibitory protein), was shown to block Fas-induced and TNF-α-induced DISC formation and subsequent activation of caspase-8.[61]

Cytotoxic T cells play a major role in vertebrate defense against viral infection.[62] They induce cell death in infected cells to prevent viral multiplication and spread of infection.[63] Cytotoxic T cells can kill their targets either by activating the Fas ligand/Fas pathway or by injecting granzyme B, a serine protease, into target cells. Cytotoxic T cells carry Fas ligand on their surface but also carry granules containing the channel-forming protein perforin and granzyme B. Upon recognizing the infected cells, the lymphocytes bind and secrete granules onto the surface of infected cells. Perforin then assembles into transmembrane channels to allow the entry of granzyme B into the target cell. Upon entry, granzyme B, which cleaves after aspartate residues in proteins ("aspase"), activates one or more of the apoptotic proteases (caspase-2, -3, -7, -8, and -10) to trigger the proteolytic death cascade (see Fig. 4.2). Fas ligand/Fas and perforin/granzyme B systems are the main apoptotic machinery that regulates homeostasis in immune cell populations.

Intrinsic Death Pathway

Cells can respond to various stressful stimuli and metabolic disturbances by triggering apoptosis. Drugs, toxins, heat, radiation, hypoxia, and viral infections are some of the stimuli known to activate death pathways. Cell death, however, is not necessarily inevitable after exposure to these agents, and the mechanisms determining the outcome of the injury are a topic of active interest. The current consensus appears to be that it is the intensity and the duration of the stimulus that determine the outcome. The stimulus must go beyond a threshold to commit cells to apoptosis. Although the exact mechanism used by each stimulus may be unique and different, a few broad patterns can be identified. For example, agents that damage DNA, such as ionizing radiation and certain xenobiotics, lead to activation of p53-mediated mechanisms that commit cells to apoptosis, at least in part through transcriptional upregulation of proapoptotic proteins.[64] Other stresses induce increased activity of stress-activated protein kinases, which results ultimately in apoptotic commitment.[65] These different mechanisms converge in the activation of caspases.

All stimuli described above change the outer mitochondrial membrane permeability and release two main groups of normally sequestered proapoptotic proteins from the intermembrane space into the cytosol.[4] The first group consists of cytochrome *c*, Smac/DIABLO, and the serine protease HtrA2/Omi.[66–68] Cytochrome *c* binds the adapter protein, Apaf-1, and procaspase-9 to form an "apoptosome"[69] complex and activates caspases. The clustering of procaspase-9 in the apoptosome complex leads to caspase-9 activation. IAPs (inhibitors of apoptosis proteins) in the cell block activate caspase-9 and cause the activation of downstream effector caspases. Smac/DIABLO and HtrA2/Omi are reported to promote apoptosis by inhibiting the activity of IAPs.[70] The second group of proapoptotic proteins, AIF, endonuclease G, and CAD (caspase-activated DNAse), are released from the mitochondria during apoptosis, but this is a late event that occurs after the cell dies. The control and regulation of these mitochondrial events during apoptosis occurs through members of the Bcl-2 family of proteins,[71] as discussed next.

In *C. elegans*, biochemical and genetic studies have indicated a role for CED-4 upstream of CED-3.[72] Upon receiving death commitment signals, CED-4 binds to pro-CED-3 and releases active CED-3.[72] However, when overexpressed,

CED-9 can inhibit the activation of pro-CED-3 by binding to CED-4 and sequestering it away from pro-CED-3. Therefore, CED-3 and CED-4 are involved in activation of apoptosis, and CED-9 inhibits apoptosis. After the discovery of caspases as CED-3 homologues, a search for activators and inhibitors analogous to CED-4 and CED-9 led to the discovery of diverse mammalian regulators of apoptosis. The plethora of these molecules and their functional diversity allowed them to be classified into four broad categories: (1) adapter proteins, (2) the Bcl-2 family of regulators, (3) inhibitors of apoptosis (IAPs), and (4) other regulators.

Adapter Proteins

As stated earlier, two major pathways of apoptosis, involving either the initiator caspase-8 or the initiator caspase-9 (see Fig. 4.2), have been recognized. Signaling by death receptors (CD95, TNFRI) occurs through a well-defined process of recruitment of caspase-8 to the death receptor by adapter proteins such as FADD. Recruitment occurs through interactions between the death domains that are present on both receptor and adapter proteins. Receptor-bound FADD then recruits caspase-8 through interactions between DEDs common to both caspase-8 and FADD, forming a DISC. In the DISC, caspase-8 activation occurs through oligomerization and autocatalysis. Activated caspase-8 then activates downstream caspase-3, culminating in apoptosis. The inhibitory protein FLIP was shown to block Fas-induced and TNF-α-induced DISC formation and subsequent activation of caspase 8.[61] Of particular interest is cellular FLIP, which stimulates caspase-8 activation at physiologically relevant levels and inhibited apoptosis upon high ectopic expression.[73] Cellular FLIP contains two DEDs that can compete with caspase-8 for recruitment to the DISC; this limits the degree of association of caspase-8 with FADD and thus limits activation of the caspase cascade. It also forms a heterodimer with caspase-8 and caspase-10 through interactions between both the DEDs and the caspase-like domains of the proteins, thus activating both caspase-8 and caspase-10.[74] Apoptotic protease activating factor-1 (Apaf-1), a CED-4 homologue in mammalian cells, affects the activation of initiator caspase-9.[75] This factor binds to procaspase-9 in the presence of cytochrome *c* and 2′-deoxyadenosine 5′-triphosphate (dATP) or adenosine triphosphate (ATP) and activates this protease, which in turn activates a downstream cascade of proteases (see Fig. 4.2).[69] By and large, Apaf-1 deficiency is embryonically lethal, and the embryos exhibit brain abnormalities similar to those seen in caspase-9 knockout mice.[76] These genetic findings support the idea that Apaf-1 is coupled to caspase-9 in the death pathway. Unlike CED-4 in nematodes, Apaf-1 requires the binding of ATP and cytochrome *c* to activate procaspase-9. The multiple WD40 repeats in the C-terminal end of Apaf-1 have a regulatory role in the activation of caspase-9.[77]

The Bcl-2 Family of Proteins

The CED-9 homologue in mammals is the Bcl-2 protein. Bcl-2 was first discovered in B-cell lymphoma as a proto-oncogene. Overexpression of Bcl-2 was shown to offer protection against a variety of death stimuli.[78] The Bcl-2 protein family includes both proapoptotic (Bcl-2, Bcl-xL, Bcl-w, Mcl-1, Nr13, and A1/Bfl1) and antiapoptotic proteins (Bax, Bak, Bok, Diva, Bcl-Xs, Bik, Bim, Hrk, Nip3, Nix, Bad, and Bid).[71,79] These proteins are characterized by the presence of Bcl-2 homology (BH) domains: BH1, BH2, BH3, and BH4 (Fig. 4.3). The proapoptotic members have two subfamilies: a multidomain and a BH3-only group (see Fig. 4.3). The relative ratio of pro- and antiapoptotic proteins determines the sensitivity of cells to various apoptotic stimuli. The best studied proapoptotic members are Bax and Bid. Exposure to various apoptotic stimuli leads to translocation of cytosolic Bax from the cytosol to the mitochondrial membrane.[80] Bax oligomerizes on the mitochondrial membrane along with another proapoptotic protein, Bak, leading to the release of cytochrome *c* from the mitochondrial membrane into the cytosol.[81]

Other proapoptotic proteins, mainly the BH3-only proteins, are thought to aid in Bax–Bak oligomerization on the mitochondrial membrane. The antiapoptotic Bcl-2 family members are known to block Bax–Bak oligomerization on the mitochondrial membrane and subsequent release of cytochrome *c* into the cytosol.[81,82] After release from the mitochondria, cytochrome *c* is known to interact with the WD40 repeats of the adapter protein Apaf-1, resulting in the formation of the apoptosome complex. Seven molecules of Apaf-1, interacting through their N-terminal caspase activation and recruitment domain, form the central hub region of the symmetrical wheel-like structure, the apoptosome. Binding of ATP/dATP to Apaf-1 triggers the formation of the apoptosome, which subsequently recruits procaspase-9 into the apoptosome complex, resulting in its activation.[83] Activated caspase-9 then activates executioner caspases, such as caspase-3 and caspase-7, eventually leading to programmed cell death. However, for the caspase-9 to be active, it must reside on the apoptosome complex.[84]

Inhibitors of Apoptosis Proteins

The IAPs, first discovered in baculoviruses and then in insects and *Drosophila*, inhibit activated caspases by directly binding to the active enzymes.[85] These proteins contain one or more baculovirus inhibitors of apoptosis repeat domains, which are responsible for the caspase inhibitory activity.[86] To date, eight mammalian IAPs have been identified: they include X-linked IAP (XIAP), c-IAP1, c-IAP2, Melanoma IAP (ML-IAP)/Livin, IAP-like protein-2 (ILP-2), neuronal apoptosis-inhibitory protein (NAIP), Bruce/Apollon, and Survivin. In mammals, caspase-3, -7, and -9 are inhibited by IAPs.[83]. There are reports suggesting aberrant expression of IAPs in many cancer tissues. For example, cIAP1

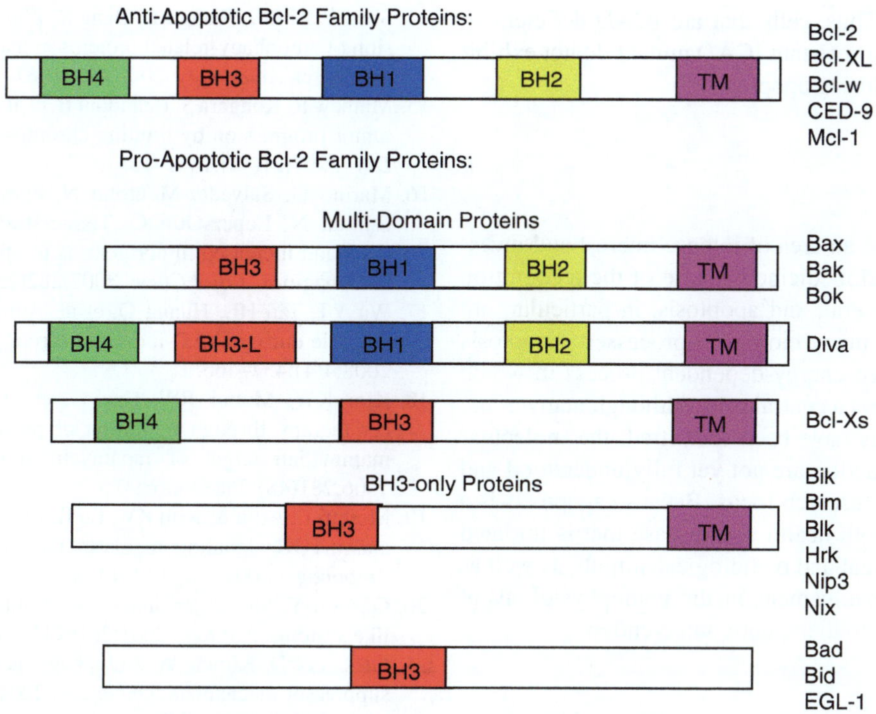

FIG. 4.3. Structural homologies in anti- and proapoptotic proteins of the Bcl-2 family. Anti- and proapoptotic proteins of the Bcl-2 family proteins are depicted, indicating Bcl-2 homologous (BH) regions BH1, BH2, BH3, and BH4 and transmembrane (TM) domains.

is overexpressed in esophageal squamous cell sarcoma[87]; the cIAP2 locus is translocated in mucosa-associate lymphoid lymphoma[88]; and Survivin has been shown to be upregulated in many cancer cells.[89]

Other Regulators

The caspase inhibitory activity of IAPs is inhibited by proteins containing an IAP-binding tetrapeptide motif.[83] The founding member of this family is Smac/DIABLO, which is released from the mitochondrial intermembrane space into the cytosol during apoptosis. In the cytosol, it interacts with several IAPs and inhibits their function. The other mitochondrial protein, Omi/HtrA2, is also known to antagonize XIAP-mediated inhibition of caspase-9 at high concentrations.[90] A serine protease, Omi/HtrA2, can proteolytically cleave and inactivate IAP proteins and thus is considered to be a more potent suppressor of IAPs than Smac.[91]

It has been reported that the heat shock proteins Hsp90, Hsp70, and Hsp27 can inhibit caspase activation by cytochrome *c* by interacting with either Apaf-1 or other players in the pathway.[92–94] A high-throughput screen identified a compound called PETCM [α-(trichloromethyl)-4-pyridineethanol] as a caspase-3 activator. Further work with PETCM revealed its involvement in apoptosome regulation.[95] This pathway also includes oncoprotein prothymosin-α and tumor suppressor putative HLA-DR-associated proteins. These proteins were shown to promote caspase-9 activation after apoptosome formation, whereas prothymosin-α inhibited caspase-9 activation by inhibiting apoptosome formation.

Protein Targets of Caspases

In an apoptotic cell, the regulatory, structural, and housekeeping proteins are the main targets of the caspases. The regulatory proteins mitogen-activated protein/extracellular signal-regulated kinase kinase-1, p21-activated kinase-2, and Mst-1 are activated upon cleavage by caspases.[96] Caspase-mediated protein hydrolysis inactivates other proteins, including focal adhesion kinase, phosphatidylinositol-3 kinase, Akt, Raf-1, IAPs, and inhibitors of caspase-activated DNAse (ICAD). Caspases also convert the antiapoptotic protein Bcl-2 into a proapoptotic protein such as Bax upon cleavage. There are many structural protein targets of caspases, which include nuclear lamins, actin, and regulatory proteins such as spectrin, gelsolin, and fodrins.[47,97]

Degradation of nuclear DNA into internucleosomal chromatin fragments is one of the hallmarks of apoptotic cell death that occurs in response to various apoptotic stimuli in a wide variety of cells. A specific caspase-activated DNase (CAD), that cleaves chromosomal DNA in a caspase-dependent manner, is synthesized with the help of ICAD. In proliferating cells, CAD is always found to be associated with ICAD in the cytosol. When cells are undergoing apoptosis, caspases (particularly caspase-3) cleave ICAD to release CAD and allow its translocation to the nucleus to cleave chromosomal DNA. Thus, cells that are ICAD deficient or which express caspase-resistant ICAD mutant do not exhibit DNA fragmentation during apoptosis.

Conclusion

Cell death has become an area of intense interest and investigation in science and medicine because of the recognition that cell death, in general, and apoptosis, in particular, are important features of many biological processes. Apoptosis is a carefully regulated energy-dependent process in which caspase activation plays a central role. Although many of the key apoptotic proteins have been identified, the molecular mechanisms of their action are not yet fully understood and thus continue to be a research focus. Because apoptosis is a vital component of both health and disease that is initiated by various physiological and pathological stimuli, as well as having widespread involvement in the pathophysiology of disease, it lends itself to therapeutic intervention.

References

1. Kerr JF, Wyllie AH, Currie AR. Apoptosis: a basic biological phenomenon with wide-ranging implications in tissue kinetics. *Br J Cancer*. 1972;26(4):239–257.
2. Horvitz HR. Worms, life, and death (Nobel lecture). *ChemBioChem*. 2003;4(8):697–711.
3. Ellis HM, Horvitz HR. Genetic control of programmed cell death in the nematode *C. elegans*. *Cell*. 1986;44(6): 817–829.
4. Saikumar P, Dong Z, Mikhailov V, Denton M, Weinberg JM, Venkatachalam MA. Apoptosis: definition, mechanisms, and relevance to disease. *Am J Med*. 1999;107(5):489–506.
5. Leist M, Jaattela M. Four deaths and a funeral: from caspases to alternative mechanisms. *Nat Rev*. 2001;2(8):589–598.
6. Nicotera P, Leist M, Ferrando-May E. Intracellular ATP, a switch in the decision between apoptosis and necrosis. *Toxicol Lett*. 1998;102-103:139–142.
7. Zong WX, Ditsworth D, Bauer DE, Wang ZQ, Thompson CB. Alkylating DNA damage stimulates a regulated form of necrotic cell death. *Genes Dev*. 2004;18(11):1272–1282.
8. Formigli L, Papucci L, Tani A, et al. Aponecrosis: morphological and biochemical exploration of a syncretic process of cell death sharing apoptosis and necrosis. *J Cell Physiol*. 2000;182(1): 41–49.
9. Klionsky DJ, Emr SD. Autophagy as a regulated pathway of cellular degradation. *Science*. 2000;290(5497):1717–1721.
10. Paglin S, Hollister T, Delohery T, et al. A novel response of cancer cells to radiation involves autophagy and formation of acidic vesicles. *Cancer Res*. 2001;61(2):439–444.
11. Cuervo AM. Autophagy: in sickness and in health. *Trends Cell Biol*. 2004;14(2):70–77.
12. Qu X, Yu J, Bhagat G, et al. Promotion of tumorigenesis by heterozygous disruption of the beclin 1 autophagy gene. *J Clin Invest*. 2003;112(12):1809–1820.
13. Karantza-Wadsworth V, White E. Role of autophagy in breast cancer. *Autophagy*. 2007;3(6):610–613.
14. Shen Y, Li DD, Wang LL, Deng R, Zhu XF. Decreased expression of autophagy-related proteins in malignant epithelial ovarian cancer. *Autophagy*. 2008;4(8):1067–1068.
15. Mathew R, Kongara S, Beaudoin B, et al. Autophagy suppresses tumor progression by limiting chromosomal instability. *Genes Dev*. 2007;21(11):1367–1381.
16. Marino G, Salvador-Montoliu N, Fueyo A, Knecht E, Mizushima N, Lopez-Otin C. Tissue-specific autophagy alterations and increased tumorigenesis in mice deficient in Atg4C/autophagin-3. *J Biol Chem*. 2007;282(25):18573–18583.
17. Wu YT, Tan HL, Huang Q, et al. Autophagy plays a protective role during zVAD-induced necrotic cell death. *Autophagy*. 2008;4(4):457–466.
18. Kim KW, Mutter RW, Cao C, et al. Autophagy for cancer therapy through inhibition of pro-apoptotic proteins and mammalian target of rapamycin signaling. *J Biol Chem*. 2006;281(48):36883–36890.
19. Moretti L, Attia A, Kim KW, Lu B. Crosstalk between Bak/Bax and mTOR signaling regulates radiation-induced autophagy. *Autophagy*. 2007;3(2):142–144.
20. Ohsumi Y. Molecular dissection of autophagy: two ubiquitin-like systems. *Nat Rev*. 2001;2(3):211–216.
21. Gozuacik D, Kimchi A. Autophagy as a cell death and tumor suppressor mechanism. *Oncogene*. 2004;23(16):2891–2906.
22. Meijer AJ, Codogno P. Regulation and role of autophagy in mammalian cells. *Int J Biochem Cell Biol*. 2004;36(12):2445–2462.
23. Ogier-Denis E, Pattingre S, El Benna J, Codogno P. Erk1/2-dependent phosphorylation of G-alpha-interacting protein stimulates its GTPase accelerating activity and autophagy in human colon cancer cells. *J Biol Chem*. 2000;275(50):39090–39095.
24. Sperandio S, de Belle I, Bredesen DE. An alternative, nonapoptotic form of programmed cell death. *Proc Natl Acad Sci USA*. 2000;97(26):14376–14381.
25. Wang Y, Li X, Wang L, et al. An alternative form of paraptosis-like cell death, triggered by TAJ/TROY and enhanced by PDCD5 overexpression. *J Cell Sci*. 2004;117(pt 8):1525–1532.
26. Sperandio S, Poksay K, de Belle I, et al. Paraptosis: mediation by MAP kinases and inhibition by AIP-1/Alix. *Cell Death Differ*. 2004;11(10):1066–1075.
27. Schneider D, Gerhardt E, Bock J, et al. Intracellular acidification by inhibition of the Na+/H+-exchanger leads to caspase-independent death of cerebellar granule neurons resembling paraptosis. *Cell Death Differ*. 2004;11(7):760–770.
28. Valamanesh F, Torriglia A, Savoldelli M, et al. Glucocorticoids induce retinal toxicity through mechanisms mainly associated with paraptosis. *Mol Vis*. 2007;13:1746–1757.
29. Henics T, Wheatley DN. Cytoplasmic vacuolation, adaptation and cell death: a view on new perspectives and features. *Biol Cell*. 1999;91(7):485–498.
30. Bowen ID, Morgan SM, Mullarkey K. Cell death in the salivary glands of metamorphosing *Calliphora vomitoria*. *Cell Biol Int*. 1993;17(1):13–33.
31. Bowen ID, Mullarkey K, Morgan SM. Programmed cell death during metamorphosis in the blow-fly *Calliphora vomitoria*. *Microsc Res Tech*. 1996;34(3):202–217.
32. Cornillon S, Foa C, Davoust J, Buonavista N, Gross JD, Golstein P. Programmed cell death in *Dictyostelium*. *J Cell Sci*. 1994;107(pt 10):2691–2704.
33. Nijhawan R, Rajwanshi A, Gautam U, Gupta SK. Cytoplasmic vacuolation, intracytoplasmic lumina, and DPAS stain-

ing in ductal carcinoma of the breast. *Diagn Cytopathol* 2003;28(6):291–294.
34. Gilloteaux J, Jamison JM, Arnold D, Summers JL. Autoschizis: another cell death for cancer cells induced by oxidative stress. *Ital J Anat Embryol.* 2001;106(2 suppl 1):79–92.
35. Jamison JM, Gilloteaux J, Taper HS, Calderon PB, Summers JL. Autoschizis: a novel cell death. *Biochem Pharmacol.* 2002;63(10): 1773–1783.
36. Gilloteaux J, Jamison JM, Lorimer HE, et al. Autoschizis: a new form of cell death for human ovarian carcinoma cells following ascorbate:menadione treatment. Nuclear and DNA degradation. *Tissue Cell* 2004;36(3):197–209.
37. Platt N, da Silva RP, Gordon S. Recognizing death: the phagocytosis of apoptotic cells. *Trends Cell Biol.* 1998;8(9):365–372.
38. Yuan J. Evolutionary conservation of a genetic pathway of programmed cell death. *J Cell Biochem.* 1996;60(1):4–11.
39. Spector MS, Desnoyers S, Hoeppner DJ, Hengartner MO. Interaction between the *C. elegans* cell-death regulators CED-9 and CED-4. *Nature.* 1997;385(6617):653–656.
40. Thornberry NA, Lazebnik Y. Caspases: enemies within. *Science.* 1998;281(5381):1312–1316.
41. Kumar S. Caspase function in programmed cell death. *Cell Death Differ.* 2007;14(1):32–43.
42. Shiozaki EN, Chai J, Rigotti DJ, et al. Mechanism of XIAP-mediated inhibition of caspase-9. *Mol Cell.* 2003;11(2):519–527.
43. Donepudi M, Mac Sweeney A, Briand C, Grutter MG. Insights into the regulatory mechanism for caspase-8 activation. *Mol Cell.* 2003;11(2):543–549.
44. Li P, Allen H, Banerjee S, Seshadri T. Characterization of mice deficient in interleukin-1 beta converting enzyme. *J Cell Biochem.* 1997;64(1):27–32.
45. Schielke GP, Yang GY, Shivers BD, Betz AL. Reduced ischemic brain injury in interleukin-1 beta converting enzyme-deficient mice. *J Cereb Blood Flow Metab.* 1998;18(2):180–185.
46. Janicke RU, Sprengart ML, Wati MR, Porter AG. Caspase-3 is required for DNA fragmentation and morphological changes associated with apoptosis. *J Biol Chem.* 1998;273(16):9357–9360.
47. Varfolomeev EE, Schuchmann M, Luria V, et al. Targeted disruption of the mouse caspase 8 gene ablates cell death induction by the TNF receptors, Fas/Apo1, and DR3 and is lethal prenatally. *Immunity.* 1998;9(2):267–276.
48. Yeh WC, Pompa JL, McCurrach ME, et al. FADD: essential for embryo development and signaling from some, but not all, inducers of apoptosis. *Science.* 1998;279(5358):1954–1958.
49. Kuida K, Haydar TF, Kuan CY, et al. Reduced apoptosis and cytochrome c-mediated caspase activation in mice lacking caspase 9. *Cell.* 1998;94(3):325–337.
50. Magnusson C, Vaux DL. Signalling by CD95 and TNF receptors: not only life and death. *Immunol Cell Biol.* 1999;77(1):41–46.
51. Ashkenazi A, Dixit VM. Death receptors: signaling and modulation. *Science.* 1998;281(5381):1305–1308.
52. Hsu H, Xiong J, Goeddel DV. The TNF receptor 1-associated protein TRADD signals cell death and NF-kappa B activation. *Cell.* 1995;81(4):495–504.
53. Chinnaiyan AM, O'Rourke K, Tewari M, Dixit VM. FADD, a novel death domain-containing protein, interacts with the death domain of Fas and initiates apoptosis. *Cell.* 1995;81(4):505–512.
54. Cowling V, Downward J. Caspase-6 is the direct activator of caspase-8 in the cytochrome c-induced apoptosis pathway: absolute requirement for removal of caspase-6 prodomain. *Cell Death Differ.* 2002;9(10):1046–1056.
55. Scaffidi C, Fulda S, Srinivasan A, et al. Two CD95 (APO-1/Fas) signaling pathways. *EMBO J.* 1998;17(6):1675–1687.
56. Grell M, Zimmermann G, Gottfried E, et al. Induction of cell death by tumour necrosis factor (TNF) receptor 2, CD40 and CD30: a role for TNF-R1 activation by endogenous membrane-anchored TNF. *EMBO J.* 1999;18(11):3034–3043.
57. Natoli G, Costanzo A, Moretti F, Fulco M, Balsano C, Levrero M. Tumor necrosis factor (TNF) receptor 1 signaling downstream of TNF receptor-associated factor 2. Nuclear factor kappaB (NFkappaB)-inducing kinase requirement for activation of activating protein 1 and NFkappaB but not of c-Jun N-terminal kinase/stress-activated protein kinase. *J Biol Chem.* 1997;272(42):26079–26082.
58. Boldin MP, Goncharov TM, Goltsev YV, Wallach D. Involvement of MACH, a novel MORT1/FADD-interacting protease, in Fas/APO-1- and TNF receptor-induced cell death. *Cell.* 1996;85(6): 803–815.
59. Imai Y, Kimura T, Murakami A, Yajima N, Sakamaki K, Yonehara S. The CED-4-homologous protein FLASH is involved in Fas-mediated activation of caspase-8 during apoptosis. *Nature.* 1999;398(6730):777–785.
60. Choi YH, Kim KB, Kim HH, et al. FLASH coordinates NF-kappa B activity via TRAF2. *J Biol Chem.* 2001;276(27): 25073–25077.
61. Thome M, Schneider P, Hofmann K, et al. Viral FLICE-inhibitory proteins (FLIPs) prevent apoptosis induced by death receptors. *Nature.* 1997;386(6624):517–521.
62. Offit PA, Cunningham SL, Dudzik KI. Memory and distribution of virus-specific cytotoxic T lymphocytes (CTLs) and CTL precursors after rotavirus infection. *J Virol.* 1991;65(3):1318–1324.
63. Zajac AJ, Quinn DG, Cohen PL, Frelinger JA. Fas-dependent CD4+ cytotoxic T-cell-mediated pathogenesis during virus infection. *Proc Natl Acad Sci USA.* 1996;93(25): 14730–14735.
64. Wu Q, Kirschmeier P, Hockenberry T, et al. Transcriptional regulation during p21WAF1/CIP1-induced apoptosis in human ovarian cancer cells. *J Biol Chem.* 2002;277(39):36329–36337.
65. Kunz M, Ibrahim S, Koczan D, et al. Activation of c-Jun NH2-terminal kinase/stress-activated protein kinase (JNK/SAPK) is critical for hypoxia-induced apoptosis of human malignant melanoma. *Cell Growth Differ.* 2001;12(3):137–145.
66. Cai J, Yang J, Jones DP. Mitochondrial control of apoptosis: the role of cytochrome c. *Biochim Biophys Acta.* 1998;1366(1–2):139–149.
67. Du C, Fang M, Li Y, Li L, Wang X. Smac, a mitochondrial protein that promotes cytochrome c-dependent caspase activation by eliminating IAP inhibition. *Cell.* 2000;102(1):33–42.
68. van Loo G, van Gurp M, Depuydt B, et al. The serine protease Omi/HtrA2 is released from mitochondria during apoptosis. Omi interacts with caspase-inhibitor XIAP and induces enhanced caspase activity. *Cell Death Differ.* 2002;9(1):20–26.
69. Zou H, Li Y, Liu X, Wang X. An APAF-1.cytochrome c multimeric complex is a functional apoptosome that activates procaspase-9. *J Biol Chem.* 1999;274(17):11549–11556.

70. Schimmer AD. Inhibitor of apoptosis proteins: translating basic knowledge into clinical practice. *Cancer Res.* 2004;64(20):7183–7190.
71. Cory S, Adams JM. The Bcl2 family: regulators of the cellular life-or-death switch. *Nat Rev Cancer.* 2002;2(9):647–656.
72. Hengartner MO, Horvitz HR. The ins and outs of programmed cell death during *C. elegans* development. *Philos Trans R Soc Lond B Biol Sci.* 1994;345(1313):243–246.
73. Chang DW, Xing Z, Pan Y, et al. c-FLIP(L) is a dual function regulator for caspase-8 activation and CD95-mediated apoptosis. *EMBO J.* 2002;21(14):3704–3714.
74. Peter ME. The flip side of FLIP. *Biochem J.* 2004; 382(pt 2):e1–e3.
75. Zou H, Henzel WJ, Liu X, Lutschg A, Wang X. Apaf-1, a human protein homologous to *C. elegans* CED-4, participates in cytochrome c-dependent activation of caspase-3. *Cell.* 1997;90(3):405–413.
76. Cecconi F, Alvarez-Bolado G, Meyer BI, Roth KA, Gruss P. Apaf1 (CED-4 homolog) regulates programmed cell death in mammalian development. *Cell.* 1998;94(6): 727–737.
77. Srinivasula SM, Ahmad M, Fernandes-Alnemri T, Alnemri ES. Autoactivation of procaspase-9 by Apaf-1-mediated oligomerization. *Mol Cell.* 1998;1(7):949–957.
78. Zhong LT, Sarafian T, Kane DJ, et al. bcl-2 inhibits death of central neural cells induced by multiple agents. *Proc Natl Acad Sci USA.* 1993;90(10):4533–4537.
79. Reed JC. Bcl-2 family proteins. *Oncogene.* 1998;17(25): 3225–3236.
80. Saikumar P, Dong Z, Patel Y, et al. Role of hypoxia-induced Bax translocation and cytochrome c release in reoxygenation injury. *Oncogene.* 1998;17(26):3401–3415.
81. Mikhailov V, Mikhailova M, Degenhardt K, Venkatachalam MA, White E, Saikumar P. Association of Bax and Bak homo-oligomers in mitochondria. Bax requirement for Bak reorganization and cytochrome c release. *J Biol Chem.* 2003;278(7):5367–5376.
82. Mikhailov V, Mikhailova M, Pulkrabek DJ, Dong Z, Venkatachalam MA, Saikumar P. Bcl-2 prevents Bax oligomerization in the mitochondrial outer membrane. *J Biol Chem.* 2001;276(21):18361–18374.
83. Shi Y. Mechanisms of caspase activation and inhibition during apoptosis. *Mol Cell.* 2002;9(3):459–470.
84. Saikumar P, Mikhailova M, Pandeswara SL. Regulation of caspase-9 activity by differential binding to the apoptosome complex. *Front Biosci.* 2007;12:3343–3354.
85. Ekert PG, Silke J, Hawkins CJ, Verhagen AM, Vaux DL. DIABLO promotes apoptosis by removing MIHA/XIAP from processed caspase 9. *J Cell Biol.* 2001;152(3):483–490.
86. Deveraux QL, Reed JC. IAP family proteins: suppressors of apoptosis. *Genes Dev.* 1999;13(3):239–252.
87. Imoto I, Yang ZQ, Pimkhaokham A, et al. Identification of cIAP1 as a candidate target gene within an amplicon at 11q22 in esophageal squamous cell carcinomas. *Cancer Res.* 2001;61(18):6629–6634.
88. Dierlamm J, Baens M, Wlodarska I, et al. The apoptosis inhibitor gene API2 and a novel 18q gene, MLT, are recurrently rearranged in the t(11;18)(q21;q21) associated with mucosa-associated lymphoid tissue lymphomas. *Blood.* 1999;93(11):3601–3609.
89. Ambrosini G, Adida C, Altieri DC. A novel anti-apoptosis gene, survivin, expressed in cancer and lymphoma. *Nat Med.* 1997;3(8):917–921.
90. Suzuki Y, Imai Y, Nakayama H, Takahashi K, Takio K, Takahashi R. A serine protease, HtrA2, is released from the mitochondria and interacts with XIAP, inducing cell death. *Mol Cell.* 2001;8(3): 613–621.
91. Yang QH, Church-Hajduk R, Ren J, Newton ML, Du C. Omi/HtrA2 catalytic cleavage of inhibitor of apoptosis (IAP) irreversibly inactivates IAPs and facilitates caspase activity in apoptosis. *Genes Dev.* 2003;17(12):1487–1496.
92. Beere HM, Wolf BB, Cain K, et al. Heat-shock protein 70 inhibits apoptosis by preventing recruitment of procaspase-9 to the Apaf-1 apoptosome. *Nat Cell Biol.* 2000;2(8):469–475.
93. Pandey P, Farber R, Nakazawa A, et al. Hsp27 functions as a negative regulator of cytochrome c-dependent activation of procaspase-3. *Oncogene.* 2000;19(16):1975–1981.
94. Pandey P, Saleh A, Nakazawa A, et al. Negative regulation of cytochrome c-mediated oligomerization of Apaf-1 and activation of procaspase-9 by heat shock protein 90. *EMBO J.* 2000;19(16):4310–4322.
95. Jiang X, Kim HE, Shu H, et al. Distinctive roles of PHAP proteins and prothymosin-alpha in a death regulatory pathway. *Science.* 2003;299(5604):223–226.
96. Widmann C, Gerwins P, Johnson NL, Jarpe MB, Johnson GL. MEK kinase 1, a substrate for DEVD-directed caspases, is involved in genotoxin-induced apoptosis. *Mol Cell Biol.* 1998;18(4):2416–2429.
97. Sanghavi DM, Thelen M, Thornberry NA, Casciola-Rosen L, Rosen A. Caspase-mediated proteolysis during apoptosis: insights from apoptotic neutrophils. *FEBS Lett.* 1998;422(2):179–184.

5 The Role of Mutation and Epimutation in the Development of Human Disease

Ashley G. Rivenbark and William B. Coleman

Introduction

The paradigm of disease causation holds that disease represents the manifestation or several manifestations of an underlying process that has one or more root causes. Thus, human diseases reflect a spectrum of pathologies and mechanisms of disease pathogenesis. The general categories of disease affecting humans include (a) hereditary diseases, (b) infectious diseases, (c) inflammatory diseases, and (d) neoplastic diseases. Pathological conditions representing each of these general categories have been described for every tissue in the body. Despite the grouping of diseases by the common features of the general disease type, the pathogenesis of each of the various diseases is unique, and in some cases multiple mechanisms can give rise to a similar pathology (disease manifestation). Disease causation may be related to intrinsic factors or extrinsic factors, but many or most diseases are multifactorial, involving a combination of intrinsic and extrinsic factors. It is now well recognized that most major diseases are ultimately the result of aberrant gene expression and that susceptibility to disease is significantly influenced by patterns of gene expression in target cells or tissues for a particular type of pathology. It follows that gene mutations and other genetic alterations are important in the pathogenesis of many human diseases. Similarly, nongenetic alterations affecting the expression of key genes, called epimutations, may also contribute to the genesis of disease at many tissue sites. In this chapter, general concepts related to the molecular basis for the major disease types are reviewed. This review is not intended to be comprehensive. Rather, the current state of understanding related to the genes and molecular mechanisms (genetic and epigenetic) that contribute to illustrative diseases is described.

Mutations and Epimutations

Mutation refers to changes in the genome that are characterized by alteration in the nucleotide sequence of a specific gene and/or other alterations at the level of the primary structure of DNA. Point mutations, insertions, deletions, and chromosomal abnormalities are all classified as mutations. In contrast, epimutation refers to alterations in the genome that do not involve changes in the primary sequence of the DNA. Aberrant DNA hypermethylation or hypomethylation and/or abnormal histone modifications resulting in alterations in chromatin structure are considered epimutations. Despite the differences between mutation and epimutation, the consequences of these molecular processes on the normal expression or function of critical genes and proteins may be the same: alteration of normal gene expression and/or normal protein function. These alterations may reflect (a) loss or reduction of normal levels of gene expression with consequent loss of protein function, (b) loss of function caused by loss of protein or synthesis of defective protein, (c) increased levels of gene expression with consequent overexpression of protein, or (d) gain-of-function mutation with consequent altered protein function. Although many human diseases can be attributed to genetic alteration or epimutation affecting a single gene (or a few genes), the actual molecular consequences of these changes can be very dramatic, resulting in major alterations in gene expression patterns secondary to the primary genetic or epigenetic gene defect.

Genetic Alterations

Disease-related genetic alterations can be categorized into two major groups: nucleotide sequence abnormalities and chromosomal abnormalities. Examples of both these forms

P.T. Cagle and T.C. Allen (eds.), *Basic Concepts of Molecular Pathology*, Molecular Pathology Library 2,
DOI: 10.1007/978-0-387-89626-7_5, © Springer Science+Business Media, LLC 2009

of molecular lesion have been characterized in familial and acquired diseases affecting various human tissues.

Nucleotide sequence alterations include changes in individual genes involving single nucleotide changes (missense and nonsense) and small insertions or deletions (some of which result in frameshift mutations). Single nucleotide alterations that involve a change in the normal coding sequence of the gene (point mutations) can give rise to an alteration in the amino acid sequence of the encoded protein. Missense mutations alter the translation of the affected codon, whereas nonsense mutations alter codons that encode amino acids to produce stop codons. This action results in premature termination of translation and the synthesis of a truncated protein product. Small deletions and insertions are typically classified as frameshift mutations because deletion or insertion of a single nucleotide (for instance) will alter the reading frame of the gene on the 3′-side of the affected site. This alteration can result in the synthesis of a protein that bears very little resemblance to the normal gene product or production of an abnormal/truncated protein because of the presence of a stop codon in the altered reading frame. In addition, deletion or insertion of one or more groups of three nucleotides will not alter the reading frame of the gene, but will alter the resulting polypeptide product, which will exhibit either loss of specific amino acids or the presence of additional amino acids within its primary structure.

Chromosomal alterations include the gain or loss of one or more chromosomes (aneuploidy), chromosomal rearrangements resulting from DNA strand breakage (translocations, inversions, and other rearrangements), and gain or loss of portions of chromosomes (amplification, large-scale deletion). The direct result of chromosomal translocation is the movement of a segment of DNA from its natural location into a new location within the genome, which can result in altered expression of the genes that are contained within the translocated region. If the chromosomal breakpoints utilized in a translocation are located within structural genes, then hybrid (chimeric) genes can be generated. The major consequence of a chromosomal deletion (involving a whole chromosome or a large chromosomal region) is the loss of specific genes that are localized to the deleted chromosomal segment, resulting in changes in the copy number of the affected genes. Similarly, gain of chromosome number or amplification of chromosomal regions results in an increase in the copy numbers of genes found in these chromosomal locations.

Epigenetic Alterations

In contemporary terms, epigenetics refers to modifications of the genome that are heritable during cell division but do not involve a change in the DNA sequence.[1] Therefore, epigenetics describes heritable changes in gene expression that are not simply attributable to nucleotide sequence variation.[2] It is now recognized that epigenetic regulation of gene expression reflects contributions from both DNA methylation as well as complex modifications of histone proteins and chromatin structure.[3] Nonetheless, DNA methylation plays a central role in nongenomic inheritance and in the preservation of epigenetic states and remains the most accessible epigenomic feature due to its inherent stability. Thus, DNA methylation represents a target of fundamental importance in the characterization of the epigenome and for defining the role of epigenetics in disease pathogenesis.

Diseases Associated with Chromosomal Alterations

Chromosomal Alterations in Cancer

Chromosomal alterations have been widely documented in the various forms of cancer, and some of these chromosomal aberrations are recognized to be causally related to cancer induction and/or progression. In some cases, these alterations represent loss of individual chromosomes, chromosome arms, or specific chromosomal segments, consistent with deletion of a tumor suppressor locus. In other cases, these alterations represent gain of individual chromosomes or specific chromosomal segments, consistent with activation of a positive mediator of cell proliferation. In addition, numerous complex chromosomal rearrangements (such as translocation) have been characterized in certain forms of cancer. In each of these cases, the result is alteration of gene dose (either as gene loss or gene amplification) and function (altered product or altered expression). Cytogenetic methods to characterize chromosomal alterations have been applied for many years, but recent developments have automated this process and opened the analysis to difficult samples (solid tumors). For instance, array-based comparative genomic hybridization (CGH) is a technique[4] that analyzes global genomic changes by documenting gains and losses of chromosomal regions in diseases such as cancer.[5–7] Recent studies have shown the potential of array CGH in detecting copy number alterations in gastric cancer,[8,9] chronic lymphocytic leukemia,[10] fallopian tube carcinomas,[11] oral squamous carcinoma,[12] bladder cancer,[13] and pancreatic cancer.[14] A few well-characterized examples from major human cancers are given in the sections that follow.

Gene Amplifications in Cancer

Several chromosomal regions have been characterized to be overrepresented in lung cancer,[15,16] suggestive of amplification of chromosomal regions. One such frequently overrepresented region is chromosome 8q, which harbors the c-myc proto-oncogene.[17] The c-myc proto-oncogene is a member of the basic helix-loop-helix superfamily of nuclear transcription factors.[18] The myc protein heterodimerizes with max, and the resulting myc–max protein complex binds to and transcriptionally activates genes that contain a CAGCTG consensus binding sequence.[19–22] Increased expression of c-myc has been reported for both small cell lung carcinoma (SCLC) and non-small cell lung carcinoma (NSCLC).[23–27] Similarly, amplification of the c-myc locus has been observed in both SCLC and NSCLC,[28] but it may be more prevalent in SCLC.[29] Thus, overexpression of c-myc in lung cancer is the frequent consequence of gene amplification at 8q24.[30–32] Gene amplification and overexpression of c-myc occur more frequently in advanced neoplasms and metastatic lesions, suggesting a role for this event in tumor progression, and partially explaining the significant correlation between c-myc amplification and poor prognosis.[31] In contrast to other cancers (such as lymphoma), point mutation of c-myc and c-myc gene translocation related to specific chromosomal alterations have not been reported in lung cancer.

Several studies have examined chromosomal alterations in breast cancer using comparative genomic hybridization approaches.[33–35] These studies (as well as those using cytogenetic techniques) have identified a number of gene amplifications in breast cancer. For instance, the TM4SF1 gene is located at 3q25.1 and is amplified frequently in BRCA1-related breast cancers.[36] Similarly, the CYP24 gene that is located at 20q13.2 is subject to recurrent amplification in breast cancer.[6] Amplification of the Her2/neu gene on 17q was first detected using Southern blotting[37,38] and is thought to occur in approximately 20–25% of breast cancer patients.[39,40] This gene amplification event results in multiple copies of the Her2/neu gene (>5 copies per cell, but perhaps as many as 50–100 copies per cell), and the increase in gene dose correlates with increased mRNA expression and overexpression of the p185/erbB2 protein. Her2/neu is a member of the epidermal growth factor receptor family that also includes Her1, Her3, and Her4. Breast cancers that are positive for Her2/neu amplification and overexpression of p185/erbB2 protein have a poor prognosis and are inherently resistant to common chemotherapeutic regimens.[41] However, the p185/erbB2 protein product of the Her2/neu gene is the target for the monoclonal antibody trastuzumab (Herceptin), and breast cancer patients who have Her2/neu amplification significantly benefit from combination therapies that include trasuzumab, resulting in improved survival.[42,43]

Common Chromosomal Deletions in Cancer

Allelotype studies of lung cancer have identified several recurring chromosomal deletions. In SCLC, frequent loss of heterozygosity (LOH) occurs at 3p (91%), 5q (71%), 13q (96%), 17p (88%), and 22q (73%).[44] NSCLCs display frequent LOH at 2q (68%), 3p (82%), 5q (60%), 9p (79%), 12q (63%), 13q (67%), 17p (89%), 18q (86%), and 22q (75%).[45] However, distinct differences in the patterns of chromosomal deletion have been noted for the histological subtypes of NSCLC. Among squamous cell carcinomas, frequent LOH was noted for 3p (82%), 9q (67%), 13q (60%), and 17p (88%).[46] In contrast, fewer chromosomal losses were noted among adenocarcinomas, with 51% LOH at 17p representing the most frequent alteration.[46] In other studies, similar findings have been reported.[47–49] Deletions affecting a specific chromosomal region (as measured by LOH) may be indicative of the presence of a tumor suppressor gene (or other negative mediator of cell proliferation) at that chromosomal location. Among lung cancers, frequent LOH affecting 3p, 5q, 13q, 17p, and 22q occur in both SCLC and NSCLC. Several regions of chromosome 3p have been implicated in lung cancer, including 3p12–p14, 3p21, and 3p25.[50–52] These observations suggest that there may be three (or more) tumor suppressor genes on human chromosome 3p.[52] Candidate tumor suppressor genes from chromosome 3p include FHIT at 3p12–p14[53] and RASSF1 at 3p21.[54] LOH at chromosome 5q typically corresponds to loss at 5q13–q21.[55] A number of genes map to this chromosomal region, including MCC (for mutated in colorectal cancer) and APC (for adenomatous polyposis coli).[56–59] Although neither of these genes has been shown to be mutated in lung cancer,[60] frequent LOH at this chromosomal region suggests the involvement of one of these or other candidate genes localized to this region in the molecular pathogenesis of lung cancer. The tumor suppressor gene Rb1 localizes to chromosome 13q14.1.[61,62] The expression of Rb1 is altered in a significant percentage of primary lung cancer.[63,64] The p53 tumor suppressor gene is located at 17p13.1[65] and is lost as a result of chromosomal deletion of this region in all lung cancer types.[66] The precise nature of the putative lung cancer tumor suppressor locus at chromosome 22q is not yet defined. However, a candidate gene, termed SEZ6L, has been localized to 22q12.1 and shown to be mutated in a SCLC cell line.[67]

A number of chromosomal regions demonstrate LOH in breast cancer, most commonly affecting 1p, 3p, 6q, 7q, 9p, 11p, 13q, 16q, 17p, 17q, 18q, and 22q.[68–75] Many of these chromosomal locations harbor known or candidate tumor suppressor genes. BRCA1 localizes to 17q21[76] and BRCA2 localizes to 13q12–q13.[77] Other regions of 17p may also contain tumor suppressor loci that are important in breast cancer.[78] For instance, the p53 tumor suppressor gene is located at 17p13.1[65] and has been implicated in the genesis of some breast cancers. Similarly, genes located in other regions of 13q (including the Rb1 locus at 13q14.1[61,62]) may be important in breast carcinogenesis. Chromosome 16q harbors the CDH1 gene, which encodes E-cadherin.[79] Loss of E-cadherin has been implicated in breast cancer metastasis.[80] Other known tumor suppressor genes that localize to chromosomal regions which demonstrate LOH in breast cancer include the DCC gene at 18q.[81] Multiple regions of chromosome 1 and 11 are affected by LOH in breast cancers, suggesting the possibility of multiple distinct tumor suppressor loci on each chromosome.[82–84] Several studies have examined chromosomal alterations in ductal carcinoma in situ (DCIS) in an attempt to identify genetic changes occurring in the preneoplastic breast that may contribute to malignant conversion. These studies found frequent LOH affecting various chromosomal regions including 8p, 13q, 16q, 17p, and 17q in DCIS.[85,86] These results are consistent with the idea that known tumor suppressor genes (such as BRCA1, BRCA2, and p53) may be involved in the early stages of breast carcinogenesis.

Complex Chromosomal Rearrangements in Cancer

Chromosomal translocations have been studied from many years in hematopoietic malignancy. In fact, the first consistent chromosomal aberration observed in human cancer was the so-called Philadelphia chromosome, which was described in chronic myelogenous leukemia.[87,88] This chromosomal aberration was later found to represent a balanced translocation between chromosomes 9 and 22 [t(9;22)(q34.1;q11.2)].[89] Since then, numerous chromosomal translocations have been described in leukemia and lymphoma.

A major chromosomal translocation in acute myelogenous leukemia (AML) involves chromosomes 15 and 17. The resulting chromosomal abnormality [t(15;17)(q21;q21)], which occurs exclusively in acute promyelocytic leukemia (APL), is a balanced and reciprocal translocation in which the PML (for promyelocytic leukemia) gene on chromosome 15 and the RARα gene on chromosome 17 are disrupted and fused to form a hybrid gene.[90,91] The PML–RARα fusion gene (located on chromosome 15) encodes a chimeric mRNA and a novel protein. On the derivative chromosome 15, both the PML and RARα genes are oriented in a head-to-tail orientation. There are three major forms of the PML–RARα fusion gene, corresponding to different breakpoints in the PML gene.[92–94] The breakpoint in the RARα gene occurs in the same general location in all cases, involving the sequences within intron 2. Approximately 40–50% of cases have a PML breakpoint in exon 6 (the long form, termed bcr1), 40–50% of cases have the PML breakpoint in exon 3 (the short form, termed bcr3), and 5–10% of cases have a breakpoint in PML exon 6 that is variable (the variable form, termed bcr2). In each form of the translocation, the PML–RARα fusion protein retains the 5′-DNA-binding and dimerization domains of PML and the 3′-DNA-binding, heterodimerization, and ligand (retinoic acid)-binding domains of RARα. Recent studies indicate that the different forms of PML–RARα fusion mRNA correlate with clinical presentation or prognosis. In particular, the bcr3 type of PML–RARα correlates with higher leukocyte counts at time of presentation.[93,94] Both higher leukocyte counts and variant morphology are adverse prognostic findings, and the bcr3 type of PML–RARα does not independently predict poorer disease-free survival.[94] The presence of the t(15;17) translocation consistently predicts responsiveness to a specific treatment utilizing all-trans-retinoic acid (ATRA). Retinoic acid is a ligand for the retinoic acid receptor (RAR). The gene encoding the RAR is involved in the t(15;17) chromosomal abnormality. ATRA has been suggested to function by overcoming the blockade of myeloid cell maturation, allowing the neoplastic cells to mature (differentiate).[95,96]

Progress toward characterization of complex chromosomal rearrangements in solid tumors was hindered by technical limitations until recently, when spectral karyotyping became available. A number of studies using spectral karyotyping of various human cancers, including lung cancer, have now appeared.[48,97,98] These studies identified a number of unbalanced chromosomal translocations, in many cases involving some of the same chromosomal regions that are frequently deleted in lung cancer. These complex chromosomal rearrangements may alter the structure or expression of genes localized to the affected chromosomal regions. However, additional investigation will be required to characterize the molecular consequences associated with specific chromosomal rearrangements cancers of solid tissues, such as breast, prostate, colon, and lung.

Chromosomal Alterations in Nonneoplastic Diseases

Chromosomal alterations are not limited to neoplastic diseases. Various forms of chromosomal abnormality have been documented in a wide variety of human genetic diseases. A few examples of diseases that are associated with gene inversions, gene deletions, and gene duplications are provided here.

Structural variants affecting specific chromosomes have been identified in the general population to be the cause of genetic disease in the offspring of parents who exhibit certain DNA inversions. Williams–Beuren syndrome has an incidence of 1 in 20,000–50,000 births and is associated with a 1.5-Mb inversion at 7q11.23.[99] This chromosomal abnormality occurs with a 5% frequency in the general population.[100] Angelman syndrome has an incidence of 1 in 10,000 to 20,000 births and is associated with a 4-Mb inversion at 15q12[101]. This chromosomal abnormality occurs in 9% of the general population.[102] The genetic inversions associated with some disease states are not detected in the general population.[103] Patients with hemophilia A carry a 400 Kb inversion in intron 22 of the *factor VIII* gene, and two copies are found in an inverted orientation.[104] This genetic inversion is the product of a nonallelic homologous recombination event that results in the inactivation of the *factor VIII* gene.[105] A smaller genetic inversion in the *emerin* gene has been identified associated with Emery–Dreifuss muscular dystrophy.[106] Hunter syndrome is an X-linked dominant disorder in which nonallelic homologous recombination between the idunorate 2-sulfatase gene (*IDS*) and an *IDS* pseudogene generates a genomic inversion, resulting in a disruption of the functional *IDS* gene. This genetic inversion occurs in approximately 13% of Hunter syndrome patients.[107,108] Soto syndrome is a microdeletion syndrome. Fathers of Soto syndrome patients carry a 1.9-Mb inversion at 5q35 that predisposes their offspring to this disease.[109] Polymorphic inversion at olfactory receptor gene clusters found at 4p16 and 8p23 occur at frequencies of 12.5% and 26%, respectively.[110] These inversion events predispose affected individuals to a form of genetic instability characterized by constitutional chromosomal translocations. Heterozygous carriers of these translocations exhibit no phenotypic characteristics, whereas their offspring who inherit these translocations show phenotypes from mild dysmorphic features to Wolf–Hirschhorn syndrome, characterized by growth defects and severe mental retardation.[111].

α-Thalassemia affects 5–40% of the population in Africa and 40–80% in South Asia. This disorder results from a homologous deletion of an approximately 4-kb segment of DNA that is flanked by two α-globin genes on chromosome 16q13.3. A nonallelic homologous recombination event between these two copies of the α-globin genes results in the deletion of one functional copy, producing the disease phenotype.[101,112,113] Red and green pigment genes are located on Xq23, and individuals that have normal color vision have one copy of the red pigment gene and one or more copies of the green pigment gene.[114] In red-green color blindness, which affects 4–5% of males, genetic deletions or fusions caused by nonallelic homologous recombination occur.[115] In patients with incontinentia pigmenti, an 11-kb deletion occurs secondary to nonallelic homologous recombination between two low-copy repeats with one in the diseased gene (*NEMO*) and another DNA segment that is 4 kb downstream of the gene.[116,117] Hereditary neuropathy with liability to pressure palsy (HNPP) is a common autosomal dominant neurological disorder that is caused by a 1.4-Mb deletion on 17p12.[118] The gene *NF1* that encodes for neurofibromatosis type 1 is located on 17q11.2, and a 1.5-Mb deletion encompassing this gene accounts for 5–22% of patients with this disease.[119,120] Patients with DiGeorge syndrome/velocardiofacial syndrome (DGS/VCFS) can exhibit a 3-Mb deletion within a region-specific repeat unit, LCR22, that is flanked by LCR22A and LCR22D, or a 1.5-Mb deletion that is flanked by LCR22A and LCR22B, located on chromosome 22q11.2.[121] Patients with this congenital disease experience recurrent infection, have heart defects, and exhibit distinct facial features. Smith–Magenis syndrome (SMS) affects 1 in 25,000 individuals and is caused by a 4-Mb deletion that differentially affects several loci on chromosome 17, and the severity of mental retardation exhibited by the patient is determined by the nature of the specific genes involved in the deletion.[122]

Charcot–Marie–Tooth disease (CMT) is an inherited autosomal dominant disorder that occurs in about 1 in 25,000 individuals and is characterized by atrophy of the muscles in the legs, progressing over time to the hands, forearms, and feet. There are two clinical classes of CMT, type I and type II. In CMT type I (CMT1A), 75% of individuals have a duplication in one of the peripheral myelin protein 22 (*PMP22*) genes.[123] Duplication events affecting *PMP22* on both copies of chromosome 17p12 produce a more severe form of CMT1A, reflecting the presence of four copies of the *PMP22* gene.[124] Pelizaeus–Merzbacher disease is a central nervous system disorder affecting the myelin sheath covering the nerve fibers in the brain. The majority of patients with this disease have a duplication of the proteolipid protein gene (*PLP1*), which is found on Xq21–22.[125]

Diseases Associated with Gene Mutations

Gene Mutations in Cancer

Mutations affecting a variety of genes have been characterized in cancer. Some of these mutations represent activating mutations of proto-oncogenes (or other positive mediators of cell proliferation) and others represent inactivating mutations of tumor suppressor genes (or other negative mediators of cell proliferation). Mutations in these genes synergize with other genetic (chromosomal) and epigenetic abnormalities to drive neoplastic transformation and tumorigenesis in affected cells and tissues. This review focuses on the mutation of *ras* and *p53* in cancer.

Mutation of the ras Gene Family in Cancer

Three genes constitute the ras gene family: H-ras, K-ras, and N-ras. The ras protein functions in cell signaling as a hete-rotrimeric G protein, toggling between inactive and active forms. In response to a specific cell stimulus, inactive ras protein releases bound guanosine diphosphate (GDP) and binds a guanosine triphosphate (GTP) molecule.[126,127] In this GTP-bound active configuration, cell signaling occurs until the intrinsic GTPase activity of the ras protein itself cleaves the GTP to GDP, resulting in reacquisition of the inactive configuration.[126,127] Mutant K-ras proteins lack intrinsic GTPase activity and remain in a continuously active form.[128] The constitutive activation of K-ras through this mutational mechanism leads to the induction of multiple signaling pathways involved with cell proliferation and cell survival.[129]

Each of these ras family of oncogenes have been found to be mutated with varying frequencies in major forms of human cancer.[130] K-ras is mutated in cancers of the lung, colon, and pancreas, as well as in myelodysplastic syndrome and some other cancers (such as seminoma). The frequency of K-ras mutation in pancreatic cancer is estimated to be 90%,[130] suggesting that this gene mutation is very important in the malignant conversion of this tissue. H-ras is mutated in cancers of the bladder, kidney, and thyroid, and N-ras is mutated in seminoma, thyroid, and acute myelogenous leukemia.[130] In lung cancer, the majority of ras mutations occur in the K-ras gene.[131,132] K-ras is mutated in 15–20% of all NSCLC and in 30–50% of lung adenocarcinomas,[29,131,133–137] but it is infrequently mutated in other lung cancer types.[29,134] In lung adenocarcinomas, 85% of K-ras mutations affect codon 12.[132] Certain carcinogens found in cigarette smoke, such as benzo[a]pyrene, have been shown to preferentially adduct codon 12 of K-ras, and this adduct is not effectively repaired.[138,139] No K-ras mutations are found in adenocarcinomas of nonsmokers, supporting a specific role of tobacco carcinogens in the mutation of K-ras.[140] The majority of K-ras codon 12 mutations are G to T transversions,[141] resulting in either glycine to cysteine (GGT to TGT) or glycine to valine (GGT to GTT) amino acid substitutions in the mutant protein.

Mutation of the p53 Tumor Suppressor Gene in Cancer

The p53 tumor suppressor gene is one of the most frequently mutated genes in cancer.[142,143] The p53 protein functions as a transcriptional regulator and mediator of cellular responses to DNA damage and stress.[144] Point mutation of p53 leads to synthesis of mutant forms of the protein that do not fold properly,[145] resulting in a nonfunctional protein that will not bind DNA.[146,147] When this mutant protein oligomerizes with other p53 molecules (normal or mutant), the resultant tetramers (or higher-order oligomers) are nonfunctional.[148] Thus, cells with mutant p53 (or p53 deficiency due to chromosomal deletion) become susceptible to progressive genomic instability and accumulation of additional genetic damage.[149,150] Mutational inactivation of p53 has been intensely studied since the discovery of the gene two decades ago. The IARC TP53 Mutation Database (http://www-p53.iarc.fr)[151,152] catalogues reported mutations in the p53 gene. The most recent compilation (Release 12, November 2007) contained 24,819 mutations, including germline and somatic mutations. Of these mutations, 23,544 were characterized as somatic mutations, the majority (17,358, or 74%) of which were missense mutations. Although missense mutations represent the prevalent mutation type affecting p53, numerous other mutation types have been documented, including frameshift, nonsense, and splice site mutations among others. p53 mutations have been documented in subsets of most major forms of human cancer, including cancers of the colorectal, lung, breast, head and neck, esophagus, brain, liver, skin, pancreas, and bladder, as well as in hematological malignancies.

Inactivation of the p53 protein can result from mutations of the p53 gene affecting codons encompassing the majority of the gene, with most of the mutations falling within the highly conserved region of the gene that includes exons 5–9.[143,153–155] Nevertheless, a few hotspots for mutation have been identified. These mutational hotspots correspond to the codons that encode amino acids which are important for the functional interaction of the p53 protein with the DNA molecule.[156] The mutational spectrum of the p53 gene is also known to reflect the carcinogenic insults encountered by various tissues. For example, consider p53 mutations in liver cancers and lung cancers. In hepatocellular carcinoma, p53 is frequently the target of chromosomal deletion involving 17p13.1, perhaps occurring in as many as 41% of liver cancers examined.[157] However, p53 mutation occurs in conjunction with chromosomal deletion of the p53 locus in hepatocellular carcinoma. A summary of numerous studies of p53 mutation in hepatocellular carcinoma suggests that mutations occur in approximately 27% (547/2029) of these tumors.[157] Point mutation of codon 249 of the p53 gene represents a hotspot in hepatocellular carcinoma, accounting for approximately 30% of p53 mutations in hepatocellular carcinoma.[157] This mutation, which results in a G to T transversion (AGG to AGT, arginine to serine), was first recognized in patients from Qidong, China,[158–160] but it has also been recognized in hepatocellular carcinomas from Africa and North America.[161–163] The p53 codon 249 mutation has been attributed to exposure to aflatoxin B_1.[164] Hence, the mutational spectrum of the p53 gene in hepatocellular carcinoma differs significantly between geographic areas with high aflatoxin exposure and those with low exposures.[162,165,166] Similar to liver cancer, the p53 gene may be involved in the molecular pathogenesis of lung cancer through chromosomal deletion of 17p13.1, as well as through gene mutation. Approximately 50% of NSCLCs and 90% of SCLCs harbor mutations in the p53 gene.[167–169] The p53 mutational spectrum in lung cancer indicates that G to T transversions dominate and that specific hotspot codons are frequently mutated (including codons 157, 158, 175, 245, 248, 249, and 273).[170] The types of mutations detected may reflect the interaction of the DNA with specific carcinogens found in cigarette smoke.[171,172]

Gene Mutations in Nonneoplastic Diseases

The phenylalanine hydroxylase gene (*PAH*) maps to chromosome 12q23.2[173,174] and is expressed in the liver and kidney.[175] Mutations in the *PAH* gene, which encodes the enzyme l-phenylalanine hydroxylase, cause phenylketonuria (PKU).[176] PKU is an autosomal recessive inherited disease, causing mental retardation, a mousy odor, light pigmentation, and peculiarities of sitting, standing, and walking, as well as eczema and epilepsy.[176] The average incidence of PKU in the United States is 1 in 8000 individuals.[177] The metabolic defect in PKU reflects the inability of these patients to hydrolyze phenylalanine to tyrosine, resulting in hyperphenylalaninemia. Untreated hyperphenylalaninemia is toxic to the brain and leads to symptoms of mental retardation.[176] The *PAH* locus encompasses 1.5 Mb of DNA with single nucleotide polymorphisms (SNPs), repeat sequences, polymorphisms, and *cis* control elements embedded in the sequence, and harbors five other genes. Thus, a wide range of disease-causing mutations can affect the genes contained in this region, resulting in variable disease manifestations.[176,178] PKU is one of the first genetic disorders to benefit from an effective rational therapy.[176] PKU can be identified with a biochemical test in newborns and can be treated by a phenylalanine-free, tyrosine-supplemented diet, which permits normal or near-normal cognitive development.[176]

Cystic fibrosis (CF), an autosomal recessive disease that occurs in approximately 1 in 3,500 newborns, is the most common lethal inherited genetic disease among the Caucasian population, affecting almost 30,000 Americans.[179–181] Treatment advances for CF patients have increased life expectancy to more than 36 years today.[179] The molecular basis for the development of CF is mutation of the *cystic fibrosis transmembrane conductance regulator* (*CFTR*).[182,183] The diagnosis of CF is based on several clinical characteristics, a family history of CF or a positive CF newborn screening test, and mutation in the *CFTR* gene and/or protein.[184] The symptomatology of CF is highly variable from patient to patient, possibly owing to differences in CFTR function caused by specific mutations.[185–187] Given that epithelia in CF patients behave as though they are impermeable to chloride ions,[188] it was expected that the *CFTR* gene would encode a chloride channel. However, it was found that the *CFTR* gene encodes an ABC transporter homologue that actually functions as a chloride channel which is directly activated by phosphorylation.[183] The most common mutation in the *CFTR* gene, designated ΔF508, results in deletion of phenylalanine at position 508 within nucleotide binding domain 1 of the CFTR protein.[183] The ΔF508 mutation accounts for more than 65% of *CFTR* mutations, although more than 1,000 other *CFTR* mutations have been documented. The ΔF508-CFTR protein is not properly folded during posttranslational processing and is rapidly degraded, resulting in deficiency for the CFTR protein.[189]

The best characterized genetic risk factor for emphysema (and COPD) is α1-antitrypsin deficiency.[190] However, α1-antitrypsin deficiency is a rare condition and only accounts for 1–2% of patients with emphysema,[191] typically those with panacinar pathology.[192] α1-Antitrypsin is a member of the serine protease inhibitor superfamily of proteins, which function to inactivate neutrophil elastase and other proteases to maintain the protease-antiprotease balance.[193] Normally, α1-antitrypsin is synthesized in the liver and secreted into the blood, producing serum concentrations of 20–53 μmol/l.[192] In deficient individuals, α1-antitrypsin levels in the serum fall below 20 μmol/l.[192] A number of different mutant alleles for α1-antitrypsin have been identified and characterized.[194] Most clinically recognized cases of α1-antitrypsin deficiency are caused by the presence of the Z allele, which is characterized by the substitution of lysine for glutamic acid at position 342 (Glu342Lys).[192] The Glu342Lys mutation leads to polymerization of the Z form of α1-antitrypsin within hepatocytes, resulting in impaired secretion and reduction of the plasma levels of α1-antitrypsin to 15% of normal.[193,195] To compound the α1-antitrypsin deficiency in these patients, the Z form of α1-antitrypsin does not function properly.[196] Thus, deficiency in the antielastase activity of α1-antitrypsin caused by diminished serum levels and decreased activity of the defective protein (Z form) sets up a chronic condition in the patient characterized by excess elastase activity and destruction of the lung.

Diseases Associated with Epimutation

Epimutations in Cancer

Neoplastic transformation is associated with alterations in DNA methylation, including both global hypomethylation and gene-specific hypermethylation.[197–199] Gains of DNA methylation in cancer cells typically reflect hypermethylation of CpG islands in gene promoter regions, which contributes to gene silencing.[197] Methylation-dependent gene silencing is a normal mechanism for regulation of gene expression.[200] However, in cancer cells methylation-dependent epigenetic gene silencing represents a mutation-independent mechanism for inactivation of tumor suppressor genes or other negative mediators of neoplastic transformation.[201] A significant number of cancer-related genes have been identified across all chromosome locations that are subject to methylation-dependent silencing,[202] and many of these genes contribute to the hallmarks of cancer.[203] It is likely that more genes are silenced by hypermethylation than through genetic alterations.[204] Additionally, some genes that are silenced though epigenetic mechanisms (including DNA methylation) in many tumor types are not subject to genetic mutation.[204] Similarly, genetic and epigenetic alterations can function together to alter gene expression in cancer cells. There are several examples of genes (*MLH1* and *p16*^{INK4A}) in which one allele is subject to mutation, and subsequently the remaining normal allele becomes silenced through promoter hypermethylation.[204] Thus, epigenetic modifications complement and collaborate with genetic changes to transform normal cells into neoplastic cells. These observations combine to strongly suggest that epigenetic events, and specifically those involving DNA methylation, represent fundamental aspects of cancer and play key roles in neoplastic transformation and progression.

Epimutations in Lung Cancer

Several genes are known to be hypermethylated in lung cancer, resulting in gene silencing.[205] Among these genes, p16^{INK4A} is found to be methylated in more than 40% of lung cancers examined.[206,207] Although it is recognized that hypermethylation of p16^{INK4A} occurs frequently in lung cancer, there is debate regarding the importance of this mechanism of inactivation among lung cancer patients who are smokers versus nonsmokers (or never smokers). In one study, deletion or mutation of the p16^{INK4A} gene was only observed in the lung cancers of smokers, while hypermethylation was found only among nonsmokers.[208] In contrast, other studies have found hypermethylation of p16^{INK4A} as the prevalent mechanism of inactivation among lung cancers from smokers.[209] Other epigenetically regulated genes in lung cancer include CDH1,[210] CDKN1A,[211] DAPK1,[212] ESR1,[213] GJB2,[214] GSTP1,[215] HS3ST2,[216] PRDM2,[217] PRKCDBP,[218] RASSF1,[219] and SFN.[220] Similar to p16^{INK4A}, RASSF1 may be subject to inactivation through multiple molecular mechanisms:

hypermethylation[219] and deletion of chromosome 3p21.[54] Using recently developed array-based methods, a number of putative targets for promoter methylation in lung cancer (including some new gene targets) have been identified, including HIC1, IRF7, ASC, RIPK3, FABP3, and PAX3.[221]

Epimutations in Colorectal Cancer

Methylation of multiple gene targets occurs during colorectal carcinogenesis and some of these methylation events occur early in the multistep process. HLTF, hMLH1, MGMT, APC, and ATM are frequently methylated in colorectal adenomas (42–78%), at least one of these genes was methylated in 100% of the adenomas tested, and 61% contained at least three methylated genes.[222] Similarly, another study found that p16^{INK4A}, MGMT, and hMLH1 are methylated in 34%, 49%, and 7% of colorectal adenomas, respectively.[223] These genes were not methylated in normal colonic mucosa and were infrequently methylated in hyperplastic polyps (5%, 10%, and 7.5%, respectively).[223] These observations suggest that methylation events that occur early in colorectal tumorigenesis may significantly contribute to progression of preneoplastic lesions toward malignant conversion. In fact, the methylation patterns observed in colorectal adenomas are preserved with progression to colorectal carcinoma. HLTF, hMLH1, MGMT, APC, and ATM are methylated in colorectal carcinomas with similar or greater frequency (47–77%) compared to that observed in adenomas, and 68% contain at least three methylated genes.[222]

Epimutations in Breast and Prostate Cancer

A number of studies have shown the presence of aberrant DNA methylation of genes in breast carcinogenesis, affecting genes including p16^{INK4A}, p14^{ARF}, Cyclin D2, and Slit2.[224] The majority (86%) of breast tumors showed hypermethylation of at least one gene, and there was a significant correlation between methylation in tumor and serum DNA of all genes.[224] In other studies, nipple aspiration was applied as a noninvasive technique to obtain DNA samples for examination of tumor-associated gene methylation.[225] Hypermethylation of one or more genes, including GSTP1, RARβ2, p16^{INK4A}, p14^{ARF}, RASSF1, and DAP-kinase, was detected in DNA from nipple aspirate, and identical gene hypermethylation was found in 82% of matched breast tumor tissue samples.[225] However, no aberrant levels of gene methylation were seen in the normal and benign breast tissue or in nipple aspirates from healthy women.[225] Several studies have examined methylation of GSTP1 in prostate cancers.[226] Hypermethylation of GSTP1 has been analyzed in more than 1,000 prostate tumor samples and has been shown to be methylated in the majority (81%) of these tumors. In addition, a recent study of GSTP1, ARF, p16^{INK4A}, and MGMT methylation levels in urine sediments reported that these four genes were able to identify the majority (87%) of prostate cancer patients with extremely high specificity (100%) compared to control samples.[227] These findings suggest that DNA from urine samples could be utilized as a screening tool to reduce the number of biopsies performed on men with a high risk of developing prostate cancer.[226]

Epimutations in Other Cancers

In addition to the major cancer sites described here, aberrant epigenetic regulation of gene expression has been documented in various other human cancers, including cancers of the oral cavity,[228] cervix,[229] ovary,[230] skin,[231] pancreas,[232] and brain,[233] and leukemia[234] and lymphoma.[235]

Epimutations in Nonneoplastic Disease

Evidence in the literature strongly suggests that loss of CFTR function causes cystic fibrosis (CF), and a large number of mutations in the *CFTR* gene have been characterized.[183] However, there is some evidence that the *CFTR* gene may be subject to methylation-dependent epigenetic regulation. The promoter of the *CFTR* gene is remarkably GC rich.[236,237] Early studies found hypermethylation of the *CFTR* promoter in various cell lines analyzed, which correlated with low levels of expression.[238,239] However, more detailed bisulfite sequencing analysis of the *CFTR* promoter has not been reported. Nevertheless, these results suggest that the *CFTR* gene may be subject to epigenetic regulation through promoter hypermethylation, possibly representing a nonmutational mechanism for inactivation of *CFTR*, leading to cystic fibrosis.

References

1. Coleman WB, Rivenbark AG. Quantitative DNA methylation analysis: the promise of high-throughput epigenomic diagnostic testing in human neoplastic disease. *J Mol Diagn.* 2006;8:152–156.
2. Murrell A, Rakyan VK, Beck S. From genome to epigenome. *Hum Mol Genet.* 2005;14(Special No. 1):R3-R10.
3. Fuks F. DNA methylation and histone modifications: teaming up to silence genes. *Curr Opin Genet Dev.* 2005;15:490-495.
4. Kallioniemi A, Kallioniemi OP, Sudar D, et al. Comparative genomic hybridization for molecular cytogenetic analysis of solid tumors. *Science.* 1992;258:818-821.
5. Bignell GR, Huang J, Greshock J, et al. High-resolution analysis of DNA copy number using oligonucleotide microarrays. *Genome Res.* 2006;14:287-295.
6. Albertson DG, Ylstra B, Segraves R, et al. Quantitative mapping of amplicon structure by array CGH identifies CYP24 as a candidate oncogene. *Nat Genet.* 2000;25:144-146.
7. Jain AN, Chin K, Borresen-Dale AL, et al. Quantitative analysis of chromosomal CGH in human breast tumors associates copy number abnormalities with p53 status and patient survival. *Proc Natl Acad Sci USA.* 2001;98:7952-7957.
8. Weiss MM, Kuipers EJ, Postma C, et al. Genomic profiling of gastric cancer predicts lymph node status and survival. *Oncogene.* 2003;22:1872-1879.
9. Weiss MM, Snijders AM, Kuipers EJ, et al. Determination of amplicon boundaries at 20q13.2 in tissue samples of human

gastric adenocarcinomas by high-resolution microarray comparative genomic hybridization. *J Pathol.* 2003;200:320-326.
10. Schwaenen C, Nessling M, Wessendorf S, et al. Automated array-based genomic profiling in chronic lymphocytic leukemia: development of a clinical tool and discovery of recurrent genomic alterations. *Proc Natl Acad Sci USA.* 2004;101:1039-1044.
11. Snijders AM, Nowee ME, Fridlyand J, et al. Genome-wide-array-based comparative genomic hybridization reveals genetic homogeneity and frequent copy number increases encompassing CCNE1 in fallopian tube carcinoma. *Oncogene.* 2003;22:4281-4286.
12. Garnis C, Coe BP, Zhang L, et al. Overexpression of LRP12, a gene contained within an 8q22 amplicon identified by high-resolution array CGH analysis of oral squamous cell carcinomas. *Oncogene.* 2004;23:2582-2586.
13. Veltman JA, Fridlyand J, Pejavar S, et al. Array-based comparative genomic hybridization for genome-wide screening of DNA copy number in bladder tumors. *Cancer Res.* 2003;63:2872-2880.
14. Holzmann K, Kohlhammer H, Schwaenen C, et al. Genomic DNA-chip hybridization reveals a higher incidence of genomic amplifications in pancreatic cancer than conventional comparative genomic hybridization and leads to the identification of novel candidate genes. *Cancer Res.* 2004;64:4428-4433.
15. Berrieman HK, Ashman JN, Cowen ME, et al. Chromosomal analysis of non-small-cell lung cancer by multicolour fluorescent in situ hybridisation. *Br J Cancer.* 2004;90:900-905.
16. Balsara BR, Sonoda G, du Manoir S, et al. Comparative genomic hybridization analysis detects frequent, often high-level, overrepresentation of DNA sequences at 3q, 5p, 7p, and 8q in human non-small cell lung carcinomas. *Cancer Res.* 1997;57:2116-2120.
17. Nowell PC, Croce CM. Chromosomal approaches to the molecular basis of neoplasia. *Symp Fundam Cancer Res.* 1986;39:17-29.
18. Grandori C, Eisenman RN. Myc target genes. *Trends Biochem Sci.* 1997;22:177-181.
19. Hurlin PJ, Huang J. The MAX-interacting transcription factor network. *Semin Cancer Biol* 2006;16:265-274.
20. Blackwood EM, Eisenman RN. Max: a helix-loop-helix zipper protein that forms a sequence-specific DNA-binding complex with Myc. *Science.* 1991;251:1211-1217.
21. Blackwood EM, Kretzner L, Eisenman RN. Myc and Max function as a nucleoprotein complex. *Curr Opin Genet Dev.* 1992;2:227-235.
22. Blackwood EM, Luscher B, Kretzner L, et al. The Myc:Max protein complex and cell growth regulation. *Cold Spring Harb Symp Quant Biol.* 1991;56:109-117.
23. Gazzeri S, Brambilla E, Caron de Fromentel C, et al. p53 genetic abnormalities and myc activation in human lung carcinoma. *Int J Cancer.* 1994;58:24-32.
24. Broers JL, Viallet J, Jensen SM, et al. Expression of c-myc in progenitor cells of the bronchopulmonary epithelium and in a large number of non-small cell lung cancers. *Am J Respir Cell Mol Biol.* 1993;9:33-43.
25. Lorenz J, Friedberg T, Paulus R, et al. Oncogene overexpression in non-small-cell lung cancer tissue: prevalence and clinicopathological significance. *Clin Invest.* 1994;72:156-163.
26. Volm M, Drings P, Wodrich W, et al. Expression of oncoproteins in primary human non-small cell lung cancer and incidence of metastases. *Clin Exp Metastasis.* 1993;11:325-329.
27. Wodrich W, Volm M. Overexpression of oncoproteins in non-small cell lung carcinomas of smokers. *Carcinogenesis.* 1993;14:1121-1124.
28. Mitani S, Kamata H, Fujiwara M, et al. Analysis of c-myc DNA amplification in non-small cell lung carcinoma in comparison with small cell lung carcinoma using polymerase chain reaction. *Clin Exp Med.* 2001;1:105-111.
29. Richardson GE, Johnson BE. The biology of lung cancer. *Semin Oncol.* 1993;20:105-127.
30. Krystal G, Birrer M, Way J, et al. Multiple mechanisms for transcriptional regulation of the myc gene family in small-cell lung cancer. *Mol Cell Biol.* 1988;8:3373-3381.
31. Johnson BE, Russell E, Simmons AM, et al. MYC family DNA amplification in 126 tumor cell lines from patients with small cell lung cancer. *J Cell Biochem Suppl.* 1996;24:210-217.
32. Levin NA, Brzoska P, Gupta N, et al. Identification of frequent novel genetic alterations in small cell lung carcinoma. *Cancer Res.* 1994;54:5086-5091.
33. van Beers EH, Nederlof PM. Array-CGH and breast cancer. *Breast Cancer Res.* 2006;8:210.
34. Albertson DG. Profiling breast cancer by array CGH. *Breast Cancer Res Treat.* 2003;78:289-298.
35. Climent J, Garcia JL, Mao JH, et al. Characterization of breast cancer by array comparative genomic hybridization. *Biochem Cell Biol.* 2007;85:497-508.
36. Jonsson G, Naylor TL, Vallon-Christersson J, et al. Distinct genomic profiles in hereditary breast tumors identified by array-based comparative genomic hybridization. *Cancer Res.* 2005;65: 7612-7621.
37. Slamon DJ, Clark GM, Wong SG, et al. Human breast cancer: correlation of relapse and survival with amplification of the HER-2/neu oncogene. *Science.* 1987;235:177-182.
38. Slamon DJ, Godolphin W, Jones LA, et al. Studies of the HER-2/neu proto-oncogene in human breast and ovarian cancer. *Science.* 1989;244:707-712.
39. Pauletti G, Dandekar S, Rong H, et al. Assessment of methods for tissue-based detection of the HER-2/neu alteration in human breast cancer: a direct comparison of fluorescence in situ hybridization and immunohistochemistry. *J Clin Oncol.* 2000;18:3651-3664.
40. Perez EA, Baweja M. HER2-positive breast cancer: current treatment strategies. *Cancer Invest.* 2008;26:545-552.
41. Oldenhuis CN, Oosting SF, Gietema JA, et al. Prognostic versus predictive value of biomarkers in oncology. *Eur J Cancer.* 2008;44: 946-953.
42. Jahanzeb M. Adjuvant trastuzumab therapy for HER2-positive breast cancer. *Clin Breast Cancer.* 2008;8:324-333.
43. Widakowich C, Dinh P, de Azambuja E, et al. HER-2 positive breast cancer: what else beyond trastuzumab-based therapy? *Anticancer Agents Med Chem.* 2008;8:488-496.
44. Kawanishi M, Kohno T, Otsuka T, et al. Allelotype and replication error phenotype of small cell lung carcinoma. *Carcinogenesis.* 1997;18:2057-2062.
45. Shiseki M, Kohno T, Adachi J, et al. Comparative allelotype of early and advanced stage non-small cell lung carcinomas. *Genes Chromosomes Cancer.* 1996;17:71-77.
46. Sato S, Nakamura Y, Tsuchiya E. Difference of allelotype between squamous cell carcinoma and adenocarcinoma of the lung. *Cancer Res.* 1994;54:5652-5655.
47. Tsuchiya E, Nakamura Y, Weng SY, et al. Allelotype of non-small cell lung carcinoma: comparison between loss of

heterozygosity in squamous cell carcinoma and adenocarcinoma. *Cancer Res*. 1992;52:2478-2481.

48. Luk C, Tsao MS, Bayani J, et al. Molecular cytogenetic analysis of non-small cell lung carcinoma by spectral karyotyping and comparative genomic hybridization. *Cancer Genet Cytogenet*. 2001; 125:87-99.
49. Lui WO, Tanenbaum DM, Larsson C. High level amplification of 1p32-33 and 2p22-24 in small cell lung carcinomas. *Int J Oncol*. 2001;19:451-457.
50. Hibi K, Takahashi T, Yamakawa K, et al. Three distinct regions involved in 3p deletion in human lung cancer. *Oncogene*. 1992;7: 445-449.
51. Brauch H, Tory K, Kotler F, et al. Molecular mapping of deletion sites in the short arm of chromosome 3 in human lung cancer. *Genes Chromosomes Cancer*. 1990;1:240-246.
52. Wistuba II, Behrens C, Virmani AK, et al. High resolution chromosome 3p allelotyping of human lung cancer and preneoplastic/preinvasive bronchial epithelium reveals multiple, discontinuous sites of 3p allele loss and three regions of frequent breakpoints. *Cancer Res*. 2000;60:1949-1960.
53. Zochbauer-Muller S, Wistuba II, Minna JD, et al. Fragile histidine triad (FHIT) gene abnormalities in lung cancer. *Clin Lung Cancer*. 2000;2:141-145.
54. Dammann R, Schagdarsurengin U, Seidel C, et al. The tumor suppressor RASSF1A in human carcinogenesis: an update. *Histol Histopathol*. 2005;20:645-663.
55. Hosoe S, Ueno K, Shigedo Y, et al. A frequent deletion of chromosome 5q21 in advanced small cell and non-small cell carcinoma of the lung. *Cancer Res*. 1994;54:1787-1790.
56. Kinzler KW, Nilbert MC, Su LK, et al. Identification of FAP locus genes from chromosome 5q21. *Science*. 1991;253:661-665.
57. Kinzler KW, Nilbert MC, Vogelstein B, et al. Identification of a gene located at chromosome 5q21 that is mutated in colorectal cancers. *Science*. 1991;251:1366-1370.
58. Groden J, Thliveris A, Samowitz W, et al. Identification and characterization of the familial adenomatous polyposis coli gene. *Cell*. 1991;66:589-600.
59. Joslyn G, Carlson M, Thliveris A, et al. Identification of deletion mutations and three new genes at the familial polyposis locus. *Cell*. 1991;66:601-613.
60. Horii A, Nakatsuru S, Miyoshi Y, et al. Frequent somatic mutations of the APC gene in human pancreatic cancer. *Cancer Res*. 1992;52:6696-6698.
61. Fung YK, Murphree AL, T'Ang A, et al. Structural evidence for the authenticity of the human retinoblastoma gene. *Science*. 1987;236:1657-1661.
62. Lee WH, Bookstein R, Hong F, et al. Human retinoblastoma susceptibility gene: cloning, identification, and sequence. *Science*. 1987;235:1394-1399.
63. Hensel CH, Hsieh CL, Gazdar AF, et al. Altered structure and expression of the human retinoblastoma susceptibility gene in small cell lung cancer. *Cancer Res*. 1990;50:3067-3072.
64. Xu HJ, Hu SX, Cagle PT, et al. Absence of retinoblastoma protein expression in primary non-small cell lung carcinomas. *Cancer Res*. 1991;51:2735-2739.
65. McBride OW, Merry D, Givol D. The gene for human p53 cellular tumor antigen is located on chromosome 17 short arm (17p13). *Proc Natl Acad Sci USA*. 1986;83:130-134.
66. Yokota J, Wada M, Shimosato Y, et al. Loss of heterozygosity on chromosomes 3, 13, and 17 in small-cell carcinoma and on chromosome 3 in adenocarcinoma of the lung. *Proc Natl Acad Sci USA*. 1987;84:9252-9256.
67. Nishioka M, Kohno T, Takahashi M, et al. Identification of a 428-kb homozygously deleted region disrupting the SEZ6L gene at 22q12.1 in a lung cancer cell line. *Oncogene*. 2000;19:6251-6260.
68. Callahan R, Cropp C, Merlo GR, et al. Genetic and molecular heterogeneity of breast cancer cells. *Clin Chim Acta*. 1993;217:63-73.
69. Cleton-Jansen AM, Moerland EW, Kuipers-Dijkshoorn NJ, et al. At least two different regions are involved in allelic imbalance on chromosome arm 16q in breast cancer. *Genes Chromosomes Cancer*. 1994;9:101-107.
70. Eiriksdottir G, Sigurdsson A, Jonasson JG, et al. Loss of heterozygosity on chromosome 9 in human breast cancer: association with clinical variables and genetic changes at other chromosome regions. *Int J Cancer*. 1995;64:378-382.
71. Gudmundsson J, Barkardottir RB, Eiriksdottir G, et al. Loss of heterozygosity at chromosome 11 in breast cancer: association of prognostic factors with genetic alterations. *Br J Cancer*. 1995;72:696-701.
72. Morelli C, Sherratt T, Trabanelli C, et al. Characterization of a 4-Mb region at chromosome 6q21 harboring a replicative senescence gene. *Cancer Res*. 1997;57:4153-4157.
73. Huang H, Qian C, Jenkins RB, et al. Fish mapping of YAC clones at human chromosomal band 7q31.2: identification of YACS spanning FRA7G within the common region of LOH in breast and prostate cancer. *Genes Chromosomes Cancer*. 1998;21:152-159.
74. Mars WM, Saunders GF. Chromosomal abnormalities in human breast cancer. *Cancer Metastasis Rev*. 1990;9:35-43.
75. Kerangueven F, Noguchi T, Coulier F, et al. Genome-wide search for loss of heterozygosity shows extensive genetic diversity of human breast carcinomas. *Cancer Res*. 1997;57:5469-5474.
76. Miki Y, Swensen J, Shattuck-Eidens D, et al. A strong candidate for the breast and ovarian cancer susceptibility gene BRCA1. *Science*. 1994;266:66-71.
77. Wooster R, Neuhausen SL, Mangion J, et al. Localization of a breast cancer susceptibility gene, BRCA2, to chromosome 13q12-13. *Science*. 1994;265:2088-2090.
78. Schultz DC, Vanderveer L, Berman DB, et al. Identification of two candidate tumor suppressor genes on chromosome 17p13.3. *Cancer Res*. 1996;56:1997-2002.
79. Cleton-Jansen AM. E-cadherin and loss of heterozygosity at chromosome 16 in breast carcinogenesis: different genetic pathways in ductal and lobular breast cancer? *Breast Cancer Res*. 2002;4:5-8.
80. Debies MT, Welch DR. Genetic basis of human breast cancer metastasis. *J Mammary Gland Biol Neoplasia*. 2001;6:441-451.
81. Fearon ER, Cho KR, Nigro JM, et al. Identification of a chromosome 18q gene that is altered in colorectal cancers. *Science*. 1990;247:49-56.
82. Bieche I, Champeme MH, Lidereau R. Loss and gain of distinct regions of chromosome 1q in primary breast cancer. *Clin Cancer Res*. 1995;1:123-127.
83. Bieche I, Khodja A, Lidereau R. Deletion mapping in breast tumor cell lines points to two distinct tumor-suppressor genes in the 1p32-pter region, one of deleted regions (1p36.2) being located within the consensus region of LOH in neuroblastoma. *Oncol Rep*. 1998;5:267-272.

84. Kerangueven F, Eisinger F, Noguchi T, et al. Loss of heterozygosity in human breast carcinomas in the ataxia telangiectasia, Cowden disease and BRCA1 gene regions. *Oncogene*. 1997;14:339-347.
85. Aldaz CM, Chen T, Sahin A, et al. Comparative allelotype of in situ and invasive human breast cancer: high frequency of microsatellite instability in lobular breast carcinomas. *Cancer Res*. 1995;55:3976-3981.
86. Radford DM, Fair KL, Phillips NJ, et al. Allelotyping of ductal carcinoma in situ of the breast: deletion of loci on 8p, 13q, 16q, 17p and 17q. *Cancer Res*. 1995;55:3399-3405.
87. Nowell PC, Hungerford DA. A minute chromosome in human chronic granulocytic leukemia [abstract]. *Science*. 1960;132:1497.
88. Nowell PC, Hungerford DA. Chromosome studies on normal and leukemic human leukocytes. *J Natl Cancer Inst*. 1960;25:85-109.
89. Rowley JD. Letter: A new consistent chromosomal abnormality in chronic myelogenous leukaemia identified by quinacrine fluorescence and Giemsa staining. *Nature*. 1973;243:290-293.
90. Kakizuka A, Miller WH Jr, Umesono K, et al. Chromosomal translocation t(15;17) in human acute promyelocytic leukemia fuses RAR alpha with a novel putative transcription factor, PML. *Cell*. 1991;66:663-674.
91. de The H, Lavau C, Marchio A, et al. The PML-RAR alpha fusion mRNA generated by the t(15;17) translocation in acute promyelocytic leukemia encodes a functionally altered RAR. *Cell*. 1991;66:675-684.
92. Miller WH, Jr., Kakizuka A, Frankel SR, et al. Reverse transcription polymerase chain reaction for the rearranged retinoic acid receptor alpha clarifies diagnosis and detects minimal residual disease in acute promyelocytic leukemia. *Proc Natl Acad Sci USA*. 1992;89:2694-2698.
93. Huang W, Sun GL, Li XS, et al. Acute promyelocytic leukemia: clinical relevance of two major PML-RAR alpha isoforms and detection of minimal residual disease by retrotranscriptase/polymerase chain reaction to predict relapse. *Blood*. 1993; 82:1264-1269.
94. Gallagher RE, Willman CL, Slack JL, et al. Association of PML-RAR alpha fusion mRNA type with pretreatment hematologic characteristics but not treatment outcome in acute promyelocytic leukemia: an intergroup molecular study. *Blood*. 1997;90:1656-1663.
95. Vyas RC, Frankel SR, Agbor P, et al. Probing the pathobiology of response to all-trans retinoic acid in acute promyelocytic leukemia: premature chromosome condensation/fluorescence in situ hybridization analysis. *Blood*. 1996;87:218-226.
96. Grignani F, Fagioli M, Alcalay M, et al. Acute promyelocytic leukemia: from genetics to treatment. *Blood*. 1994;83:10-25.
97. Sy SM, Fan B, Lee TW, et al. Spectral karyotyping indicates complex rearrangements in lung adenocarcinoma of nonsmokers. *Cancer Genet Cytogenet*. 2004;153:57-59.
98. Sy SM, Wong N, Lee TW, et al. Distinct patterns of genetic alterations in adenocarcinoma and squamous cell carcinoma of the lung. *Eur J Cancer*. 2004;40:1082-1094.
99. Greenberg F. Williams syndrome. *Pediatrics*. 1989;84:922-923.
100. Osborne LR, Li M, Pober B, et al. A 1.5 million-base pair inversion polymorphism in families with Williams–Beuren syndrome. *Nat Genet*. 2001;29:321-325.
101. Inoue K, Lupski JR. Molecular mechanisms for genomic disorders. *Annu Rev Genomic Hum Genet*. 2002;3:199-242.
102. Gimelli G, Pujana MA, Patricelli MG, et al. Genomic inversions of human chromosome 15q11–q13 in mothers of Angelman syndrome patients with class II (BP2/3) deletions. *Hum Mol Genet*. 2003;12:849-858.
103. Feuk L, Carson AR, Scherer SW. Structural variation in the human genome. *Nat Rev Genet*. 2006;7:85-97.
104. Lakich D, Kazazian HH Jr, Antonarakis SE, et al. Inversions disrupting the factor VIII gene are a common cause of severe haemophilia A. *Nat Genet*. 1993;5:236-241.
105. Naylor J, Brinke A, Hassock S, et al. Characteristic mRNA abnormality found in half the patients with severe haemophilia A is due to large DNA inversions. *Hum Mol Genet*. 1993;2: 1773-1778.
106. Small K, Iber J, Warren ST. Emerin deletion reveals a common X-chromosome inversion mediated by inverted repeats. *Nat Genet*. 1997;16:96-99.
107. Timms KM, Bondeson ML, Ansari-Lari MA, et al. Molecular and phenotypic variation in patients with severe Hunter syndrome. *Hum Mol Genet*. 1997;6:479-486.
108. Bondeson ML, Dahl N, Malmgren H, et al. Inversion of the IDS gene resulting from recombination with IDS-related sequences is a common cause of the Hunter syndrome. *Hum Mol Genet*. 1995;4:615-621.
109. Visser R, Shimokawa O, Harada N, et al. Identification of a 3.0-kb major recombination hotspot in patients with Sotos syndrome who carry a common 1.9-Mb microdeletion. *Am J Hum Genet*. 2005;76:52-67.
110. Giglio S, Broman KW, Matsumoto N, et al. Olfactory receptor-gene clusters, genomic-inversion polymorphisms, and common chromosome rearrangements. *Am J Hum Genet*. 2001;68:874-883.
111. Giglio S, Calvari V, Gregato G, et al. Heterozygous submicroscopic inversions involving olfactory receptor-gene clusters mediate the recurrent t(4;8)(p16;p23) translocation. *Am J Hum Genet*. 2002;71:276-285.
112. Higgs DR, Old JM, Pressley L, et al. A novel alpha-globin gene arrangement in man. *Nature*. 1980;284:632-635.
113. Lauer J, Shen CK, Maniatis T. The chromosomal arrangement of human alpha-like globin genes: sequence homology and alpha-globin gene deletions. *Cell*. 1980;20:119-130.
114. Vollrath D, Nathans J, Davis RW. Tandem array of human visual pigment genes at Xq28. *Science*. 1988;240:1669-1672.
115. Nathans J, Piantanida TP, Eddy RL, et al. Molecular genetics of inherited variation in human color vision. *Science*. 1986; 232:203-210.
116. Aradhya S, Woffendin H, Jakins T, et al. A recurrent deletion in the ubiquitously expressed NEMO (IKK-gamma) gene accounts for the vast majority of incontinentia pigmenti mutations. *Hum Mol Genet*. 2001;10:2171-2179.
117. Smahi A, Courtois G, Vabres P, et al. Genomic rearrangement in NEMO impairs NF-kappaB activation and is a cause of incontinentia pigmenti. The International Incontinentia Pigmenti (IP) Consortium. *Nature*. 2000;405:466-472.
118. Chance PF, Alderson MK, Leppig KA, et al. DNA deletion associated with hereditary neuropathy with liability to pressure palsies. *Cell*. 1993;72:143-151.
119. Upadhyaya M, Ruggieri M, Maynard J, et al. Gross deletions of the neurofibromatosis type 1 (NF1) gene are predominantly

of maternal origin and commonly associated with a learning disability, dysmorphic features and developmental delay. *Hum Genet.* 1998;102:591-597.
120. Valero MC, Pascual-Castroviejo I, Velasco E, et al. Identification of de novo deletions at the NF1 gene: no preferential paternal origin and phenotypic analysis of patients. *Hum Genet.* 1997;99: 720-726.
121. Edelmann L, Pandita RK, Spiteri E, et al. A common molecular basis for rearrangement disorders on chromosome 22q11. *Hum Mol Genet.* 1999;8:1157-1167.
122. Chen KS, Manian P, Koeuth T, et al. Homologous recombination of a flanking repeat gene cluster is a mechanism for a common contiguous gene deletion syndrome. *Nat Genet.* 1997;17:154-163.
123. Pasternak JJ. Molecular genetics of neurological disorders. In: Pasternak JJ, ed. An Introduction to Human Molecular Genetics: Mechanisms of Inherited Diseases. Bethesda, MD: Fitzgerald Science Press, 1999:286-287.
124. Lupski JR, de Oca-Luna RM, Slaugenhaupt S, et al. DNA duplication associated with Charcot-Marie-Tooth disease type 1A. *Cell.* 1991;66:219-232.
125. Inoue K, Osaka H, Imaizumi K, et al. Proteolipid protein gene duplications causing Pelizaeus–Merzbacher disease: molecular mechanism and phenotypic manifestations. *Ann Neurol.* 1999;45:624-632.
126. Barbacid M. Ras genes. *Annu Rev Biochem.* 1987;56:779-827.
127. Barbacid M. Ras oncogenes: their role in neoplasia. *Eur J Clin Invest.* 1990;20:225-235.
128. Wittinghofer A, Scheffzek K, Ahmadian MR. The interaction of Ras with GTPase-activating proteins. *FEBS Lett.* 1997;410:63-67.
129. Hancock JF. Ras proteins: different signals from different locations. *Nat Rev Mol Cell Biol.* 2003;4:373-384.
130. Downward J. Targeting RAS signalling pathways in cancer therapy. *Nat Rev Cancer.* 2003;3:11-22.
131. Mills NE, Fishman CL, Rom WN, et al. Increased prevalence of K-ras oncogene mutations in lung adenocarcinoma. *Cancer Res.* 1995;55:1444-1447.
132. Slebos RJ, Kibbelaar RE, Dalesio O, et al. K-ras oncogene activation as a prognostic marker in adenocarcinoma of the lung. *N Engl J Med.* 1990;323:561-565.
133. Rodenhuis S, Slebos RJ, Boot AJ, et al. Incidence and possible clinical significance of K-ras oncogene activation in adenocarcinoma of the human lung. *Cancer Res.* 1988;48:5738-5741.
134. Rodenhuis S, Slebos RJ. Clinical significance of ras oncogene activation in human lung cancer. *Cancer Res.* 1992;52:2665s-2669s.
135. Reynolds SH, Anna CK, Brown KC, et al. Activated protooncogenes in human lung tumors from smokers. *Proc Natl Acad Sci USA.* 1991;88:1085-1089.
136. Suzuki Y, Orita M, Shiraishi M, et al. Detection of ras gene mutations in human lung cancers by single-strand conformation polymorphism analysis of polymerase chain reaction products. *Oncogene.* 1990;5:1037-1043.
137. Li S, Rosell R, Urban A, et al. K-ras gene point mutation: a stable tumor marker in non-small cell lung carcinoma. *Lung Cancer.* 1994;11:19-27.
138. Feng Z, Hu W, Chen JX, et al. Preferential DNA damage and poor repair determine *ras* gene mutational hotspot in human cancer. *J Natl Cancer Inst.* 2002;94:1527-1536.
139. Hu W, Feng Z, Tang MS. Preferential carcinogen-DNA adduct formation at codons 12 and 14 in the human K-ras gene and their possible mechanisms. *Biochemistry.* 2003;42:10012-10023.
140. Ahrendt SA, Decker PA, Alawi EA, et al. Cigarette smoking is strongly associated with mutation of the K-ras gene in patients with primary adenocarcinoma of the lung. *Cancer.* 2001;92:1525-1530.
141. Slebos RJ, Hruban RH, Dalesio O, et al. Relationship between K-ras oncogene activation and smoking in adenocarcinoma of the human lung. *J Natl Cancer Inst.* 1991;83:1024-1027.
142. Hollstein M, Sidransky D, Vogelstein B, et al. p53 mutations in human cancers. *Science.* 1991;253:49-53.
143. Hollstein M, Shomer B, Greenblatt M, et al. Somatic point mutations in the p53 gene of human tumors and cell lines: updated compilation. *Nucleic Acids Res.* 1996;24:141-146.
144. Kastan MB, Onyekwere O, Sidransky D, et al. Participation of p53 protein in the cellular response to DNA damage. *Cancer Res.* 1991;51:6304-6311.
145. Gannon JV, Greaves R, Iggo R, et al. Activating mutations in p53 produce a common conformational effect. A monoclonal antibody specific for the mutant form. *EMBO J.* 1990;9:1595-1602.
146. Kern SE, Kinzler KW, Baker SJ, et al. Mutant p53 proteins bind DNA abnormally in vitro. *Oncogene.* 1991;6:131-136.
147. Kern SE, Kinzler KW, Bruskin A, et al. Identification of p53 as a sequence-specific DNA-binding protein. *Science.* 1991;252: 1708-1711.
148. Kern SE, Pietenpol JA, Thiagalingam S, et al. Oncogenic forms of p53 inhibit p53-regulated gene expression. *Science.* 1992;256:827-830.
149. Kramer A, Neben K, Ho AD. Centrosome replication, genomic instability and cancer. *Leukemia.* 2002;16:767-775.
150. Wahl GM, Linke SP, Paulson TG, et al. Maintaining genetic stability through TP53 mediated checkpoint control. *Cancer Surv.* 1997;29:183-219.
151. Olivier M, Eeles R, Hollstein M, et al. The IARC TP53 database: new online mutation analysis and recommendations to users. *Hum Mutat.* 2002;19:607-614.
152. Petitjean A, Mathe E, Kato S, et al. Impact of mutant p53 functional properties on TP53 mutation patterns and tumor phenotype: lessons from recent developments in the IARC TP53 database. *Hum Mutat.* 2007;28:622-629.
153. Hainaut P, Hernandez T, Robinson A, et al. IARC Database of p53 gene mutations in human tumors and cell lines: updated compilation, revised formats and new visualisation tools. *Nucleic Acids Res.* 1998;26:205-213.
154. Hainaut P, Soussi T, Shomer B, et al. Database of p53 gene somatic mutations in human tumors and cell lines: updated compilation and future prospects. *Nucleic Acids Res.* 1997;25:151-157.
155. Hollstein M, Rice K, Greenblatt MS, et al. Database of p53 gene somatic mutations in human tumors and cell lines. *Nucleic Acids Res.* 1994;22:3551-3555.
156. Petitjean A, Achatz MI, Borresen-Dale AL, et al. TP53 mutations in human cancers: functional selection and impact on cancer prognosis and outcomes. *Oncogene.* 2007;26:2157-2165.
157. Grisham JW. Molecular genetic alterations in primary hepatocellular neoplasms: Hepatocellular adenoma, hepatocellular carcinoma, and hepatoblastoma. In: Coleman WB, Tsongalis GJ, eds. The Molecular Basis of Human Cancer. Totowa: Humana Press, 2002:269-346.

158. Hsu IC, Metcalf RA, Sun T, et al. Mutational hotspot in the p53 gene in human hepatocellular carcinomas. *Nature*. 1991;350: 427-428.
159. Scorsone KA, Zhou YZ, Butel JS, et al. p53 mutations cluster at codon 249 in hepatitis B virus-positive hepatocellular carcinomas from China. *Cancer Res*. 1992;52:1635-1638.
160. Li D, Cao Y, He L, et al. Aberrations of p53 gene in human hepatocellular carcinoma from China. *Carcinogenesis*. 1993;14: 169-173.
161. Bressac B, Kew M, Wands J, et al. Selective G to T mutations of p53 gene in hepatocellular carcinoma from southern Africa. *Nature*. 1991;350:429-431.
162. Ozturk M. p53 mutation in hepatocellular carcinoma after aflatoxin exposure. *Lancet*. 1991;338:1356-1359.
163. Soini Y, Chia SC, Bennett WP, et al. An aflatoxin-associated mutational hotspot at codon 249 in the p53 tumor suppressor gene occurs in hepatocellular carcinomas from Mexico. *Carcinogenesis*. 1996;17:1007-1012.
164. Puisieux A, Lim S, Groopman J, et al. Selective targeting of p53 gene mutational hotspots in human cancers by etiologically defined carcinogens. *Cancer Res*. 1991;51:6185-6189.
165. Oda T, Tsuda H, Scarpa A, et al. p53 gene mutation spectrum in hepatocellular carcinoma. *Cancer Res*. 1992;52:6358-6364.
166. Kress S, Jahn UR, Buchmann A, et al. p53 mutations in human hepatocellular carcinomas from Germany. *Cancer Res*. 1992;52: 3220-3223.
167. Chiba I, Takahashi T, Nau MM, et al. Mutations in the p53 gene are frequent in primary, resected non-small cell lung cancer. Lung Cancer Study Group. *Oncogene*. 1990;5:1603-1610.
168. Curiel DT, Buchhagen DL, Chiba I, et al. A chemical mismatch cleavage method useful for the detection of point mutations in the p53 gene in lung cancer. *Am J Respir Cell Mol Biol*. 1990;3: 405-411.
169. D'Amico D, Carbone D, Mitsudomi T, et al. High frequency of somatically acquired p53 mutations in small-cell lung cancer cell lines and tumors. *Oncogene*. 1992;7:339-346.
170. Robles AI, Linke SP, Harris CC. The p53 network in lung carcinogenesis. *Oncogene*. 2002;21:6898-6907.
171. Greenblatt MS, Bennett WP, Hollstein M, et al. Mutations in the p53 tumor suppressor gene: clues to cancer etiology and molecular pathogenesis. *Cancer Res*. 1994;54:4855-4878.
172. Denissenko MF, Pao A, Tang M, et al. Preferential formation of benzo[a]pyrene adducts at lung cancer mutational hotspots in P53. *Science*. 1996;274:430-432.
173. Woo SL, Lidsky AS, Guttler F, et al. Cloned human phenylalanine hydroxylase gene allows prenatal diagnosis and carrier detection of classical phenylketonuria. *Nature*. 1983;306:151-155.
174. Kwok SC, Ledley FD, DiLella AG, et al. Nucleotide sequence of a full-length complementary DNA clone and amino acid sequence of human phenylalanine hydroxylase. *Biochemistry*. 1985;24:556-561.
175. Wang Y, DeMayo JL, Hahn TM, et al. Tissue- and development-specific expression of the human phenylalanine hydroxylase/chloramphenicol acetyltransferase fusion gene in transgenic mice. *J Biol Chem*. 1992;267:15105-15110.
176. Scriver CR. The PAH gene, phenylketonuria, and a paradigm shift. *Hum Mutat*. 2007;28:831-845.
177. DiLella AG, Kwok SC, Ledley FD, et al. Molecular structure and polymorphic map of the human phenylalanine hydroxylase gene. *Biochemistry*. 1986;25:743-749.
178. Konecki DS, Wang Y, Trefz FK, et al. Structural characterization of the 5′ regions of the human phenylalanine hydroxylase gene. *Biochemistry*. 1992;31:8363-8368.
179. Farrell PM, Rosenstein BJ, White TB, et al. Guidelines for diagnosis of cystic fibrosis in newborns through older adults: Cystic Fibrosis Foundation consensus report. *J Pediatr*. 2008;153:S4-S14.
180. Davis PB. Cystic fibrosis: new perceptions, new strategies. *Hosp Pract (Off Ed)*. 1992;27:79-83, 87-78, 93-74 passim.
181. Davies JC, Alton EW, Bush A. Cystic fibrosis. *BMJ*. 2007;335: 1255-1259.
182. Riordan JR, Rommens JM, Kerem B, et al. Identification of the cystic fibrosis gene: cloning and characterization of complementary DNA. *Science*. 1989;245:1066-1073.
183. Gadsby DC, Vergani P, Csanady L. The ABC protein turned chloride channel whose failure causes cystic fibrosis. *Nature*. 2006;440:477-483.
184. Rosenstein BJ, Cutting GR. The diagnosis of cystic fibrosis: a consensus statement. Cystic Fibrosis Foundation Consensus Panel. *J Pediatr*. 1998;132:589-595.
185. Zielenski J, Tsui LC. Cystic fibrosis: genotypic and phenotypic variations. *Annu Rev Genet*. 1995;29:777-807.
186. Mickle JE, Cutting GR. Genotype-phenotype relationships in cystic fibrosis. *Med Clin North Am*. 2000;84:597-607.
187. Zielenski J. Genotype and phenotype in cystic fibrosis. *Respiration*. 2000;67:117-133.
188. Quinton PM. Chloride impermeability in cystic fibrosis. *Nature*. 1983;301:421-422.
189. Du K, Sharma M, Lukacs GL. The DeltaF508 cystic fibrosis mutation impairs domain–domain interactions and arrests post-translational folding of CFTR. *Nat Struct Mol Biol*. 2005;12:17-25.
190. Tomashefski JF Jr, Crystal RG, Wiedemann HP, et al. The bronchopulmonary pathology of alpha-1 antitrypsin (AAT) deficiency: findings of the Death Review Committee of the national registry for individuals with Severe Deficiency of Alpha-1 Antitrypsin. *Hum Pathol*. 2004;35:1452-1461.
191. Lieberman J, Winter B, Sastre A. Alpha 1-antitrypsin Pi-types in 965 COPD patients. *Chest*. 1986;89:370-373.
192. Stoller JK, Aboussouan LS. Alpha 1-antitrypsin deficiency. *Lancet*. 2005;365:2225-2236.
193. Lomas DA, Mahadeva R. Alpha 1-antitrypsin polymerization and the serpinopathies: pathobiology and prospects for therapy. *J Clin Invest*. 2002;110:1585-1590.
194. DeMeo DL, Silverman EK. Alpha 1-antitrypsin deficiency. 2: Genetic aspects of alpha(1)-antitrypsin deficiency: phenotypes and genetic modifiers of emphysema risk. *Thorax*. 2004;59:259-264.
195. Lomas DA, Evans DL, Finch JT, et al. The mechanism of Z alpha 1-antitrypsin accumulation in the liver. *Nature*. 1992;357:605-607.
196. Ogushi F, Fells GA, Hubbard RC, et al. Z-type alpha 1-antitrypsin is less competent than M1-type alpha 1-antitrypsin as an inhibitor of neutrophil elastase. *J Clin Invest*. 1987;80:1366-1374.
197. Herman JG, Baylin SB. Gene silencing in cancer in association with promoter hypermethylation. *N Engl J Med*. 2003;349:2042-2054.
198. Baylin S. DNA methylation and epigenetic mechanisms of carcinogenesis. *Dev Biol*. 2001;106:85-87.
199. Baylin SB, Herman JG, Graff JR, et al. Alterations in DNA methylation: a fundamental aspect of neoplasia. *Adv Cancer Res*. 1998;72:141-196.

200. Momparler RL. Cancer epigenetics. *Oncogene*. 2003;22: 6479-6483.
201. Jones PA, Laird PW. Cancer epigenetics comes of age. *Nat Genet*. 1999;21:163-167.
202. Tsou JA, Hagen JA, Carpenter CL, et al. DNA methylation analysis: a powerful new tool for lung cancer diagnosis. *Oncogene*. 2002;21:5450-5461.
203. Hanahan D, Weinberg RA. The hallmarks of cancer. *Cell*. 2000;100:57-70.
204. Baylin SB, Ohm JE. Epigenetic gene silencing in cancer: a mechanism for early oncogenic pathway addiction. *Nat Rev Cancer*. 2006;6:107-116.
205. Toyooka S, Toyooka KO, Maruyama R, et al. DNA methylation profiles of lung tumors. *Mol Cancer Ther*. 2001;1:61-67.
206. Tanaka R, Wang D, Morishita Y, et al. Loss of function of p16 gene and prognosis of pulmonary adenocarcinoma. *Cancer*. 2005;103:608-615.
207. Shapiro GI, Park JE, Edwards CD, et al. Multiple mechanisms of p16INK4A inactivation in non-small cell lung cancer cell lines. *Cancer Res*. 1995;55:6200-6209.
208. Sanchez-Cespedes M, Decker PA, Doffek KM, et al. Increased loss of chromosome 9p21 but not p16 inactivation in primary non-small cell lung cancer from smokers. *Cancer Res*. 2001;61: 2092-2096.
209. Liu Y, Lan Q, Siegfried JM, et al. Aberrant promoter methylation of p16 and MGMT genes in lung tumors from smoking and never-smoking lung cancer patients. *Neoplasia*. 2006;8:46-51.
210. Tsou JA, Shen LY, Siegmund KD, et al. Distinct DNA methylation profiles in malignant mesothelioma, lung adenocarcinoma, and non-tumor lung. *Lung Cancer*. 2005;47:193-204.
211. Zhu WG, Srinivasan K, Dai Z, et al. Methylation of adjacent CpG sites affects Sp1/Sp3 binding and activity in the p21(Cip1) promoter. *Mol Cell Biol*. 2003;23:4056-4065.
212. Soria JC, Rodriguez M, Liu DD, et al. Aberrant promoter methylation of multiple genes in bronchial brush samples from former cigarette smokers. *Cancer Res*. 2002;62:351-355.
213. Lai JC, Cheng YW, Chiou HL, et al. Gender difference in estrogen receptor alpha promoter hypermethylation and its prognostic value in non-small cell lung cancer. *Int J Cancer*. 2005;117:974-980.
214. Chen Y, Huhn D, Knosel T, et al. Downregulation of connexin 26 in human lung cancer is related to promoter methylation. *Int J Cancer*. 2005;113:14-21.
215. Wistuba II, Gazdar AF, Minna JD. Molecular genetics of small cell lung carcinoma. *Semin Oncol*. 2001;28:3-13.
216. Miyamoto K, Asada K, Fukutomi T, et al. Methylation-associated silencing of heparan sulfate d-glucosaminyl 3-*O*-sulfotransferase-2 (3-OST-2) in human breast, colon, lung and pancreatic cancers. *Oncogene*. 2003;22:274-280.
217. Du Y, Carling T, Fang W, et al. Hypermethylation in human cancers of the RIZ1 tumor suppressor gene, a member of a histone/protein methyltransferase superfamily. *Cancer Res*. 2001;61:8094-8099.
218. Xu XL, Wu LC, Du F, et al. Inactivation of human SRBC, located within the 11p15.5-p15.4 tumor suppressor region, in breast and lung cancers. *Cancer Res*. 2001;61:7943-7949.
219. Dammann R, Takahashi T, Pfeifer GP. The CpG island of the novel tumor suppressor gene RASSF1A is intensely methylated in primary small cell lung carcinomas. *Oncogene*. 2001;20:3563-3567.
220. Osada H, Tatematsu Y, Yatabe Y, et al. Frequent and histological type-specific inactivation of 14-3-3sigma in human lung cancers. *Oncogene*. 2002;21:2418-2424.
221. Fukasawa M, Kimura M, Morita S, et al. Microarray analysis of promoter methylation in lung cancers. *J Hum Genet*. 2006;51:368-374.
222. Bai AH, Tong JH, To KF, et al. Promoter hypermethylation of tumor-related genes in the progression of colorectal neoplasia. *Int J Cancer*. 2004;112:846-853.
223. Petko Z, Ghiassi M, Shuber A, et al. Aberrantly methylated CDKN2A, MGMT, and MLH1 in colon polyps and in fecal DNA from patients with colorectal polyps. *Clin Cancer Res*. 2005;11:1203-1209.
224. Sharma G, Mirza S, Prasad CP, et al. Promoter hypermethylation of p16(INK4A), p14(ARF), CyclinD2 and Slit2 in serum and tumor DNA from breast cancer patients. *Life Sci*. 2007;80:1873-1881.
225. Krassenstein R, Sauter E, Dulaimi E, et al. Detection of breast cancer in nipple aspirate fluid by CpG island hypermethylation. *Clin Cancer Res*. 2004;10:28-32.
226. Perry AS, Foley R, Woodson K, et al. The emerging roles of DNA methylation in the clinical management of prostate cancer. *Endocr Relat Cancer*. 2006;13:357-377.
227. Hoque MO, Topaloglu O, Begum S, et al. Quantitative methylation-specific polymerase chain reaction gene patterns in urine sediment distinguish prostate cancer patients from control subjects. *J Clin Oncol*. 2005;23:6569-6575.
228. Ha PK, Califano JA. Promoter methylation and inactivation of tumour-suppressor genes in oral squamous-cell carcinoma. *Lancet Oncol*. 2006;7:77-82.
229. Duenas-Gonzalez A, Lizano M, Candelaria M, et al. Epigenetics of cervical cancer. An overview and therapeutic perspectives. *Mol Cancer*. 2005;4:38.
230. Balch C, Huang TH, Brown R, et al. The epigenetics of ovarian cancer drug resistance and resensitization. *Am J Obstet Gynecol*. 2004;191:1552-1572.
231. van Doorn R, Gruis NA, Willemze R, et al. Aberrant DNA methylation in cutaneous malignancies. *Semin Oncol*. 2005;32:479-487.
232. Kuroki T, Tajima Y, Kanematsu T. Role of hypermethylation on carcinogenesis in the pancreas. *Surg Today*. 2004;34:981-986.
233. Debinski W, Gibo D, Mintz A. Epigenetics in high-grade astrocytomas: opportunities for prevention and detection of brain tumors. *Ann N Y Acad Sci*. 2003;983:232-242.
234. Claus R, Almstedt M, Lubbert M. Epigenetic treatment of hematopoietic malignancies: in vivo targets of demethylating agents. *Semin Oncol*. 2005;32:511-520.
235. Kay PH, Spagnolo DV, Taylor J, et al. DNA methylation and developmental genes in lymphomagenesis – more questions than answers? *Leuk Lymphoma*. 1997;24:211-220.
236. Chou JL, Rozmahel R, Tsui LC. Characterization of the promoter region of the cystic fibrosis transmembrane conductance regulator gene. *J Biol Chem*. 1991;266:24471-24476.
237. Yoshimura K, Nakamura H, Trapnell BC, et al. The cystic fibrosis gene has a "housekeeping"-type promoter and is expressed at low levels in cells of epithelial origin. *J Biol Chem*. 1991;266:9140-9144.
238. Denamur E, Chehab FF. Methylation status of CpG sites in the mouse and human CFTR promoters. *DNA Cell Biol*. 1995;14:811-815.
239. Koh J, Sferra TJ, Collins FS. Characterization of the cystic fibrosis transmembrane conductance regulator promoter region. Chromatin context and tissue-specificity. *J Biol Chem*. 1993;268: 15912-15921.

6

Bioinformatics and Omics

Timothy Craig Allen and Philip T. Cagle

Bioinformatics and Omics

The term "genomics" originated in 1920 to describe the complete set of chromosomes and their associated genes; however, it has been in the last decade that the use of omics – genomics, transcriptomics and proteomics – and bioinformatics has led to dramatic advances in the understanding of the molecular/genetic basis of disease.[1–29]

Bioinformatics

Bioinformatics has become an essential part of omics research and requires unique practical and analytical skills for appropriate results interpretation. Bioinformatics uses computers and statistics to perform extensive omics-related research by searching biological databases and comparing gene sequences and protein on a vast scale to identify sequences or proteins that differ between diseased and healthy tissues, or between different phenotypes of the same disease.[30–37] The techniques used in omics are called high throughput because they involve analysis of very large numbers of genes, gene expression, or proteins in one procedure or combination of procedures. The vast amounts of data generated by these high-throughput studies typically require computers for analysis and comparison of differences between diseased and physiological cells and tissues, a key feature of bioinformatics. Omics and bioinformatics are used not only for the study of the genes and signaling pathways involved in human diseases, but also for identifying potential targets of therapy and the design of therapeutic drugs.

Omics

Omics – a suffix signifying the measurement of the entire complement of a given level of biological molecules and information – today encompasses a variety of new technologies that can help explain normal and abnormal cell pathways, networks, and processes via the simultaneous monitoring of thousands of molecular components.[6,7]

Genomics

Genomics provides platforms for the study of genomes and their genes, including haplotyping and single nucleotide polymorphism detection, by investigating single nucleotide polymorphisms (SNPs) and mutations using high-throughput genome sequencing techniques such as high-density DNA microarrays/DNA (oligonucleotide) chips.[7,38–44] The base sequence of the genes of the human mitochondrial genome was completed in 1981, and in 2003 the base sequence of the genes of the entire human genome was completed.[45–47] The human genomic sequence data from the International Human Genome Sequencing Consortium can be mined using tools that are now publicly available.[48] Public databases can also be mined for SNPs, human mitochondrial genome, and other human DNA polymorphic markers.[49–54]

Transcriptomics

Also termed "functional genomics," transcriptomics provides information about the expression of individual genes at the messenger RNA (mRNA) level and correlates patterns of expression with biological function.[7,55–64] A variety of techniques have been developed to investigate gene expression. These techniques include serial analysis of gene expression (SAGE), suppression subtractive hybridization (SSH), differential display (DD) analysis, RNA arbitrarily primer-polymerase chain reaction (RAP-PCR), amplified restriction fragment-length polymorphism (AFLP), total gene expression analysis (TOGA), and use of internal standard competitive template primers in a quantitative multiplex reverse transcription (RT)-PCR method [StaRT-(PCR)], restriction endonucleolytic analysis of differentially expressed sequences (READS), differential screening (DS), high-density cDNA filter hybridization analysis (HDFCA), and gene expression microarrays.[65–67]

P.T. Cagle and T.C. Allen (eds.), *Basic Concepts of Molecular Pathology*, Molecular Pathology Library 2,
DOI: 10.1007/978-0-387-89626-7_6, © Springer Science+Business Media, LLC 2009

Proteomics

Proteomics investigates individual protein concentrations present in a biological system and studies the structural, functional and regulatory roles of proteins in the cell and in pathways, including how and where they are expressed.[7,68–86] Because gene function is ultimately performed by the proteins transcribed from the genes and mRNA, proteomics is essential to comprehend the actual definitive functioning of a gene or pathway.[3,4] Fortunately, proteomics can be performed on surgical specimens including needle biopsies, cytology specimens, serum, and other fluids.[3,4] Frequently, two-dimensional gel electrophoresis (two-dimensional polyacrylamide gel electrophoresis) and mass spectrometry are employed in proteomics to initially fractionate the groups of proteins in a specimen. Proteins in a given sample fraction may later be identified using fingerprinting or sequence tag techniques.[3,4]

Metabolomics

Metabolomics (also termed metabonomics) comprises study of metabolic profiles by investigating the compounds in a process and involves the characterization and quantification of small organic molecules in either circulatory or cell–tissue systems.[7,87–90] Nuclear magnetic resonance and mass spectrometry are techniques frequently used in metabolomics.[7,87–90]

Glycomics

A challengingly diverse variety of cell-surface glycoconjugates control various biological events, including, among others, cell adhesion, cell differentiation, immunological recognition, and virus–cell recognition.[91] Glycomic analysis, termed "glycomics," is the systemic study of the variations in cellular glycoconjugates, using mass spectrometry and microarray-based and computational technologies.[91,92] In mammals, the glycome includes all oligosaccharides from glycoconjugates such as glycolipids, proteoglycans, and glycoproteins.[92] As it is difficult to profile the entire glycome because of the large diversity of oligosaccharides in the structure and function of an organism, glycomics targets specific groups of oligosaccharides, glycoproteins, and glycoconjugates, as well as the characterization of glycans and their corresponding attachments to proteins or lipids, to determine the role of glycan heterogeneity in diseases.[92] Significant challenges in mammalian glycomics include the complexity of the glycans, the complex biosynthesis of the epitopes, the multivalent nature of biological glycan recognition, and the subtle phenotypes of glycan manipulation, often requiring a multicellular environment to become manifest.[93] The integration of glycomics with genomics and proteomics is important, as the relationship between the glycome and genome is uncertain, and as there is constant reevaluation of the relative impacts of genomics, proteomics, and posttranslational modulation of protein levels and functions.[93] High-throughput glycomic technologies, such as lectin microarrays, will assist in the integration of these data sets and ultimately help determine the role of carbohydrates in various biological pathways.[93]

Cytomics

Cytomes are cellular systems, subsystems, and functional components of an organism.[94] Cytomics, the study of the heterogeneity of cytomes, is the multimolecular, single-cell analysis of cell–system heterogeneity using image or flow cytometry.[94–97] Cytomics may assist in the identification of rare cells lost in the background with genomic and proteomic profiling, such as contaminant cancer cells in autologous transplant tissue that may cause relapse.[96] As technology improves, the use of cytomics, integrated with proteomics, may allow for the use of flow cytometry to sort separate heterogeneous populations of cells using specific phenotypic characteristics, ultimately identifying and validating tumor markers that advance the potential to create personalized molecular medicine.[94,98]

Physiomics

The physiome is defined as the total integration of genome, proteome, and metabolome, from cells to organisms.[99] Physiomics is a branch of omics using large-scale and experimental databases, and computer algorithms, to examine the physiological phenotypes of, and the relationships between, genes and proteins.[99] Useful in the development of drugs and biosensors and biochips, physiomics will be enhanced by the emergence of new technologies including microtechnologies to probe the functional realm of living cells, ultimately extending electroanalytical chemistry and biosensor technologies from the micro-scale to the nano-scale, allowing for the study of subcellular and organellar physiology.[99–101]

Techniques

DNA microarrays are employed to simultaneously screen for the presence or expression of large numbers of genes.[102–109] Suppression subtractive hybridization (SSH) selectively amplifies target cDNA fragments (differentially expressed genes) and suppresses nontarget DNA.[110,111] Serial analysis of gene expression (SAGE) is used for global analysis of gene expression and provides a comprehensive qualitative and the quantitative expression profile of virtually every gene in a cell population or tissue, and SAGE libraries from different cells and tissues have been created.[112–114] 2-DE, a computer software developed in the mid-1970s to separate complex mixtures of proteins, has become a powerful tool for allowing simultaneous resolution of thousands of proteins contained in an organism's proteome.[115,116] Today, a variety of softwares are available for 2-DE experiments, but one must be careful with their use.[1] 2-DE experiments generate great quantities of data that might require weeks to analyze and interpret for appropriate protein identification.[1] Poor experimental design, poor statistical treatment, and high rates of false-positive results hinder appropriate protein identification with these softwares.[1] The massive amounts of data produced by these studies have become problematic for journal article reviewers, who may find themselves faced with many pages of raw bioinformatics data when reviewing an article.[117]

Microarray technologies have become a powerful tool for simultaneously monitoring expression patterns of thousands of genes; however, a important task is to now propose analytical methods to identify groups of genes manifesting similar expression patterns that are activated by similar conditions.[118] Other concerns include the longevity of data sets and the robustness of conclusions built upon them, the desire to integrate data from diverse scientific disciplines, and the facility to productively mine accrued data.[119] Ultimately, omics technologies in biomarker discovery and cancer detection may allow for the production of diagnostic tools that are simple, fast, robust, portable, and cost-effective, yielding clinical diagnosis systems for use outside the laboratory, including the clinic and hospital bedside.[120]

Conclusion

The need for better and more accurate diagnosis, prognosis, and treatment response indicators has yielded increased utilization of omics technologies, particularly in the area of gene expression profiling.[121] U.S. Food and Drug Administration-approved proprietary testing platforms based on microarray technologies already exist and will continue to expand.[122] Despite the rapid evolution of, and widespread adoption of, DNA microarray technology technologies, there remain considerable uncertainty and skepticism regarding data obtained using these technologies.[123] Although rapid medical advances in the laboratory have not necessarily translated into rapid treatment advances, continuing omics research to understand the conceptual framework of disease – including disease progression and treatment response – along with improved and more efficient bioinformatics tools to analyze the great amounts of data originating from omics investigations may permit future diagnostic, prognostic, and therapeutic benefit to arise from these technologies.[6] The rapid progress of omics and bioinformatics has evolved into a promising new era for personalized patient care.[124]

References

1. Biron DG, Brun C, Lefevre T, et al. The pitfalls of proteonomics experiments without the correct use of bioinformatics tools. *Proteomics*. 2006;20:5577–5596.
2. McKusick VA. Genomics: structure and functional studies of genomes. *Genomics*. 1997;45:244–249.
3. Palagi PM, Hernandez P, Walther D, Appel RD. Proteome informatics I: bioinformatics tools for processing experimental data. *Proteomics*. 2006;6:5435–5444.
4. Lisacek F, Cohen-Boulakia S, Appel RD. Proteome informatics II: bioinformatics for comparative proteomics. *Proteomics*. 2006;6: 5445–5466.
5. Maojo V, Martin-Sanchez F. Bioinformatics: towards new directions for public health. *Methods Inf Med*. 2004;43:208–214.
6. Bilello JA. The agony and ecstasy of "OMIC" technologies in drug development. *Curr Mol Med*. 2005;5:39–52.
7. Morel NM, Holland JM, van der Greef P, et al. Primer on medial genomics, part. XIV Introduction to systems biology: a new approach to understanding disease and treatment. *Mayo Clin Proc*. 2004;79:651–658.
8. Provart NJ, McCourt P. Systems approaches to understanding cell signaling and gene regulation. *Curr Opin Plant Biol*. 2004;7:605–609.
9. Wheelock AM, Goto S. Effects of post-electrophoretic analysis on variance in gel-based proteomics. *Expert Rev Proteomics*. 2006;3:129–142.
10. Debouck C, Metcalf B. The impact of genomics on drug discovery. *Annu Rev Pharmacol Toxicol*. 2000;40:193–207.
11. Ghosh D. High throughput and global approaches to gene expression. *Comb Chem High Throughput Screen*. 2000;3:411–420.
12. Hanke J. Genomics and new technologies as catalysts for change in the drug discovery paradigm. *J Law Med Ethics*. 2000;28(4 suppl):15–22.
13. Harris T. Genetics, genomics, and drug discovery. *Med Res Rev*. 2000;20:203–211.
14. Rudert F. Genomics and proteomics tools for the clinic. *Curr Opin Mol Ther*. 2000;2:633–642.
15. Merrick BA, Bruno ME. Genomic and proteomic profiling for biomarkers and signature profiles of toxicity. *Curr Opin Mol Ther*. 2004;6:600–667.
16. Chalkley RJ, Hansen KC, Baldwin MA. Bioinformatic methods to exploit mass spectrometric data for proteomic applications. *Methods Enzymol*. 2005;402:289–312.
17. Dennis JL, Oien KA. Hunting the primary: novel strategies for defining the origin of tumours. *J Pathol*. 2005;205:236–247.
18. Englbrecht CC, Facius A. Bioinformatics challenges in proteomics. *Comb Chem High Throughput Screen*. 2005;8:705–715.
19. Fung ET, Weinberger SR, Gavin E, Zhang F. Bioinformatics approaches in clinical proteomics. *Expert Rev Proteomics*. 2005;2:847–862.
20. Kremer A, Schneider R, Terstappen GC. A bioinformatics perspective on proteomics: data storage, analysis, and integration. *Biosci Rep*. 2005;25:95–106.
21. Mount DW, Pandey R. Using bioinformatics and genome analysis for new therapeutic interventions. *Mol Cancer Ther*. 2005;4:1636–1643.
22. Nishio K, Arao T, Shimoyama T, et al. Translational studies for target-based drugs. *Cancer Chemother Pharmacol*. 2005;56(suppl 1):90–93.
23. Katoh M, Katoh M. Bioinformatics for cancer management in the post-genome era. *Technol Cancer Res Treat*. 2006;5:169–175.
24. Miles AK, Matharoo-Ball B, Li G, Ahmad M, Rees RC. The identification of human tumour antigens: current status and future developments. *Cancer Immunol Immunother*. 2006;55: 996–1003.

25. Quackenbush J. Microarray analysis and tumor classification. *N Engl J Med*. 2006;354:2463–2472.
26. Redfern O, Grant A, Maibaum M, Orengo C. Survey of current protein family databases and their application in comparative, structural and functional genomics. *J Chromatogr B Anal Technol Biomed Life Sci*. 2005;815:97–107.
27. Iqbal O, Fareed J. Clinical applications of bioinformatics, genomics, and pharmacogenomics. *Methods Mol Biol*. 2006;316: 159–177.
28. Reeves GA, Thornton JM. BioSapiens Network of Excellence. Integrating biological data through the genome. *Hum Mol Genet*. 2006;15(Special No. 1):R81–R87.
29. Waggoner A. Fluorescent labels for proteomics and genomics. *Curr Opin Chem Biol*. 2006;10:62–66.
30. Ritchie MD. Bioinformatics approaches for detecting gene-gene and gene-environment interactions in studies of human disease. *Neurosurg Focus*. 2005;19:E2.
31. Hanai T, Hamada H, Okamoto M. Application of bioinformatics for DNA microarray data to bioscience, bioengineering and medical fields. *J Biosci Bioeng*. 2006; 101:377–384.
32. Goodman N. Biological data becomes computer literate: new advances in bioinformatics. *Curr Opin Biotechnol*. 2002;13: 68–71.
33. Ness SA. Basic microarray analysis: strategies for successful experiments. *Methods Mol Biol*. 2006;316:13–33.
34. Perco P, Rapberger R, Siehs C, et al. Transforming omics data into context: bioinformatics on genomics and proteomics raw data. *Electrophoresis*. 2006;27:2659–2675.
35. Haoudi A, Bensmail H. Bioinformatics and data mining in proteomics. *Expert Rev Proteomics*. 2006;3:333–343.
36. Ivanov AS, Veselovsky AV, Dubanov AV, Skvortsov VS. Bioinformatics platform development: from gene to lead compound. *Methods Mol Biol*. 2006;316:389–431.
37. Teufel A, Krupp M, Weinmann A, Galle PR. Current bioinformatics tools in genomic biomedical research (review). *Int J Mol Med*. 2006;17:967–973.
38. Regnstrom K, Burgess DJ. Pharmacogenomics and its potential impact on drug and formulation development. *Crit Rev Ther Drug Carrier Syst*. 2005;22:465–492.
39. Willard HF, Angrist M, Ginsburg GS. Genomic medicine: genetic variation and its impact on the future of health care. *Philos Trans R Soc Lond B Biol Sci*. 2005;360:1543–1550.
40. Garraway LA, Seller WR. From integrated genomics to tumor lineage dependency. *Cancer Res*. 2006;66:2506–2508.
41. McDunn JE, Chung TP, Laramie JM, Townsend RR, Cobb JP. Physiologic genomics. *Surgery*. 2006;139:133–139.
42. Tost J, Gut IG. Genotyping single nucleotide polymorphisms by mass spectrometry. *Mass Spectrom Rev*. 2002;21: 388–418.
43. Thomas DC, Haile RW, Duggan D. Recent developments in genomewide association scans: a workshop summary and review. *Am J Hum Genet*. 2005;77:337–345.
44. Bernig T, Chanock SJ. Challenges of SNP genotyping and genetic variation: its future role in diagnosis and treatment of cancer. *Expert Rev Mol Diagn*. 2006;6:319–331.
45. Anderson S, Bankier AT, Barrell BG, et al. Sequence and organization of the human mitochondrial genome. *Nature*. 1981;290: 457–465.
46. Mundy C. The human genome project: a historical perspective. *Pharmacogenomics*. 2001;2:37–49.
47. International Human Genome Sequencing Consortium. Finishing the euchromatic sequence of the human genome. *Nature*. 2004;431:931–945.
48. Baxevanis AD. Using genomic databases for sequence-based biological discovery. *Mol Med*. 2003;9:185–192.
49. The International HapMap Consortium. The International HapMap Project. *Nature*. 2003;426:789–796.
50. Thorisson GA, Stein LD. The SNP Consortium website: past, present and future. *Nucleic Acids Res*. 2003;31:124–127.
51. Liu T, Johnson JA, Casella G, Wu R. Sequencing complex diseases with HapMap. *Genetics*. 2004;168:503–511.
52. Riva A, Kohane IS. A SNP-centric database for the investigation of the human genome. *BMC Bioinform*. 2004;5:33.
53. Kong X, Matise TC. MAP-O-MAT: internet-based linkage mapping. *Bioinformatics*. 2005;21:557–559.
54. Brandon MC, Lott MT, Nguyen KC, et al. MITOMAP: a human mitochondrial genome database-2004 update. *Nucleic Acids Res*. 2005;33:D611–D613.
55. Carulli JP, Artinger M, Swain PM, et al. High throughput analysis of differential gene expression. *J Cell Biochem Suppl*. 1998;30–31: 286–296.
56. Scheel J, Von Brevern MC, Horlein A, et al. Yellow pages to the transcriptome. *Pharmacogenomics*. 2002;3:791–807.
57. Hedge PS, White IR, Debouck C. Interplay of transcriptomics and proteomics. *Curr Opin Biotechnol*. 2003;14:647–651.
58. Suzuki M, Hayashizaki Y. Mouse-centric comparative transcriptomics of protein coding and non-coding RNAs. *Bioessays*. 2004;26:833–843.
59. Breitling R, Herzyk P. Biological master games: using biologists' reasoning to guide algorithm development for integrated functional genomics. *OMICS*. 2005;9:225–232.
60. Storck T, von Brevern MC, Behrens CK, Scheel J, Bach A. Transcriptomics in predictive toxicology. *Curr Opin Drug Discov Dev*. 2002;5:90–97.
61. Hu YF, Kaplow J, He Y. From traditional biomarkers to transcriptome analysis in drug development. *Curr Mol Med*. 2005;5:29–38.
62. Kralj M, Kraljevic S, Sedic M, Kurjak A, Pavelic K. Global approach to perinatal medicine: functional genomics and proteomics. *J Perinat Med*. 2005;33:5–16.
63. Morgan KT, Jayyosi Z, Hower MA, et al. The hepatic transcriptome as a window on whole-body physiology and pathophysiology. *Toxicol Pathol*. 2005;33:136–145.
64. Jansen BJ, Schalkwijk J. Transcriptomics and proteomics of human skin. *Brief Funct Genomic Proteomic*. 2003;1:326–341.
65. Liang P, Zhu W, Zhang X, et al. Differential display using one-base anchored oligo-dT primers. *Nucleic Acids Res*. 1994;22:5763–5764.
66. Ahmed FE. Molecular techniques for studying gene expression in carcinogenesis. *J Environ Sci Health C Environ Carcinog Ecotoxicol Rev*. 2002;20:77–116.
67. Muller-Hagen G, Beinert T, Sommer A. Aspects of lung cancer gene expression profiling. *Curr Opin Drug Discov Dev*. 2004;7:290–303.
68. Anderson JS, Mann M. Functional genomics by mass spectrometry. *FEBS Lett*. 2000;480:25–31
69. Liotta LA, Petricoin EF III. The promise of proteomics. *Clin Adv Hematol Oncol*. 2003;1:460–462.
70. Jain KK. Role of oncoproteomics in the personalized management of cancer. *Expert Rev Proteomics*. 2004;1:49–55.

71. Hanash S. Disease proteomics. *Nature*. 2003;422:226–232.
72. Baggerman G, Vierstraete E, De Loof A, Schoofs L. Gel-based versus gel-free proteomics: a review. *Comb Chem High Throughput Screen*. 2005;8:669–677.
73. Calvo KR, Liotta LA, Petricoin EF. Clinical proteomics: from biomarker discovery and cell signaling profiles to individualized personal therapy. *Biosci Rep*. 2005;25:107–125.
74. Brown RE. Morphoproteomics: exposing protein circuitries in tumors to identify potential therapeutic targets in cancer patients. *Expert Rev Proteomics*. 2005;2:337–348.
75. Kalia A, Gupta RP. Proteomics: a paradigm shift. *Crit Rev Biotechnol*. 2005;25:173–198.
76. Scaros O, Fisler R. Biomarker technology roundup: from discovery to clinical applications, a broad set of tools is required to translate from the lab to the clinic. Biotechniques. 2005;Suppl:30-32.
77. Clarke W, Chan DW. ProteinChips: the essential tools for proteomic biomarker discovery and future clinical diagnostics. *Clin Chem Lab Med*. 2005;43:1279–1280.
78. Kolch W, Mischak H, Pitt AR. The molecular make-up of a tumour: proteomics in cancer research. *Clin Sci (Lond)*. 2005;108: 369–383.
79. Patel PS, Telang SD, Rawal RM, Shah MH. A review of proteomics in cancer research. *Asian Pac J Cancer Prev*. 2005;6:113–117.
80. Roboz J. Mass spectrometry in diagnostic oncoproteomics. *Cancer Invest*. 2005;23:465–478.
81. Waldburg N, Kahne T, Reisenauer A, et al. Clinical proteomics in lung diseases. *Pathol Res Pract*. 2004;200:147–154.
82. Stroncek DF, Burns C, Martin BM, et al. Advancing cancer biotherapy with proteomics. *J Immunother*. 2005;28:183–192..
83. Fleming K, Kelley LA, Islam SA, et al. The proteome: structure, function and evolution. *Philos Trans R Soc Lond B Biol Sci*. 2006;361:441–451.
84. Domon B, Aebersold R. Mass spectrometry and protein analysis. *Science*. 2006;312:212–217.
85. Gulmann C, Sheehan KM, Kay EW, Liotta LA, Petricoin EF 3rd. Array-based proteomics: mapping of protein circuitries for diagnostics, prognostics, and therapy guidance in cancer. *J Pathol*. 2006;208:595–606.
86. Kingsmore SF. Multiplexed protein measurement: technologies and applications of protein and antibody arrays. *Nat Rev Drug Discov*. 2006;5:310–320.
87. Davis CD, Milner J. Frontiers in nutrigenomics, proteomics, metabolomics and cancer prevention. *Mutat Res*. 2004;551:51–64.
88. Griffin JL, Bollard ME. Metabonomics: its potential as a tool in toxicology for safety assessment and data integration. *Curr Drug Metab*. 2004;5:389–398.
89. Rochfort S. Metabolomics reviewed: a new "omics" platform technology for systems biology and implications for natural products research. *J Nat Prod*. 2005;68:1813–1820.
90. Griffin JL. The Cinderella story of metabolic profiling: does metabolomics get to go to the functional genomics ball? *Philos Trans R Soc Lond B Biol Sci*. 2006;361:147–161.
91. Mahal LK. Glycomics: towards bioinformatics approaches to understanding glycosylation. *Anticancer Agents Med Chem*. 2008;8:37–51.
92. Packer NH, von der Lieth C-W, Aoki-Kinoshita KF, et al. Frontiers in glycomics: bioinformatics and biomarkers in disease. *Proteomics*. 2008;8:8–20.
93. Pilobello KT, Mahal LK. Deciphering the glycocode: the complexity and analytical challenge of glycomics. *Curr Opin Chem Biol*. 2007;11:300–305.
94. Bernas T, Gregori G, Asem EK, Robinson JP. Integrating cytomics and proteomics. *Mol Cell Protemics*. 2006;5:2–13.
95. Gomase VS, Tagore S. Cytomics. *Curr Drug Metab*. 2008;9:263–266.
96. Tarnok A. Slide-based cytometry for cytomics: a minireview. *Cytometry A*. 2006;69A:555–562.
97. Valet G. Cytomics as a new potential for drug discovery. *Drug Discov Today*. 2006;11:785–791.
98. Tarnok A, Bosci J, Brockhoff G. Cytomics: importance of multimodal analysis of cell function and proliferation in oncology. *Cell Prolif*. 2006;39:495–505.
99. Gomase VS, Tagore S. Physiomics. *Curr Drug Metab*. 2008;9: 259–262.
100. Porterfield DM. Measuring metabolism and biophysical flux in the tissue, cellular and sub-cellular domains: recent developments in self-referencing amperometry for physiological sensing. *Biosens Bioelectron*. 2007;15:1186–1196.
101. ul Haque A, Chatni MR, Li G, Porterfield DM. Biochips and other microtechnologies for physiomics. *Expert Rev Proteomics*. 2007;4:553–463.
102. Ramsay G. DNA chips: state-of-the art. *Nat Biotechnol*. 1997;16: 40–44.
103. Duggan DJ, Bittner M, Chen Y, Meltzer P, Trent JM. Expression profiling using cDNA microarrays. *Nat Genet*. 1999;21(suppl 1):10–14.
104. Chen l, Ren J. High-throughput DNA analysis by microchip electrophoresis. *Comb Chem High Throughput Screen*. 2004;7:29–43.
105. Heller MJ. DNA microarray technology: devices, systems, and applications. *Annu Rev Biomed Eng*. 2002;4:129–153.
106. Obeid PJ, Christopoulos TK. Microfabricated systems for nucleic acid analysis.*Crit Rev Clin Lab Sci*. 2004;41:429–465.
107. Shi L, Tong W, Goodsaid F, et al. QA/QC: challenges and pitfalls facing the microarray community and regulatory agencies. *Expert Rev Mol Diagn*. 2004;4:761–777.
108. Zhumabayeva B, Chenchik A, Siebert PD, Herrler M. Disease profiling arrays: reverse format cDNA arrays complimentary to microarrays. *Adv Biochem Eng Biotechnol*. 2004; 86:191–213.
109. Brentani RR, Carraro DM, Verjovski-Almeida S, et al. Gene expression arrays in cancer research: methods and applications. *Crit Rev Oncol Hematol*. 2005;54:95–105.
110. Diatchenko L, Lau YF, Campbell AP, et al. Suppression subtractive hybridization: a method for generating differentially regulated or tissue-specific cDNA probes and libraries. *Proc Natl Acad Sci USA*. 1996;93:6025–6030.
111. Wang X, Feuerstein GZ. Suppression subtractive hybridization: application in the discovery of novel pharmacological targets. *Pharmacogenomics*. 2000;1:101–108.
112. Velculescu VE, Vogelstein B, Kinzler KW. Analyzing uncharted transcriptomes with SAGE. *Trends Genet*. 2000;16:423–425.
113. Polyak K, Riggins GJ. Gene discovery using the serial analysis of gene expression technique: implications for cancer research. *J Clin Oncol*. 2001;19:2948–2958.
114. Riggins GJ. Using serial analysis of gene expression to identify tumor markers and antigens. *Dis Markers*. 2001;17:41–48.
115. Vihinen M. Bioinformatics in proteomics. *Biomol Eng*. 2001;18:241–248.

116. Rabilloud T. Two-dimensional gel electrophoresis in proteomics: old, old-fashioned, but it still climbs up the mountains. *Proteomics*. 2002;2:3–10.
117. Nicholson JK. Reviewers peering from under a pile of 'omics' data. *Nature*. 2006;440:992.
118. Belacel N, Wang Q, Cuperlovic-Culf M. Clustering methods for microarray gene expression data. *OMICS*. 2006;10:507–531.
119. Taylor CF. Progress in standards for reporting omics data. *Curr Opin Drug Discov Dev*. 2007;10:254–263.
120. Zhang X, Li L, Wei D, Yap Y, Chen F. Moving cancer diagnostics from bench to bedside. *Trends Biotechnol.* 2007;25:166–173.
121. Culhane AC, Howlin J. Molecular profiling of breast cancer: transcriptomic studies and beyond. *Cel Mol Life Sci.* 2007; 64;3185–3200.
122. Finn WG. Diagnostic pathology and laboratory medicine in the age of "omics": a paper from the 2006 William Beaumont Hospital Symposium on Molecular Pathology. *J Mol Diagn.* 2007;9:431–436.
123. Wilkes T, Laux H, Foy CA. Microarray data quality-a review of current developments. *OMICS*. 2007;11:1–13.
124. Tanaka H. Bioinformatics and genomics for opening new perspective for personalized care. *Stud Health Technol Inform.* 2008;134:47–58.

7
General Approach to Molecular Pathology

Gregory L. Blakey and Daniel H. Farkas

Introduction

Once a highly specialized subdiscipline of laboratory medicine, molecular diagnostics now infiltrates all of anatomic and clinical pathology. The shift from dependence on a few, relatively cumbersome methods to a wider range of technologies has facilitated this expansion. In addition, the completion of the Human Genome Project and the growing amount of sequence data related to infection, cancer, and other disease states have yielded additional applications of molecular biology for the clinical laboratory (Fig. 7.1). As the various phases of testing can be automated in many instances, molecular biological experience is no longer a prerequisite. In fact, performance of nucleic acid extraction and amplification in a tabletop unit is possible.[1,2] Increasingly, miniaturization will further move molecular testing to the point of care.[3]

Although varying from protocol to protocol, the workflow for many molecular pathological tests follows a similar scheme (Fig. 7.2). As with other laboratory tests, preanalytical variables can dramatically affect molecular assay results. Therefore, appropriate care should be paid to specimen collection, transport, processing, and storage. In particular, fresh samples submitted for analysis of RNA must be processed rapidly to prevent degradation.

Nucleic Acid Extraction

Although the traditional organic means of extracting nucleic acids, with phenol and chloroform, are useful in the clinical laboratory, less cumbersome methods now dominate. For instance, silica-based spin column technologies couple ease of use with adaptability. With these methods, a digested sample is loaded onto a column to which nucleic acids bind in the presence of chaotropic salts. Microcentrifugation of the column followed by washing removes proteins and other macromolecules. Finally, nucleic acids are eluted with buffer or water. As with more traditional techniques, conventional spin column-based nucleic acid isolation suffers from the need for manual labor, not an insignificant factor in the busy clinical laboratory.[4,5]

Automation of nucleic acid extraction techniques has increased laboratory efficiency and decreased turnaround time. Of particular importance for quantitative testing, reduced inter- and intraoperator variability also results. Commonly used automated nucleic acid extraction techniques often rely on magnetized glass beads. After cell lysis, nucleic acids bind the beads, which can be robotically manipulated with a magnet. The beads are washed, and then nucleic acids are eluted with water or an appropriate buffer. Throughput of automated extraction robots ranges from approximately 6 to 96 or more samples per run. Depending on the model, these machines are capable of processing various sample types, including plasma, whole blood, fresh tissue, and formalin-fixed tissue. Typical isolation choices include total nucleic acid, DNA, and RNA.

Although nucleic acids are most commonly purified for use in molecular protocols, crude cell lysates can be used in selected cases. Such an approach may be especially useful when only minute amounts of samples are available, as with macro- or microdissected cells. In these instances, the decrease in nucleic acid yield because of the extraction process might be unacceptable. Crude lysates have been successfully used as starting material for polymerase chain reaction (PCR)-based clinical diagnostics.[6]

Amplification Technologies

Target Amplification

Polymerase Chain Reaction

In combination with a variety of other methods (Fig. 7.3), PCR remains the core molecular pathological technique. It results in a manyfold increase in target sequences by virtue of repeated cycles in which product becomes additional template. Despite the superior sensitivity and convenience of

P.T. Cagle and T.C. Allen (eds.), *Basic Concepts of Molecular Pathology*, Molecular Pathology Library 2,
DOI: 10.1007/978-0-387-89626-7_7, © Springer Science+Business Media, LLC 2009

PCR versus Southern blotting and other venerable methods, the potential for contamination of the laboratory space and reagents with reaction products - amplicons - slowed entry of molecular diagnostics into traditional laboratory settings. The relatively recent availability of thermal cyclers that can continually monitor product formation has mitigated this

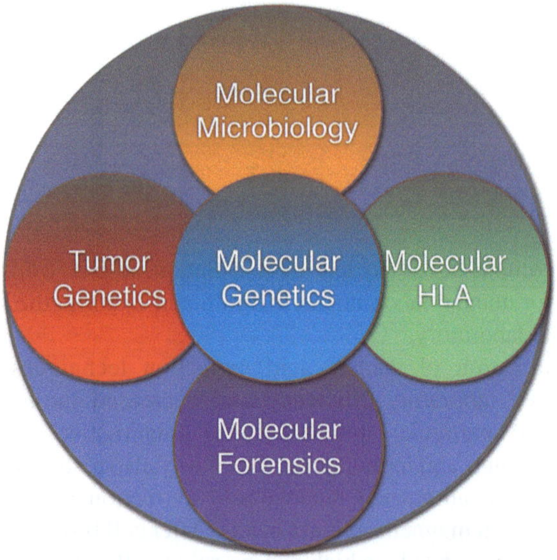

FIG. 7.1. Molecular diagnostics encompasses the use of nucleic acids to diagnose infection, malignancy (hematological and solid tumors), and genetic diseases. Other major foci include the investigation of human remains and crime scenes (forensics) and evaluation of transplant donors and recipients [human leukocyte antigen (HLA) testing].

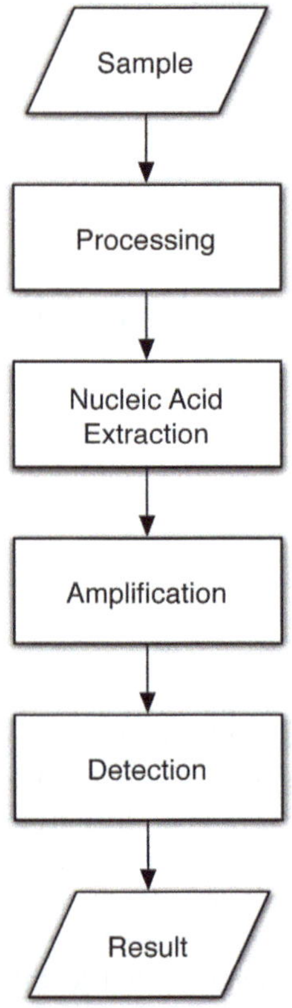

FIG. 7.2. The general steps in many molecular diagnostic protocols are similar. For example, a blood sample received for genetic testing first may be processed to isolate white blood cells. Then, DNA is extracted from these cells and used to set up polymerase chain reaction. A multitude of detection methods are available (see text for further discussion).

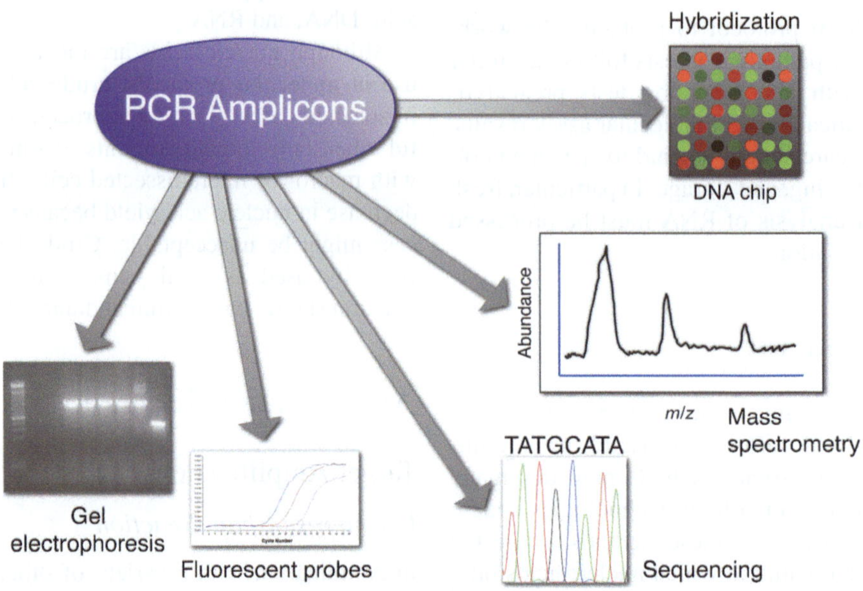

FIG. 7.3. Numerous techniques allow detection and analysis of polymerase chain reaction (PCR) products, amplicons. Amplicons can be sized by gel electrophoresis, detected in real time with fluorescent probes, and sequenced with a variety of methods. Mass spectrometry approaches allow precise fragment sizing or base composition determination, and DNA "chips" permit parallel hybridization of amplicon and thousands of probes. (Sequencing diagram modified from Blakey and Farkas,[7] by permission of the Colorado Association for Continuing Medical Laboratory Education).

problem. These instruments allow real-time PCR in which subsequent detection steps requiring opening of the reaction vessel are usually not needed. Besides lowering the risk of contamination, real-time PCR allows rapid turnaround time; the high-speed thermal cyclers used rapidly change the reaction temperature of metal blocks or air in which reaction vessels are suspended. Applications of real-time PCR include genotyping, pathogen identification, and target quantification.

Besides the standard reagents [primers flanking the target site, heat-resistant polymerase, deoxyribonucleotide triphosphates (dNTPs), and various buffer components], real-time PCR requires fluorescent molecules for ultimate product detection. Fluorescent dyes such as SYBR Green that nonspecifically bind double-stranded DNA can be used, but false-positive results may occur. More commonly, polymorphism and mutation detection are accomplished with fluorescently labeled oligonucleotide probes. Some of the most frequently used of the many such probe systems are hybridization probes, hydrolysis probes, and molecular beacons. Paired hybridization probes anneal to adjacent sites on the target sequence. One hybridization probe has an acceptor fluorophore, the other a donor fluorophore. When in close proximity, the energy from the donor excites the acceptor via fluorescence resonance energy transfer (FRET; Fig. 7.4a). The real-time thermal cycler measures this signal. A hydrolysis probe consists of a single oligonucleotide labeled

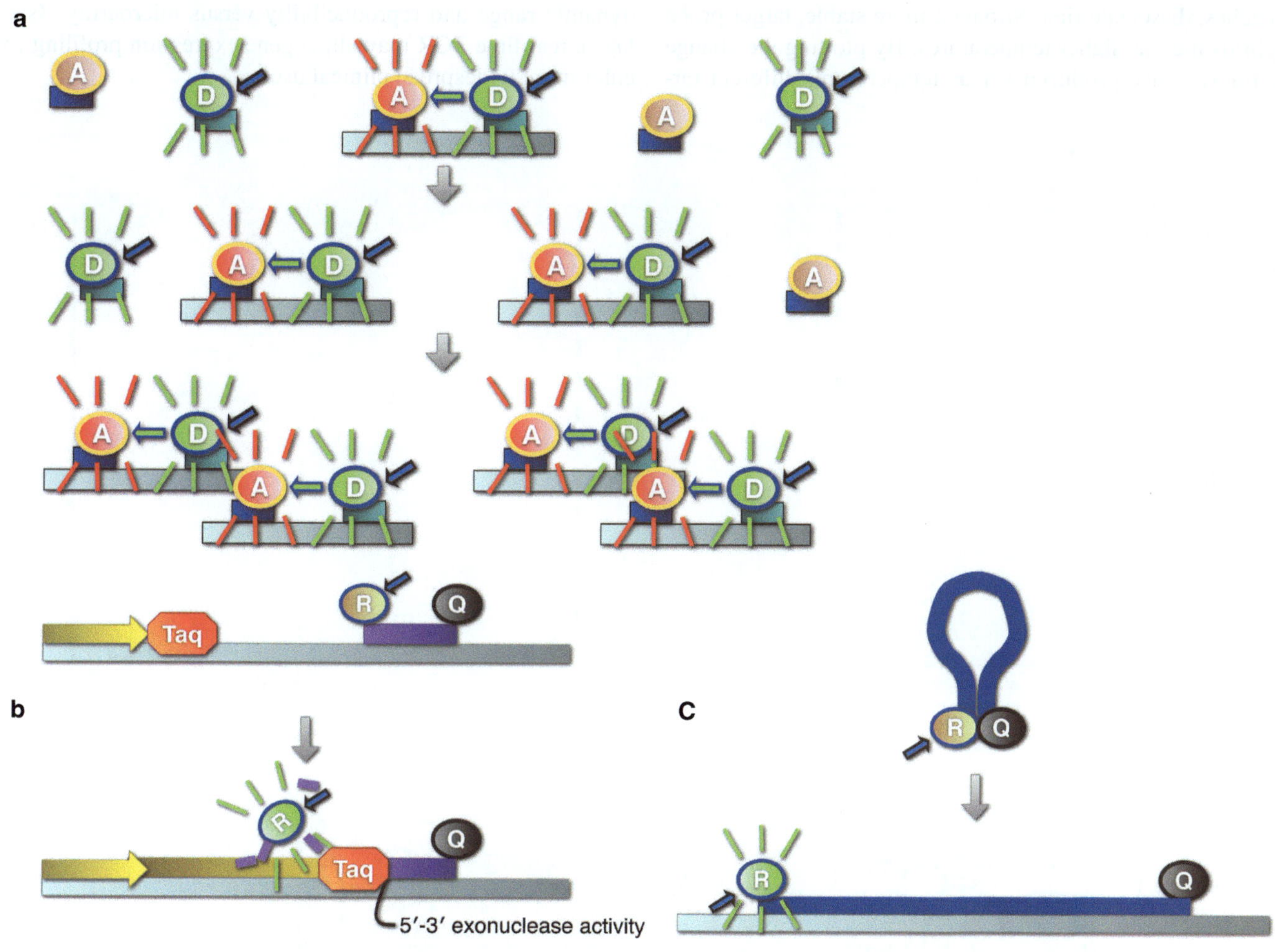

FIG. 7.4. Many different strategies allow detection of amplicons in real-time polymerase chain reaction (PCR). These three probe systems rely on fluorescence resonance energy transfer (FRET) between fluorophores. *Blue arrows* represent light emitted by the thermal cycler. (**a**) Dual hybridization probes bind the target sequence, whereupon the acceptor fluorophore (*A*) on one probe receives light from the adjacent donor fluorophore (*D*) on the other probe. As the PCR progresses, fluorescence increases. (Modified from Blakey and Farkas,[7] by permission of the Colorado Association for Continuing Medical Laboratory Education.) (**b**) A hydrolysis probe, an oligonucleotide labeled with reporter (*R*) and quencher (*Q*) fluorophores, binds the target sequence during the extension phase of PCR. *Taq* polymerase hydrolyzes bound probe, freeing the reporter from the quencher. (Modified from Blakey and Farkas,[7] by permission of the Colorado Association for Continuing Medical Laboratory Education.) (**c**) A molecular beacon probe, composed of a stem-loop oligonucleotide structure with terminal reporter (*R*) and quencher (*Q*) fluorophores. Binding of the probe to the target sequence separates quencher from reporter, allowing fluorescence of the latter.

with reporter and quencher fluorophores; fluorescence is inhibited by FRET between the two. Upon dissociation of the two fluorophores by the exonuclease activity of the DNA polymerase during the extension phase of PCR, the reporter emits light (Fig. 7.4b). Similarly, molecular beacons contain a stem-loop structure that brings together reporter and quencher fluorophores attached to the respective ends of the probe. When the probe binds to the target sequence, the reporter and quencher are physically separated (Fig. 7.4c).

Melting curve analysis with hybridization probes can be used to distinguish different target sequences. Hybridized strands separate at the melting temperature (T_m). To determine T_m, after PCR is completed, the thermal cycler slowly increases the temperature while continuously monitoring fluorescence. As the temperature rises, probes that are less tightly bound to target, either by design or because of mismatches, dissociate first. Stronger, more stable, target-probe hybrids melt at higher temperatures. By plotting the change in fluorescence produced versus temperature, different target amplicon species (e.g., normal versus mutant or different pathogens) can be distinguished (Fig. 7.5).

Real-time PCR permits relatively easy quantification of analytes such as microbial genomes or fusion transcripts. As the initial concentration of an analyte of interest increases, the reaction cycle number at which fluorescence exceeds background decreases. Therefore, external calibrators can be used to construct a standard curve, thus permitting quantification.

Polymerase chain reaction systems with 96- and 384-well plates increase throughput. Multiplexing (more than one reaction per tube or well) enhances this. However, untoward interactions between the various components of a multiplexed reaction limit the number of analytes that can be interrogated simultaneously. High-throughput gene expression profiling may be accomplished by advances such as highly parallel picoliter-scale PCR.[8] With potentially superior dynamic range and reproducibility versus microarrays (see later), real-time PCR may allow gene expression profiling to enter more widespread clinical use.

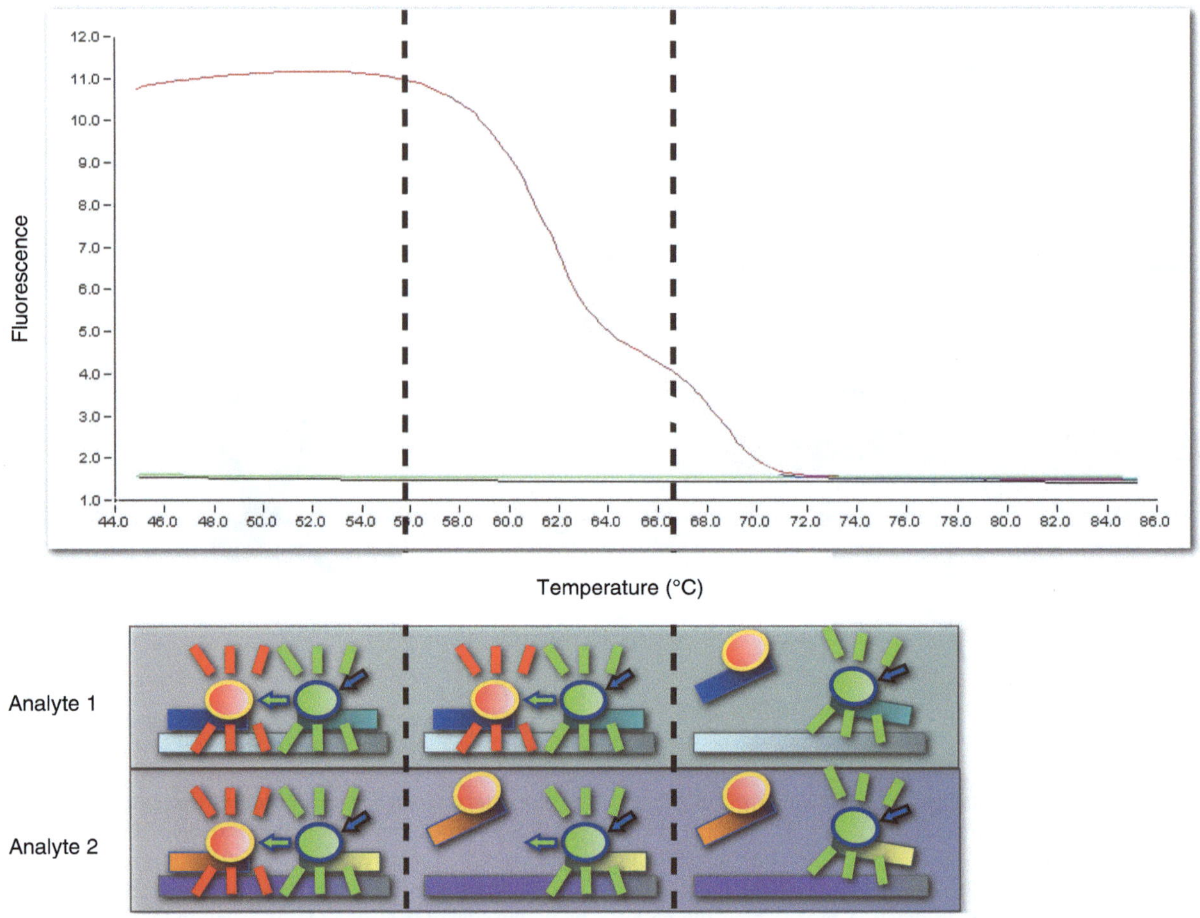

Fig. 7.5. Amplicons can be distinguished by melting curve analysis. Two hybridization probe pairs differentiate between a mixture of analytes 1 and 2 in this illustration (*left*). As the thermal cycler ramps up the temperature, one of the probes for analyte 2 dissociates first (*middle*), resulting in drop in total fluorescence. The temperature at which this loss of fluorescence occurs is the melting temperature for analyte 2. A second drop indicates the dissociation of at least one of the analyte 1 probes (*right*).

Other Target Amplification Techniques

Besides PCR, target amplification techniques include the proprietary nucleic acid sequence-based amplification and transcription-mediated amplification, both of which rely on the isothermal production of RNA intermediates.

Signal Amplification

Instead of increasing the number of targets to boost sensitivity, signal amplification techniques rely on multiplying the detection signal. Signal amplification techniques include Hybrid Capture, Invader, and branched chain DNA (bDNA), all proprietary.

With Hybrid Capture, in wide use for the identification of human papillomavirus, hybrids of viral DNA and probe RNA are bound by antibodies attached to a reaction well. Labeled antibodies complete the sandwich of the hybrids, allowing chemiluminescent detection.

Invader assays depend on recognition of the combination of the target, a probe, and a so-called invader oligonucleotide by a proprietary enzyme (Cleavase); probe cleavage results. A probe fragment in turn interacts with a looped DNA structure labeled with reporter and quencher fluorophores (FRET cassette). Cleavage again results, releasing the reporter, which fluoresces. Invader chemistry is isothermal. Thus, a thermal cycler is not required. In addition, as amplicons are not produced, the risk of contamination is minimized. Genotyping is a common application.

bDNA technology uses a combination of capture molecules and labels to amplify the amount of signal. Applications include quantification of human immunodeficiency, hepatitis B, and hepatitis C viruses.

Restriction Fragment Analysis and Southern Blotting

Restriction Fragment Length Polymorphism

Restriction fragment length polymorphism (RFLP) analysis, previously a mainstay of genotyping, involves cutting target molecules (usually after amplification with PCR) with a restriction endonuclease, followed by gel electrophoresis. The resulting fragment patterns can be compared to those of controls to deduce the genotype. Many RLFP assays have been supplanted by those based on real-time PCR.

Southern Blotting

A particularly laborious and time-consuming (turnaround times of many days) technique, the Southern blot is still used in the clinical laboratory, particularly when target sequences are too long to safely amplify with PCR. In Southern blotting, relatively large quantities (microgram) of DNA are digested with a restriction endonuclease, fractionated via gel electrophoresis, and then transferred to a solid support, such as a nylon membrane. Labeled probes then hybridize to complementary target sequences, permitting detection.

Sequencing

Sequencing is used widely in genetics laboratories, particularly for the identification of heterogeneous mutations within a given gene or exon.

Sanger Sequencing

In Sanger sequencing, fluorescently labeled dideoxyribonucleotide triphosphates are incorporated into growing nucleotide strands, causing chain termination. The resulting fragments are resolved with slab gel or, more commonly, capillary electrophoresis, to yield the sequence. Sequence read lengths are in the hundreds of base pairs. Common sequencing primers and protocols, high-throughput 384-capillary sequencers, and sophisticated analysis software form the large-scale sequencing pipeline required for the timely detection of polymorphisms and mutations in the high-volume clinical molecular genetics laboratory.

Real-Time Sequencing

Pyrosequencing, or real-time sequencing, depends on a proprietary chemical cascade that converts the pyrophosphate by-product of nucleotide incorporation to light (Fig. 7.6). With

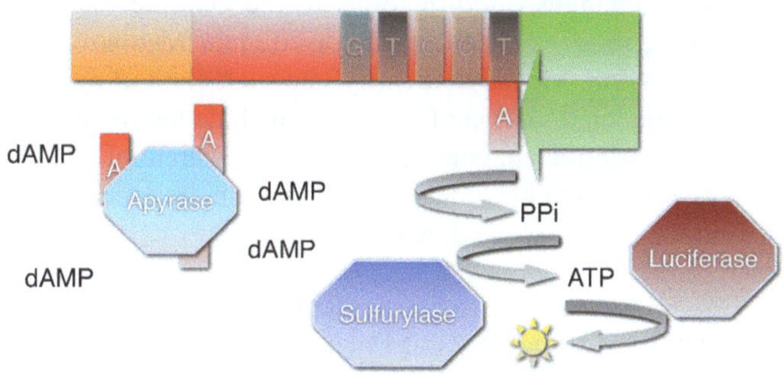

Fig. 7.6. Pyrosequencing takes advantage of the pyrophosphate byproduct of nucleotide incorporation into a growing DNA strand. Pyrophosphate is converted to light by sulfurylase and luciferase. Apyrase degrades excess deoxyribonucleotide triphosphate. As nucleotides are added individually in a predetermined order, the target sequence can be deduced as the DNA strand grows. Hence, pyrosequencing is also known as *sequencing by synthesis*. The *large arrow* represents the sequencing primer. *ATP* adenosine triphosphate; *dAMP* deoxyadenosine monophosphate; *PPi* inorganic phosphate. (Modified from Blakey and Farkas,[7] by permission of the Colorado Association for Continuing Medical Laboratory Education).

inkjet printer technology, the pyrosequencing instrument adds individual dNTPs in a specified order to reaction vessels. Only when a dNTP incorporates at a given position is light generated and detected by the instrument; therefore, the sequence of the target can be inferred.

Although Sanger sequencing remains the predominant sequencing technique, real-time sequencing offers several advantages. Sequence reads can begin immediately after the sequencing primers. In addition, detection occurs during the sequencing reaction, reducing turnaround time. Compared with real-time PCR, pyrosequencing has the advantage of quicker assay development in some instances, given that design and optimization of fluorescently labeled probes is not required. Limitations of pyrosequencing include difficulty reading homopolymeric repeats (several consecutive identical bases) and the relatively short sequence read lengths (less than approximately 100 base pairs).

Advanced Sequencing

Rapid progress in high-throughput sequencing technologies hastens the goal of quickly and inexpensively sequencing a human genome for clinical purposes. One recently introduced machine, based on pyrosequencing chemistry and the so-called shotgun sequencing approach, can sequence a bacterial genome in as little as a few hours.[9] An effort to more fully characterize genomic changes of malignancies, the National Institutes of Health's Cancer Genome Atlas project (http://cancergenome.nih.gov), could eventually generate new clinical assays based on new sequencing technologies.

Methylation Detection Methods

Epigenetic changes, modifications to the genome that do not alter the sequence itself, are involved in the regulation of gene expression. Means to identify one such class of epigenetic change, methylation of cytosines, include methylation-sensitive restriction enzymes that generate differential RFLP patterns and chemical conversion (bisulfite treatment) of methylated bases followed by sequencing.

Fluorescence In Situ Hybridization and Microarrays

Fluorescence In Situ Hybridization

Fluorescently labeled probes hybridized to histological and cytological preparations allow the detection of translocations, amplifications, and deletions. Fluorescence in situ hybridization (FISH) is used widely in the study of hematological malignancies, solid tumors, and genetic disorders. Platforms with automated hybridization and analysis will increase the practicality of FISH in the clinical laboratory. Although generally less sensitive than PCR, FISH has the advantage of preserving histological context.

Chromosomal Microarrays

Composed of hundreds or thousands of probes or clones representing every chromosome, chromosomal microarrays achieve a higher resolution than the staining techniques of classic cytogenetics. Thus, this technique, also known as array comparative genomic hybridization, may detect small deletions and duplications missed by traditional karyotyping. Before hybridization, samples may undergo whole genome amplification to increase sensitivity. Preliminary studies have shown that formalin-fixed cells and tissues may be successfully used with chromosomal microarrays,[10,11] raising the possibility of its routine use in anatomic pathology.

Oligonucleotide Microarrays

Oligonucleotide microarrays are constructed by spotting hundreds to thousands of oligonucleotide probes onto a glass slide or other substrate. Alternatively, probes can be incorporated in situ during construction of the array. By hybridizing cDNA, derived from cellular mRNA via reverse transcription, to these arrays, the expression levels of the various genes in a cell can be measured. Gene expression profiling has been repeatedly demonstrated to generate "signatures" that correlate with prognosis of different cancers, although variabilities in the means of assay performance and statistical analysis have sometimes rendered interlaboratory comparison difficult. Platform standardization and other efforts to improve reproducibility will likely pave the way for gene expression profiling to more fully enter routine clinical use. Selected example applications of microarray technologies are listed in Table 7.1.

TABLE 7.1. Applications of microarray technologies.

Application	Format	Comment
Gene expression (mRNA) profiling	Solid phase, liquid bead	Can identify disease-specific signatures
Global chromosomal abnormality detection	Solid phase	Resolutions depends on number of clones or probes; may not detect balanced translocations
MicroRNA expression profiling	Solid phase, liquid bead	Robust tumor-specific signatures are obtainable from fixed tissues
Pathogen detection	Solid phase, liquid bead	Has potential to replace many conventional virological and bacteriological methods
Polymorphism and mutation detection	Solid phase, liquid bead	A pharmacogenetic test is the first U.S. Flood and Drug Administration - approved microarray-based assay
Sequencing	Solid phase	Alternative to traditional sequencing techniques

Liquid Bead Microarrays

Liquid bead arrays consisting of sets of labeled polystyrene microbeads can be used to simultaneously interrogate numerous targets for genotyping, pathogen identification, and other purposes. As with traditional flow cytometry, beads are passed single file through a laser-based detector. One laser identifies the bead based on its unique color while another quantifies attached analyte. Besides nucleic acids, proteins, cytokines, and other biomolecules can be attached to the beads, enhancing the multiplexing power of this technology.

Mass Spectrometry

Although not "molecular pathology" - the diagnosis of disease using nucleic acids - in the strict sense, proteomic profiling with mass spectrometry is touted by some as a powerful tool for the screening and diagnosis of malignancies and the identification of pathogens. Numerous variants of mass spectrometry separate proteins by mass and charge (Fig. 7.7a). The resultant protein mass fingerprints reveal the relative quantities of the protein and peptide species present (Fig. 7.7b). Such patterns derived from blood may constitute proteomic disease signatures.[12] Confirmation of these sometimes controversial, nascent diagnostic approaches is needed before implementation in the clinical laboratory.

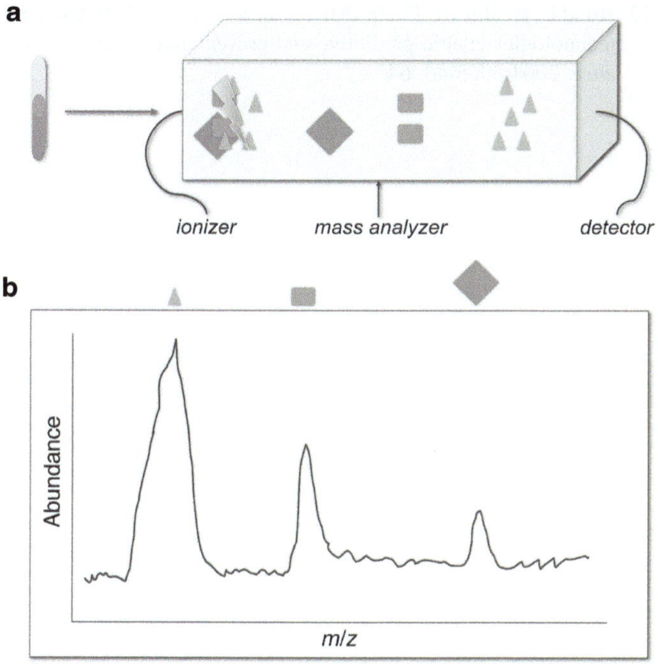

FIG. 7.7. (**a**) Mass spectrometry can be used to separate serum proteins. Following ionization or other means of dissociation, proteins move through the mass analyzer based on the ratio of mass to charge. A detector measures this time of flight (TOF). (**b**) Mass spectrometry TOF yields quantitative "proteomic profiles," as shown in this simplified diagram. The species with the lowest mass/charge ratio (*triangle*) is more abundant than the two species with higher mass:charge ratios (*rectangle and diamond*). Serum proteomic signatures may be composed of the shed proteins and peptides of cancer and other diseases.

Systems Biology Approaches

The emerging field of systems biology seeks to more fully understand life in terms of its myriad molecular networks. This outlook contrasts with the reductionist bent of molecular biology and the common clinical practice of employing one or a few biomarkers to diagnose a given disease. Much more predictive, less reactive medical care is promised by use of multiparameter, high-throughput molecular tools.[13] Although molecular diagnostic assays now serve primarily as adjuncts to other methods in surgical pathology, the use of global, systems-based approaches may challenge the dominance of histopathology.

Implementing Molecular Pathology

Laboratory Design Considerations

To prevent contamination, laboratories manipulating amplicons generated by PCR or other target amplification technologies must strictly observe a unidirectional workflow, from sample processing to detection (Fig. 7.8). Reagent preparation should ideally be done in an area of the laboratory separate from sample processing and nucleic acid extraction. Both these activities benefit from positive air pressure or "still boxes" and ultraviolet light to minimize airborne contaminants. Most importantly, the laboratory's postamplification room, where reaction vessels are opened after PCR, must be kept separate from the rest of the laboratory. Dedicated gloves and gowns should be kept in this area. In addition, negative air flow can prevent escape of amplicons.

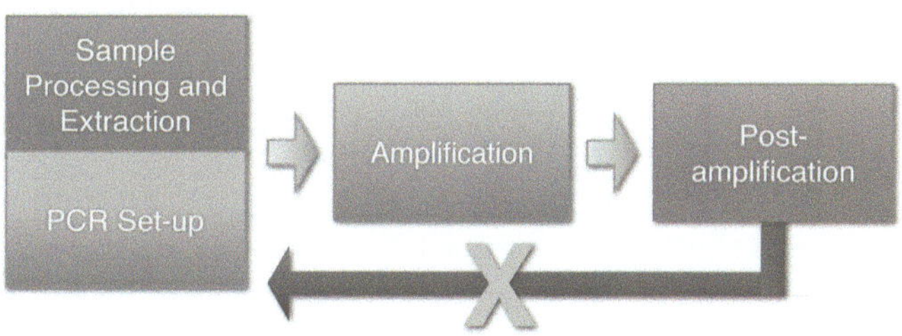

FIG. 7.8. Unidirectional workflow in the molecular pathology laboratory prevents contamination of new samples and reagents with amplicons produced in previously performed polymerase chain reactions (PCRs).

Laboratory Staffing

Early molecular pathology laboratories relied on staff trained in molecular biology to perform and troubleshoot assays. Now, those with more traditional medical technologist training increasingly perform these functions in part because of better educational opportunities. The American Society for Clinical Pathology, the American Board of Bioanalysis, and the National Credentialing Agency for Laboratory Personnel offer certification in molecular diagnostics. In addition, there are several degree-granting programs. The Training and Education section of the Association for Molecular Pathology website provides an updated list of molecular training available for technologists (http://www.amp.org/T&E/training&edu.htm).

References

1. Hughes SJ, Xi L, Raja S, et al. A rapid, fully automated, molecular-based assay accurately analyzes sentinel lymph nodes for the presence of metastatic breast cancer. *Ann Surg.* 2006;243:389–398.
2. Ulrich MP, Christensen DR, Coyne SR, et al. Evaluation of the Cepheid GeneXpert system for detecting *Bacillus anthracis*. *J Appl Microbiol.* 2006;100:1011–1016.
3. Holland CA, Kiechle FL. Point-of-care molecular diagnostic systems: past, present and future. *Curr Opin Microbiol.* 2005;8:504–509.
4. Barkham T. BioRobot EZ1 workstation compares well with manual spin kits for extraction of viral RNA from sera and saves substantial staff time. *J Clin Microbiol.* 2006;44:1598.
5. Knepp JH, Geahr MA, Forman MS, et al. Comparison of automated and manual nucleic acid extraction methods for detection of enterovirus RNA. *J Clin Microbiol.* 2003;41:3532–3536.
6. Hatanpaa KJ, Burger PC, Eshleman JR, et al. Molecular diagnosis of oligodendroglioma in paraffin sections. *Lab Invest.* 2003;83: 419–428.
7. Blakey GL, Farkas DH. Understanding Molecular Pathology: Methods and Applications. Denver: Colorado Association for Continuing Medical Laboratory Education; 2006.
8. Marcus JS, Anderson WF, Quake SR. Parallel picoliter RT-PCR assays using microfluidics. *Anal Chem.* 2006;78:956–958.
9. Margulies M, Egholm M, Altman WE, et al. Genome sequencing in microfabricated high-density picolitre reactors. *Nature (Lond).* 2005;437:367–380.
10. Ghazani AA, Arneson NC, Warren K, et al. Limited tissue fixation times and whole genomic amplification do not impact array CGH profiles. *J Clin Pathol.* 2006;59:311–315.
11. Johnson NA, Hamoudi RA, Ichimura K, et al. Application of array CGH on archival formalin-fixed paraffin-embedded tissues including small numbers of microdissected cells. *Lab Invest.* 2006;86:968–978.
12. Petricoin EF III, Ardekani AM, Hitt BA, et al. Use of proteomic patterns in serum to identify ovarian cancer. *Lancet.* 2002;359:572–577.
13. Hood L, Health JR, Phelps ME, et al. Systems biology and new technologies enable predictive and preventative medicine. *Science.* 2004;306:640–643.

8 Applications in Anatomic Pathology

Jennifer L. Hunt and Sanja Dacic

Introduction

Molecular testing in anatomic pathology is becoming increa-singly important, and molecular assays are used both diagnostically and prognostically, particularly in the workup of neoplasia and for identification or subclassification for certain infectious processes. Fresh and frozen tissues are the optimal source of DNA and RNA to serve as the template for targeted molecular analysis; however, archival formalin-fixed, paraffin-embedded tissue is a frequently used alternative source of DNA for clinical testing. Paraffin-embedded tissue is especially critical as a source of nucleic acid when pathological evaluation renders an unexpected diagnosis, and it also provides the advantage of allowing for archival analysis with correlation to outcome.

Molecular analysis can be performed at different levels of resolution, from the whole chromosome down to the specific nucleotide sequence. At the chromosomal level, classical cytogenetics discerns chromosome structure and number and can detect many major translocations and deletions. This technique requires fresh cells that are capable of entering into cell division. The cells are arrested in metaphase so that the chromosomes can be visualized individually for analysis. A newer technique that can be combined with classical cytogenetics is spectral karyotyping (SKY). For this specialized analysis, the metaphase chromosomes are subjected to fluorescent in situ hybridization with specific probes that are labeled with combinations of five different fluorescent tags. The result involves the "painting" of each chromosome with a unique fluorescent signal that can be differentiated with the assistance of computer detection system. This technique allows for resolution of complex karyotypes.

In situ hybridization [chromogenic (CISH) or fluorescent (FISH)] enables detection of known sequences of DNA and can be performed on whole cells or sections cut from paraffin or frozen tissue. The probes used in ISH can highlight amplifications, specific translocations mutations, or genomic material from infectious organisms. In paraffin-embedded tissues, FISH is a particularly powerful tool for molecular analysis, given the known difficulties in DNA extraction from fixed material.

Comparative genomic hybridization (CGH) is another relatively new technique that can detect and map changes in copy number for specific DNA sequences. The DNA from a test genome (e.g., tumor) and a reference genome (e.g., normal tissue) are differentially labeled and hybridized to normal metaphase chromosomes. In the past few years, microarray-based formats for CGH (array CGH) have been developed and are beginning to be widely used in preference to chromosome-based CGH. Overrepresentation (amplifications) or underrepresentation (deletions) of the test sample signal can be resolved with computer software. This technique is excellent for discovery of unique genetic alterations (deletions and amplifications), but it is not commonly used as a diagnostic tool.

In anatomic pathology laboratories, genetic material is often analyzed at the nucleic acid level, especially with the ubiquitous technique of polymerase chain reaction (PCR). Currently, many of the PCR-based diagnostic molecular tests yield positive or negative results to identify disease-specific genetic changes.

As molecular testing becomes more focused on quantitation of molecular targets, the need for relatively pure cell populations will increase. Tissue microdissection is an excellent method to obtain relatively pure cellular samples of morphologically confirmed cell types.[1–3] This cellular purity results in more accurate test results and will limit the contamination from normal DNA that usually reduces assay sensitivity. Microdissection can be performed in a variety of ways, all of which have different advantages and disadvantages that have been reviewed elsewhere. Basically, these methods range from simple and inexpensive manual methods to laser-capture microdissection (LCM) methods that require expensive and complex equipment. Microdissection of target tissue is followed by DNA or RNA extraction. The common sequence of tissue processing in anatomic clinical or developmental molecular pathology laboratories is illustrated in Fig. 8.1.

P.T. Cagle and T.C. Allen (eds.), *Basic Concepts of Molecular Pathology*, Molecular Pathology Library 2,
DOI: 10.1007/978-0-387-89626-7_8, © Springer Science+Business Media, LLC 2009

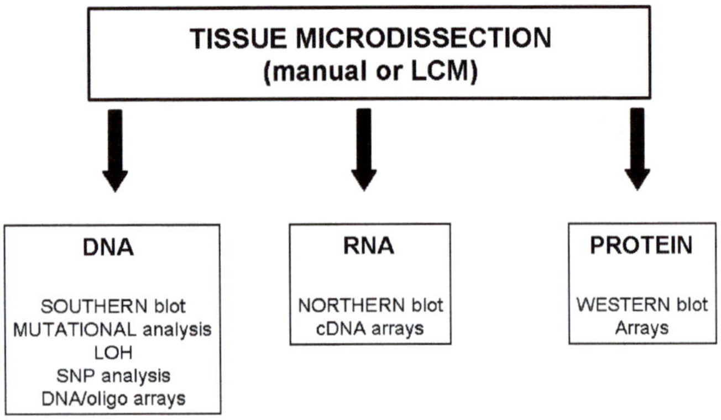

FIG. 8.1. Tissue processing in anatomic clinical or developmental molecular pathology laboratories. Manual or laser-capture microdissection of frozen or paraffin-embedded target tissue is followed by extraction of DNA, RNA, or protein, which are subsequently analyzed by appropriate molecular techniques.

Anatomic Pathology Testing to Detect or Characterize Neoplasia

Rapid advances in molecular technologies and expanding knowledge about lung tumor carcinogenesis have resulted in many studies of potential diagnostic and prognostic biomarkers. Based on current understanding, most tumors have multiple genetic and epigenetic alterations, including inactivation of tumor suppressor genes and activation of oncogenes. These molecular changes can be correlated with the sequential morphological changes of multistep carcinogenesis. The status of the many genes involved in tumorigenesis has been explored with multiple different molecular techniques, including loss of heterozygosity, array CGH, direct gene sequencing, gene expression profiling, fluorescent in situ hybridization, and proteomics. Each of these assays is discussed in the relevant chapters for specific assays throughout this book. In the following discussion, a more general overview of molecular testing in carcinomas is presented.

Oncogenes

"Proto-oncogenes" are wild-type, universally present genes that can stimulate carcinogenesis when mutated. The mutant forms are designated as "oncogenes." Oncogenes act as dominant genes in which a mutation in only one copy of the gene leads to activation, often with overexpression of a protein product. A variety of mutational events can transform a proto-oncogene to an oncogene, including point mutations, translocations, and amplifications.

Detection techniques for oncogene mutations in tumor samples depend entirely on what type of mutation is occurring. For example, point mutations can be detected using techniques for gene sequencing. Translocations and amplifications are generally best detected in paraffin-embedded tissues using FISH-based techniques. Because many of these mutations are activating mutations and cause overexpression, the aberrant protein product may be detectable by immunohistochemistry (IHC). However, overexpression of an oncogene protein product by IHC alone does not necessarily signify a mutation, because other nonmutational mechanisms can also be responsible for protein overexpression.

In surgical pathology, some of the most common oncogene assays detect translocations, particularly for sarcomas and hematological malignancies. These tumorigenic translocations often reposition an oncogene partner next to a constitutively active gene. The oncogene is then aberrantly and constitutively activated in the cells harboring the translocation. Depending on the clustering of breakpoints, translocation assays can use reverse transcription-polymerase chain reaction (RT-PCR), PCR alone, or in situ hybridization. Very few epithelial tumors are known to harbor consistent translocations.

Most oncogenes that have been associated with epithelial tumors involve specific point mutations in notorious genes. One of the most prevalent oncogene mutations occurs in the *KRAS* gene. *KRAS* gene mutations occur in many different tumor types, including lung carcinomas, pancreatic carcinomas, thyroid tumors, and colon carcinomas. *KRAS* mutations usually involve one of three important codons (12, 13, or 61). In some tumor types, such as lung carcinoma, *KRAS* mutations are known to associate with the genetic damage from tobacco.[4–7] In other organ systems, such as colon carcinoma, there is an emerging role for KRAS testing for predicting nonresponse to some targeted drugs.[8]

The clustering of mutations in *KRAS* mutations predominantly in exon 2 of the gene makes it feasible to use a simple and rapid PCR-based sequencing assay for detection of these mutations. Even in small or low cellularity samples, such as cytology samples, sequencing assays can be performed. One caveat to all traditional sequencing reactions for detecting oncogene mutations, however, is that the lower limits of sensitivity are around 20–25% in most studies. This limitation means that samples that are heavily contaminated with normal adjacent tissue may yield false-negative results using a sequencing assay for point mutations.

Unfortunately, *KRAS* mutations alone are not specific for carcinoma in most organ systems. In lung carcinoma, for example, they have also been detected in the sputum of patients whose tumors are negative for *KRAS* mutations and even in individuals with no clinically detectable carcinoma.[9–12] This finding highlights the fact that mutations in *KRAS* are only one component of the carcinogenesis pathway.

Tumor Suppressor Genes

Tumor suppressor genes (TSG) in the wild-type state have two active copies (alleles). These genes are thought to functionally suppress carcinogenesis in their routine cellular activity. Because both copies of the TSG must be mutated for tumorigenesis, TSGs are designated as recessive genes, in contrast to dominant proto-oncogenes, in which only one mutation is needed. We rarely test for both mutations when assessing the status of TSGs.

The most common assays for identifying TSGs include loss of heterozygosity (LOH), CGH, array CGH (aCGH), and FISH. LOH studies have been used to identify and define the location of the many TSGs involved in carcinogenesis. Alterations in TSGs are generally thought to be cumulative and progressive throughout tumorigenesis in many different organ systems. Advanced tumors frequently show widespread complete or partial loss of each chromosomal arm, whereas precursor lesions tend to show more focal and smaller chromosomal losses.

Molecular Anatomic Testing for Targeted Therapies

The advancement in development of novel targeted therapies, such as the inhibitors of the epidermal growth factor receptor (EGFR) and other tyrosine kinases, has opened a new chapter for the molecular testing of tumors. It is clear from the experience with targeted therapy for breast cancer that standardized assays for assessing and predicting the effects of therapeutic agents should be ideally developed in parallel with targeted therapies. The first assays used in the search for biomarkers for drug therapy with EGFR inhibitors were tests for EGFR protein expression by IHC and gene copy number by FISH.[13–15] There is a good correlation between IHC and FISH results, meaning that amplification often does correlate with increased protein expression in some tumor systems, such as lung carcinoma and glioblastoma multiforme. Unfortunately, these assays have not proven to be useful in predicting patient response to EGFR inhibitors. EGFR activating point mutations have also been described, but these appear to be exclusive to lung cancer; only extremely rare mutations have been identified in other types of cancer, particularly in colorectal carcinoma.[16,17] The same deletion in exon 19 as is seen in lung cancer has also been rarely detected in squamous cell carcinoma of the head and neck.[18]

Anatomic Pathology Testing for Infectious Agents

As microbiology has become more and more reliant on molecular technology, sophisticated assays can also be readily applied to tissues in anatomic pathology. Molecular testing for infectious agents is increasingly useful as a diagnostic tool. IHC and ISH remain the popular ancillary techniques for surgical pathologists, because they can be applied to routine diagnostic material coupled with histological evaluation. However, paraffin-embedded tissues can also serve as excellent source material for PCR-based organism identification assays; this is particularly useful when tissue was not sent directly for cultures or when suspicion is high but IHC and ISH are negative or inconclusive. Nucleic acids extracted from formalin-fixed paraffin-embedded tissue are usually more fragmented than the high molecular weight DNA that can be obtained from fresh or frozen tissue. Several laboratories are now equipped to do testing for *Mycobacteria* species and viruses, such as cytomegalovirus (CMV), from paraffin-embedded tissues, through specially designed and optimized assays for fragmented DNA.

Despite the increase in isolates of *Mycobacterium tuberculosis* in the United States since 1985, there are now more isolates of nontuberculous mycobacteria (NTB) such as *Mycobacterium avium* complex. The diagnosis of tuberculosis for many years has been based on special stains of smears for acid-fast bacilli (AFB) and mycobacterial cultures. The special stains for AFB are not sensitive and do not allow the identification of the different species of mycobacteria. Mycobacterial cultures are specific, but results are usually not available for 2–3 weeks or longer. PCR using oligonucleotide primers specific for gene fragments in *M. tuberculosis* has been used to identify tuberculosis organisms from archival formalin-fixed paraffin-embedded tissue.[19] This method may detect organism with high specificity and sensitivity and is much faster than conventional cultures. The sensitivity of PCR is better than that of the special stains, in which a few mycobacteria might be missed on routine histological sections. However, PCR results should always be interpreted in a clinical context. In diagnostic molecular pathology, sampling error is an important source of false-positive and false-negative results. The percentage of false-negative results after PCR amplification of *M. tuberculosis* from formalin-fixed, paraffin-embedded tissue varies among studies from 2 to 19%, depending on the PCR technique employed.[20] The DNA extraction used for this method is performed from several whole thick sections, as the rate of PCR positivity seems to be at least partially related to the quantity of DNA used for amplification. Therefore, PCR techniques may have some disadvantages such as possibly increased effect of tissue inhibitors or increased possibility of false-positive results caused by contamination. Because of the relatively low amount of template DNA needed for amplification, PCR-based analysis is now increasingly performed on microdissected areas from tissue sections. For surgical pathologists, a molecular identification of mycobac-teria can be the method of choice when the tissue obtained by biopsy is not sent for culture.

Several types of amplification assays are available for quantitation of CMV and other viruses, most of which use a form of real-time or quantitative PCR. Currently, most assays are designed for peripheral blood, and therefore surgical pathologists are rarely involved in ordering these tests.

References

1. Hunt JL, Finkelstein SD. Microdissection techniques for molecular testing in surgical pathology. *Arch Pathol Lab Med.* 2004;128: 1372–1378.

2. Eltoum IA, Siegal GP, Frost AR. Microdissection of histologic sections: past, present, and future. *Adv Anat Pathol.* 2002;9:316–322.
3. Simone NL, Paweletz CP, Charboneau L, et al. Laser capture microdissection: beyond functional genomics to proteomics. *Mol Diagn.* 2000;5:301–307.
4. Rodenhuis S, Slebos RJ. The ras oncogenes in human lung cancer. *Am Rev Respir Dis.* 1990;142:S27–S30.
5. Westra WH, Slebos RJ, Offerhaus GJ, et al. K-ras oncogene activation in lung adenocarcinomas from former smokers. Evidence that K-ras mutations are an early and irreversible event in the development of adenocarcinoma of the lung. *Cancer (Phila).* 1993;72:432–438.
6. Nelson HH, Christiani DC, Mark EJ, et al. Implications and prognostic value of K-ras mutation for early-stage lung cancer in women. *J Natl Cancer Inst.* 1999;91:2032–2038.
7. Chapman AD, Kerr KM. The association between atypical adenomatous hyperplasia and primary lung cancer. *Br J Cancer.* 2000;83:632–636.
8. De Reynies A, Boige V, Milano G, Faivre J, Laurent-Puig P. KRAS mutation signature in colorectal tumors significantly overlaps with the cetuximab response signature. *J Clin Oncol.* 2008;26:2228–2230.
9. Mao L, Hruban RH, Boyle JO, et al. Detection of oncogene mutations in sputum precedes diagnosis of lung cancer. *Cancer Res.* 1994;54:1634–1637.
10. Ronai Z, Yabubovskaya MS, Zhang E, et al. K-ras mutation in sputum of patients with or without lung cancer. *J Cell Biochem.* Suppl 1996;25:172–176.
11. Nakajima E, Hirano T, Konaka C, et al. K-ras mutation in sputum of primary lung cancer patients does not always reflect that of cancerous cells. *Int J Oncol.* 2001;18:105–110.
12. Destro A, Bianchi P, Alloisio M, et al. K-ras and p16(INK4A) alterations in sputum of NSCLC patients and in heavy asymptomatic chronic smokers. *Lung Cancer.* 2004;44:23–32.
13. Nakamura H, Kawasaki N, Taguchi M, et al. Survival impact of epidermal growth factor receptor overexpression in patients with non-small cell lung cancer: a meta-analysis. *Thorax.* 2006;61:140–145.
14. Suzuki S, Dobashi Y, Sakurai H, et al. Protein overexpression and gene amplification of epidermal growth factor receptor in nonsmall cell lung carcinomas. An immunohistochemical and fluorescence in situ hybridization study. *Cancer (Phila).* 2005;103:1265–1273.
15. Hirsch FR, Varella-Garcia M, Bunn PA Jr, et al. Epidermal growth factor receptor in non-small-cell lung carcinomas: correlation between gene copy number and protein expression and impact on prognosis. *J Clin Oncol.* 2003;21: 3798–3807.
16. Lynch TJ, Bell DW, Sordella R, et al. Activating mutations in the epidermal growth factor receptor underlying responsiveness of non-small-cell lung cancer to gefitinib. *N Engl J Med.* 2004;350: 2129–2139.
17. Nagahara H, Mimori K, Ohta M, et al. Somatic mutations of epidermal growth factor receptor in colorectal carcinoma. *Clin Cancer Res.* 2005;11:1368–1371.
18. Lee JW, Soung YH, Kim SY, et al. Somatic mutations of EGFR gene in squamous cell carcinoma of the head and neck. *Clin Cancer Res.* 2005;11:2879–2882.
19. Johansen IS, Thomsen VO, Forsgren A, et al. Detection of Mycobacterium tuberculosis complex in formalin-fixed, paraffin-embedded tissue specimens with necrotizing granulomatous inflammation by strand displacement amplification. *J Mol Diagn.* 2004;6:231–236.
20. Marchetti G, Gori A, Catozzi L, et al. Evaluation of PCR in detection of Mycobacterium tuberculosis from formalin-fixed, paraffin-embedded tissues: comparison of four amplification assays. *J Clin Microbiol.* 1998;6:1512–1517.

9

Polymerase Chain Reaction and Reverse Transcription-Polymerase Chain Reaction

Dwight Oliver

Polymerase Chain Reaction

Polymerase chain reaction (PCR) enables one to determine if a specific needle is present in a haystack, and it can be used as a step toward the characterization of the needle. It is a quick, powerful, inexpensive DNA amplification technique that has become a fundamental tool in molecular pathology.

Theory

The PCR is one of the most significant technical innovations in molecular biology.[1] The PCR was devised by Kary Mullis and colleagues[2,3] at Cetus Corporation in California and was first described in a 1985 paper demonstrating its application in the prenatal diagnosis of sickle cell anemia[2] and then further described in an ensuing paper.[3] These works detailed how a DNA sequence could be enzymatically amplified in vitro using specific oligonucleotide primers and bacterial (Klenow) DNA polymerase. With refinement of PCR over the next 3 years, it was found that a robust PCR using a thermostable polymerase could amplify a DNA sequence by a factor of more than 10^7, even when the target DNA made up only 1 of 100,000 DNA strands in a reaction.[4] Since then, additional improvements and variations to the original reaction have been made, affording even more efficiency, sensitivity, and utility to this tool. Its application specifically to diseases of the lungs has ranged from detection of infectious diseases[5–7] to study of inflammatory mechanisms,[8,9] to use in mutation analysis,[10–12] to detection of tumors and metastases.[13–15]

Principles

The principle of PCR is illustrated in Fig. 9.1. The target DNA to be amplified in vitro can be human genomic, bacterial, viral, plasmid, or previously PCR-amplified DNA and is represented in the figure by the target's nucleotide base letters A, C, G, and T and the sugar phosphate backbone. Other components of PCR include a thermostable DNA polymerase such as *Taq* polymerase, two oligonucleotide primers, four deoxynucleotide triphosphates, magnesium, buffer, and a thermocycler. PCR achieves amplification of the DNA by repeating a three-step cycle over and over. These three steps are denaturation, annealing, and extension.

In the denaturation step, target double-stranded DNA (dsDNA) is heated to a high temperature (94–95°C) to break the hydrogen bonds between nucleotide bases on opposing strands. The dsDNA *denatures*, splitting into two intact single strands (ssDNA) that are *complementary* to each other. In the second (annealing) step, the reaction mix is cooled, typically to 50–65°C, allowing ssDNA oligonucleotide *primers* to bind (*anneal*) to the portion of the target ssDNA to which they have specifically been designed. The investigator must know the sequence of the target DNA (at least in the region of the primers) to design the primers. Two primers are required: one that is complementary to the 3′– → 5′-oriented target strand in Fig. 9.1 (the *forward* primer), and one primer that is complementary to the 5′ → 3′ target strand in the figure (the *reverse* primer). Because these primers are in overwhelming abundance compared with the target DNA, they will anneal to the target DNA much more frequently than the full-length target DNA will anneal to its full-length complementary strand as the reaction is cooled. In the third (*extension*) step, as the temperature is raised to its working optimum (72°C), *Taq* polymerase recognizes these partially double-stranded DNAs and uses the forward and reverse primers as initiation points to begin extending the primers via polymerization in a 5′– → 3′-direction along the two new (*nascent*) strands. DNA polymerase does this by selecting high-energy deoxynucleotide triphosphates (dNTPs) from solution and placing them in the nascent strands directly across from their complementary base in the template (target) strand. During the first round (*cycle*) of PCR, the DNA polymerases extend the nascent strands for a relatively long distance before falling off. At the end of the first cycle, two double-stranded copies of a portion of the target DNA have been generated from one copy.

P.T. Cagle and T.C. Allen (eds.), *Basic Concepts of Molecular Pathology*, Molecular Pathology Library 2,
DOI: 10.1007/978-0-387-89626-7_9, © Springer Science+Business Media, LLC 2009

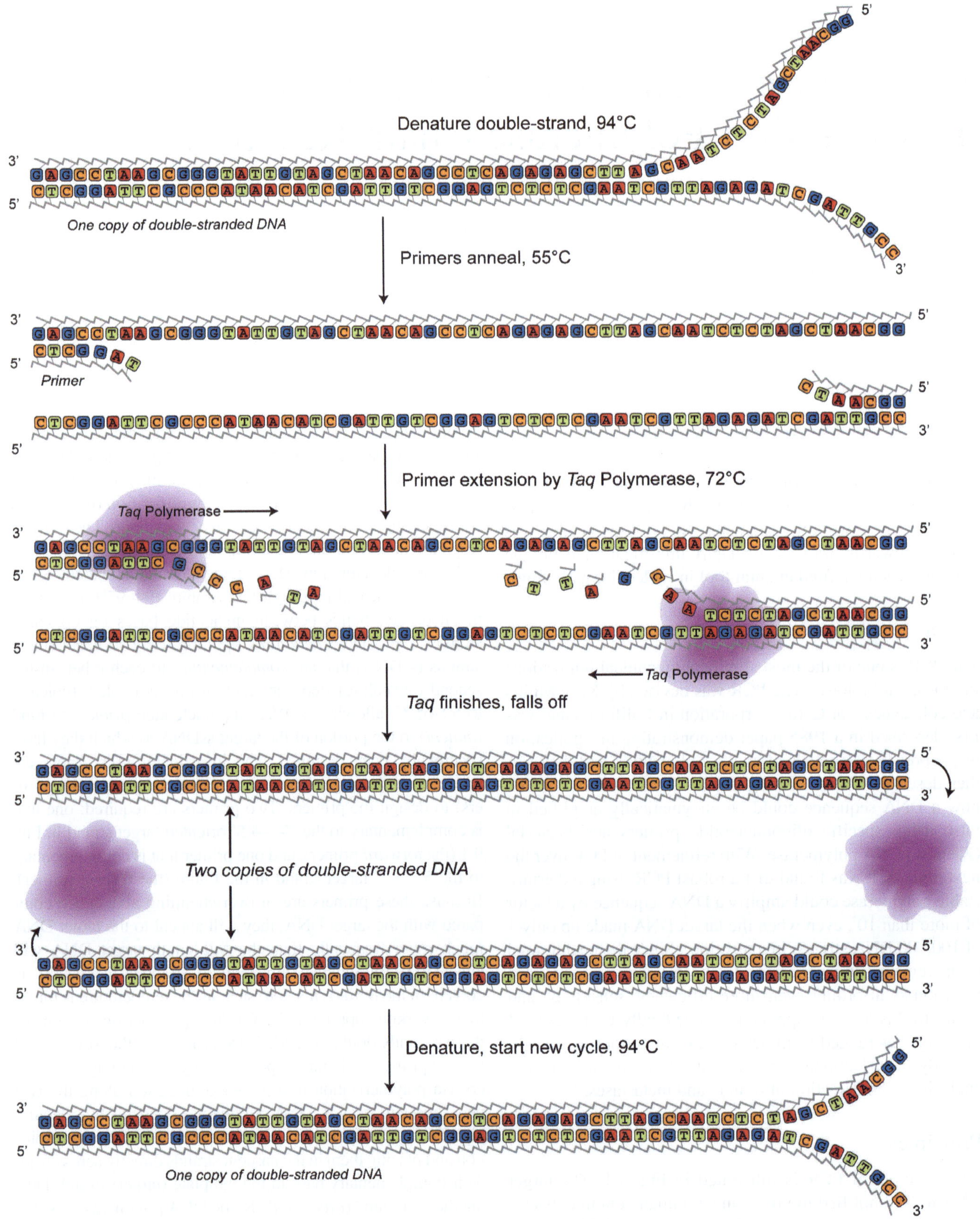

FIG. 9.1. Polymerase chain reaction steps. A single double-stranded DNA fragment is denatured and cycled through three steps – denaturation, annealing, and extension – to create an exact copy of the original. *Taq* polymerase extends primers by selecting deoxyribonucleotide triphosphates from the reaction solution based on the nucleotide sequence of the target DNA.

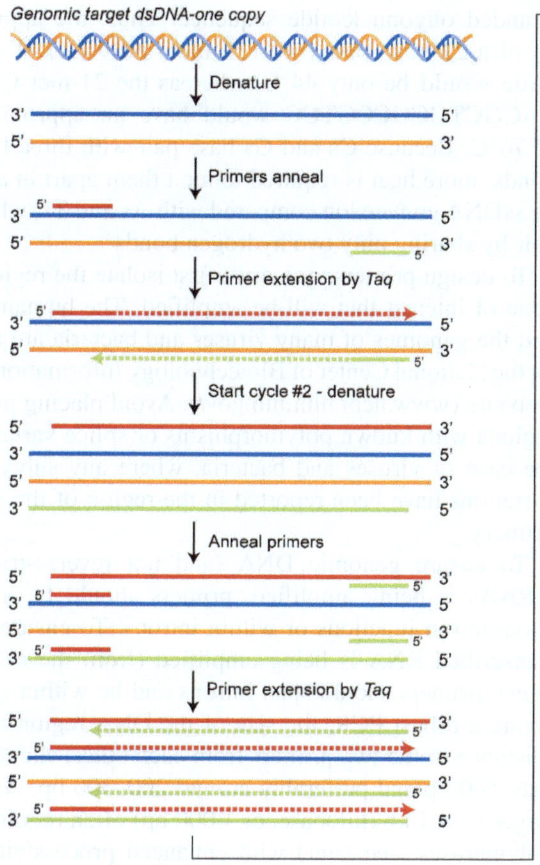

FIG. 9.2. Thirty-five cycles of polymerase chain reaction (PCR). DNA strands are abbreviated as *lines*. In the first cycle, two long PCR products of variable length (*red or green lines*) are polymerized, but with ensuing cycles the overwhelming PCR product is short, as defined by the positions of the forward (*red*) and reverse (*green*) primers. In a perfect PCR, several million amplicons are present after 35 cycles and ready for post-PCR analysis.

During the second cycle (Fig. 9.2, where DNA strands are symbolized by lines) the same process happens, but now the polymerase can only proceed as far down the DNA as the point where the opposite primer started. At the end of the second cycle the PCR products (*amplicons*) include two relatively short fragments of ssDNA whose two ends now correspond exactly to the locations of the forward and reverse primers. Both the genomic DNA and new amplicons can serve as templates in successive PCR cycles.[16] With subsequent cycles, the longer dsDNA PCR products are diluted out by the more numerous shorter dsDNA PCR products.[17,18] In a perfect PCR the amount of dsDNA doubles with each cycle so that after 30 cycles there are more than 1 billion copies of the original dsDNA (2^{30} = 1.1 billion) and more than 1 trillion copies after 40 cycles (2^{40} = 1.1 trillion). Because the primers recognize only the target DNA they are designed for, only a specific segment of DNA is amplified, even if it makes up only a fraction of all the different DNA sequences in a reaction.[4] This preferential amplification greatly facilitates post-PCR analysis of the target sequence.

PCR cycling is done in a *thermocycler*, a small automated tabletop instrument programmed by the investigator. A typical program (Table 9.1) starts with a 5–10-min denaturation at 94–95°C to ensure that the majority of the DNA (especially large chromosomal strands) is denatured. This step is followed by 30–50 cycles of brief denaturation/annealing/extension. The PCR concludes with a 5–10-min final polymerization step at 72°C to ensure nearly all amplicons are extended to their full length.

TABLE 9.1 Typical polymerase chain reaction (PCR) thermocycler program.

	Temperature (C)	Time
Initial denaturation	95	10 min
Forty cycles of		
Denaturation step	95	30 s
Annealing step	57	30 s
Extension step	72	45 s
Final polymerization	72	10 min
Post-PCR hold	4	Indefinite

Practical Polymerase Chain Reaction

Thermocyclers must be properly programmed to create an efficient PCR. Most programs are roughly similar (see Table 9.1), except for the annealing temperature and time intervals of each step. The annealing temperature is dependent on the melting temperature of the primers, and the time spent at the different steps is dictated by the size of the target DNA. The denaturation temperature is generally the same in all PCRs (94–95°C), just as the extension temperature is usually 72°C, because most polymerases used in PCR work best at this temperature (but see later discussion of polymerases and real-time PCR).

Denaturation Step Programming

The initial denaturation, which occurs before any cycling, is typically 94–95°C for 5–10 min; this gives human chromosomal DNA time to unravel and split into single strands. Smaller (e.g., viral) DNA targets may require only 3–5 min. Because ssDNA tends to reanneal while cooling, the dsDNA must be redenatured at 94°C for 10–60 s at the beginning of each cycle. GC-rich targets may require hotter/longer denaturations, while formamide or dimethylsulfoxide (DMSO) can be added to promote denaturation.[19] Temperatures above 95°C or excessive cycles should be avoided because *Taq* stability decreases under these conditions.[20]

Annealing Step Programming and Primer Design

The *annealing* temperature varies with every PCR and is a critical factor in the PCR program, as it helps dictate lower limit of detection, sensitivity, and specificity of the PCR. It must be calculated for each unique PCR and is dependent on the melting temperature of the two oligonucleotide primers. The *melting* temperature (T_m) of a DNA fragment is the temperature at which half of it has denatured into the single-stranded form (e.g., the primer is *not* annealed to its complementary target) and half is still double-stranded (primer *is* annealed to its target), assuming the number of copies of the two complementary strands is equal. Melting temperature calculation of an oligonucleotide can be very complex, with formulas that employ thermodynamics and depend on nearest neighbor nucleotides[21,22] and the salt concentrations in the reaction. Numerous websites, including those of companies that manufacture custom oligonucleotides, have free T_m calculators; after a primer sequence is typed in, the T_m is instantly calculated. Software is also available that will generate a list of potential primer pairs, with their T_ms, once the entire target sequence is entered.

Manual calculation of the approximate T_ms of short oligonucleotides can also be done using the abbreviated Wallace rule:[23]

$$T_m = 2(A + T) + 4(C + G)$$

Here T_m is in degrees centigrade, and A, T, C, and G stand for the number of each of these bases in the single-stranded oligonucleotide sequence. Thus, the approximate T_m of a 22-base pair (bp) (22-mer) poly (A) oligonucleotide would be only 44°C, whereas the 21-mer CGGCTG CACGCTGCGCCGTCC would have an approximate T_m of 76°C. Because Cs and Gs base pair with three hydrogen bonds, more heat is required to melt them apart in a dsDNA → ssDNA conversion compared with As and Ts, which base pair by sharing only two hydrogen bonds.

To design primers, one must first isolate the region of the gene of interest that will be amplified. The human genome and the genomes of many viruses and bacteria are available on the National Center of Biotechnology Information (NCBI) web site (www.ncbi.nlm.nih.gov/). Avoid placing primers in regions with known polymorphisms or splice variants or, in the case of viruses and bacteria, where any subtype/strain variations have been reported in the region of the proposed primers.

To ensure genomic DNA (and not reverse-transcribed mRNA) is being amplified, primers should be located at exon–intron junctions or within introns. To ensure reverse-transcribed RNA is being amplified (from the cDNA; see later), primers should span introns and be within exons. To create a robust PCR, the size of the DNA region amplified (distance of the two primers from each other) should be less than 500 bp and preferably around 200–300 bp. Amplicons larger than 1 kb (kilobase, or 1,000 bp) often require special polymerases (see later) with enhanced processivity. These forward and reverse primers should be only 18–25 bp long, have a 40–60% G + C content, and have T_ms in the 55–65°C range. Melting temperatures outside this range may work, and in fact are sometimes necessary, but they have a greater likelihood of resulting in a less efficient or more nonspecific PCR. In addition, the primers should have T_ms that are within 2°C of each other: add or remove bases to meet this goal. Ideally, the last five bases in the primer should include three Cs or Gs, and the 3′-end of the primers should be a C or G to promote tight base pairing at the point of *Taq* recognition and initiation of polymerization.

Just as a primer will anneal to its complementary region in the target DNA, so a pair of primers might anneal to complementary regions within themselves or within each other. Thus, **TGGCC**CATTACACTT**GGCC**ATTT is a poor primer choice because there will be some tendency for the boldface sequences to anneal to each other, forming a hairpin *stem-loop structure* within itself or to form *primer dimers* between two similar primers. Likewise, a 5′-TAGG-3′ sequence in the reverse primer could transiently anneal to a 5′-CCTA-3′ sequence in the forward primer. Many of the web sites that calculate T_ms also have a tool to check for these types of structures, as these aberrant forms can significantly reduce the yield of PCR product due to effective reduction of the available primer supply. Finally, avoid repetitive bases at the 3′-end of a primer, as this promotes slippage ("out of register") errors by *Taq* polymerase. Once all (or as many as feasible) of the foregoing rules are met and two primer sequences have been found, the primers should be checked to be certain they are not complementary to DNA sequences unrelated to the target sequence by using the NCBI BLAST (basic local alignment search tool) database at www.ncbi.nlm.nih.gov/BLAST/. This website will list all published DNA sequences that exactly or closely match the oligonucleotide sequences submitted. If one or both primers are close matches to nontarget DNA that may be present in the samples tested, new primer(s) may have to be designed.

The rules are empirical and do not guarantee a large PCR yield for reasons that are not always obvious or easily tested. It is often more expeditious to simply design more than one set of primers using the above rules and test all on a specimen to determine which pair gives the most robust PCR.

What annealing temperature should now be used in the PCR? This choice also requires trial and error. Initially the annealing temperature programmed into the thermocycler should be 5–10°C below the lowest T_m of the forward and reverse primers. At such a low temperature there will be mispriming onto nontarget sequences, resulting in amplification of nonspecific products. To enhance the specificity and perhaps even the sensitivity of the PCR, the temperature should then be systematically raised until amplicon yield drops and, hopefully, nonspecific amplicons disappear. Alternatively, the annealing temperature can be lowered, potentially increasing sensitivity, at the risk of generating nonspecific amplicons in addition to the desired PCR product.

Polymerase Chain Reaction Components

Thermostable Polymerase

The first PCRs used an *Escherichia coli* DNA polymerase[2] that was thermolabile and had to be replaced at each cycle; extension at 37°C would allow the nonspecific priming of numerous genomic sites and thus formation of nonspecific amplicons. Because the cycling of PCR repeatedly raises the reaction to 95°C, a thermostable polymerase is required. *Taq* polymerase, the most frequently used, was first purified from the bacterium *Thermus aquaticus* in 1976,[24] years before PCR. The enzyme has optimal 5′ → 3′ polymerase activity at 80°C (but will inefficiently extend primers at much lower temperatures), requires a divalent cation (Mg^{2+}), extends at a rate of 60 nucleotides/s,[25] and has a polymerization per binding event (processivity) of 50–80 bases.[26] *Taq* has double-strand-specific 5′ → 3′ exonuclease activity (see later discussion of real-time PCR) but does not have 3′ → 5′ exonuclease (proofreading) ability; its estimated error rate is 2.1×10^{-4} errors per base per duplication.[27] *Taq* is inhibited in samples containing heparin, hemoglobin, phenol,[28] urine, urea,[29] ethanol and high formamide, DMSO, or ethylenediaminetetraacetic acid (EDTA) levels.

Polymerases with fidelity superior to *Taq* include *Pfu*, *Pwo*, *Tgo* (from *Pyrococcus furiosus*, *Pyrococcus woesei*, and *Thermococcus gorgonarius*, respectively), and *Tli* (from *Thermococcus litoralis*) has superior thermostability.[30] *Thermus thermophilus* polymerase (*Tth*), unlike *Taq*, has reverse transcriptase activity (see later) as well as DNA polymerase activity. All these polymerases are readily available from different distributors.

Deoxynucleotides

The four dNTPs needed to replicate DNA are deoxyadenosine triphosphate, deoxyguanosine triphosphate, deoxycytosine triphosphate, and deoxythymidine triphosphate (dATP, dGTP, dCTP, and dTTP). They are added in equal concentration to the PCR mix, typically with a final concentration of 50–250 μM each in the reaction (200–1,000 μM total for all four dNTPs).

To prevent contamination of a new PCR by previously amplified DNA, deoxyuridine triphosphate (dUTP), instead of dTTP, is added to reactions and incorporated into the amplicons. When the bacterial enzyme uracil-*N*-glycosylase[31] is added and activated at the start of subsequent PCRs, it destroys any previously amplified PCR products that contain uracil, but it does not harm the natural TTP-containing DNA of the new sample. Uracil-*N*-glycosylase is deactivated at temperatures above 50°C and therefore does not destroy the newly polymerized DNA strands made during PCR.

Polymerase Chain Reaction Buffer

Taq requires the correct pH to function throughout the range of temperatures in PCR. The buffer Tris–HCl (10 mM) provides a pH of 8.3 at 25°C but pH 7.2 at 72°C; *Taq* has improved fidelity at this pH or lower.[32] Potassium chloride (50 mM) stabilizes the DNA and promotes primer annealing to its target. Nonionic detergents such as 0.01% Tween-20 or 0.1% Triton X-100 are often used, as well as gelatin. Fortunately, the above components (depending on manufacturer) are included in optimized 10× PCR buffer solutions provided by most *Taq* polymerase suppliers and do not have to be added individually to the reaction.

Magnesium

The magnesium concentration is important because it affects PCR specificity and efficiency through its interaction with *Taq*, whose function is dependent on the divalent cation.[33] A free [Mg^{2+}] of 1.2–1.3 mM is optimal for *Taq*, whereas much higher Mg^{2+} levels result in increased error rates caused by base substitutions and frameshift errors.[32] Magnesium is often included in 10× PCR buffer solutions at a [Mg^{2+}] of 1.5 mM, which decreases to 1.3 mM in the presence of 0.2 mM dNTPs because of equimolar binding of dNTPs and Mg^{2+} by *Taq* polymerase.[33]

Polymerase Chain Reaction Setup

Before starting experiments using PCR, accommodations must be made in the laboratory. Because of the amplification power of PCR, contamination of even minute amounts of DNA from one sample to the next must be avoided. Master mixes, containing all components of the PCR except the DNA, should be set up in a separate, dedicated room or area (hood) away from specimens or post-PCR solutions. DNA or RNA should be isolated from samples in a second room or area. DNA of the samples can then be added to the master mix in a third area, ideally. The thermocycler should also be in a separate room or area of the lab. Keep a unidirectional flow of material from pre-PCR to post-PCR: do not allow PCR-amplified material into the master mix preparation or DNA isolation areas. Label dedicated pipettors and use only aerosol-resistant tips to prevent contamination of pipettor barrels. Wear protective disposable gloves at all times, and change them and laboratory coats when going from one room or area to the next. Use ultraviolet irradiation inside hoods or on benchtops to destroy possible contaminating DNA. Use only autoclaved molecular biology-grade water in master mixes. Make certain that all plastic tubes and tips are DNAse- and RNAse free, as these enzymes will digest the target nucleic acids in specimens.

There is no universal recipe for the PCR mix, but a typical mix, including the DNA, is shown in Table 9.2. However, this is just a start. All PCRs need some adjusting – annealing temperatures, cycle step times, reagents (concentration of each primer, Mg^{2+} concentration, etc.), primer sequences – to determine the optimal conditions to give the greatest yield, fastest time, or highest specificity.

Postprocedure Analysis

PCR alone does not provide answers to an investigator's questions, but, because of the huge increase in a specific product, it makes the product's analysis much easier. To prove that a target sequence was present in a specimen, amplicons are run out on agarose or polyacrylamide electrophoresis gels to compare their lengths to DNA "ladder" markers; amplicons of the correct size are strong evidence that the target sequence was present. Southern blotting[34] with probe hybridization is an alternative. Bacterial restriction endonuclease digestion before electrophoresis or single-stranded conformation polymorphism studies can be done on PCR products to check for mutations.[33] DNA sequencing[35] of the amplicons or ligation into a plasmid for further analysis are other options, depending on the needs of the researcher.

TABLE 9.2. Typical polymerase chain reaction (PCR) reagent mix.

H_2O (nuclease-free) to final total volume of	25.0 μl
dNTPs at 5–10 mM each (20–40 mM total)	0.5 μl
Forward primer at 25 μM	0.3 μl
Reverse primer at 25 μM	0.3 μl
10× PCR buffer with Mg^{2+}	2.5 μl
Taq polymerase at 1 unit[a]/μl	1.0 μl
Target DNA 100–500 ng	1.0 μl

[a]One unit of *Taq* is defined as the amount of enzyme that will incorporate 10 nmole of deoxynucleotide triphosphates into acid-insoluble material in 30 min at 75°C

Variations

Improvements and variations in PCR have been introduced over the past 20 years to meet the needs of researchers and clinical molecular diagnostics labs.

Hot start PCR[36–39] is a technique preventing *Taq* from extending primers until a temperature of 60–80°C is reached, usually done by withholding *Taq* from the reaction until these temperatures have been reached. This step prevents *Taq* extension of primer dimers or primers that have annealed to nonspecific regions of the specimen DNA at low temperatures, such as during the preparation of the reagent mix. The result is improved specificity and yield of the PCR and is especially helpful when the target DNA is a small percentage of the total DNA. There are two common hot start methods. The first uses a wax plug to separate key reagents (e.g., dNTPs from *Taq*) in the PCR tube; the wax melts at a temperature well above the primer T_m and allows mixing of all reagents. The second method uses a modified *Taq* that is activated only after the initial 95°C 10-min denaturation step.

Nested PCR[40,41] uses two pairs of primers to improve amplicon yield and specificity. The first pair ("outer primers") is designed to amplify a larger fragment of the target DNA. These amplicons are then used, usually in a second PCR, as the template DNA for the second set ("inner primers"), which necessarily makes a smaller PCR product. Even if two sequential PCRs are run there may still be four products in the final PCR, the smallest being defined by only the inner primers. If all four primers are added at once and only one PCR is run, the T_m of the inner primers should be lower than that of the outer primers.[33]

Methylation-specific PCR[42–44] is used on genomic DNA to determine if the CpG islands within the promoter of a gene are methylated on the cytosine residue (blocking the gene expression in the cell[45]) or unmethylated (potentially allowing gene expression). For example, imprinted genes and silenced genes on X chromosomes are methylated,[46] as are the promoters of tumor suppressor genes in many cancers.[47–49] Methylation-specific PCR is based on the realization that a cytosine converts to a uracil after bisulfite treatment, whereas a methylated cytosine (5-methylcytosine) is refractory and remains as cytosine.[50] An unmethylated CpG will be converted to UpG after bisulfite, but a methylated one will remain as CpG. The change in DNA sequence of an unmethylated promoter compared with a methylated promoter after bisulfite treatment allows one to design primers that can discriminate between the two. Whichever of the two primers yields a PCR product indicates the methylation status of the promoter.

Multiplex PCR amplifies multiple different regions of DNA at one time by using multiple primer pairs in one reaction. Several targets can be analyzed in one specimen, including housekeeping genes and variably expressed genes, multiple microorganisms,[51] or multiple mutations in a genetic disorder[52] or malignancy.[53,54] Although ostensibly a timesaver, designing primers and optimizing reaction conditions so as to ensure equally robust amplification of all targets in multiplex PCR is a challenge. First, design individual primer pairs (but avoid primer dimers with every other primer), and program the thermocycler to allow optimal amplification of all targets on an individual basis. Then, combine (equimolar) primers and run the same program to see which targets are weakly amplified. Adjust primer concentrations, annealing temperatures, [Mg^{2+}], and so forth to equalize yields of targets. Primer software is also available, but in all cases trial and error are necessary to arrive at the final setup.

Other variations of PCR, each with its unique benefits, include the amplification refractory mutation system,[55] allele-specific oligonucleotide probes,[56] rapid amplification of cDNA ends,[57] and in situ PCR.[58] One variant that deserves special attention is real-time PCR.

Real-Time Polymerase Chain Reaction

Real-time PCR[59–62] is a recent innovation that has quickly become very popular in molecular biology research and molecular diagnostics. It circumvents the need for time-consuming post-PCR analysis, and it can detect DNA targets, quantify the original (before amplification) copy number of the target DNA present in a specimen, or detect specific mutations. Although very similar to conventional PCR, real-time PCR is based on two additional principles. First, *Taq* polymerase has a bonus *5′-exonuclease* activity on partially double-stranded DNA[63]; it will digest it to a single-stranded target just before polymerizing it into a double-stranded amplicon. Second, energy can be transferred between fluorescent molecules (fluorophores) attached to oligonucleotides when these small DNA fragments are used as probes against specific DNA targets. This second principle, first proposed in 1948 by Förster[64] and later supported by Stryer and Haugland,[65] indicates that electronic excitation energy can be transferred over short distances (≤50 Å) via dipole–dipole resonance between an energy donor and acceptor chromophore.[66] Transfer efficiency is proportional to the inverse sixth power of the distance separating the donor and acceptor.[65]

Fluorescence resonance energy transfer (FRET) in real-time PCR is demonstrated in the five frames in Fig. 9.3. A single oligonucleotide (TaqMan® type) probe is designed so that it specifically anneals to its complementary region of a target DNA, somewhere between the forward and reverse primers (top frame of Fig. 9.3). A donor/emitter fluorophore covalently attached to one end of this oligonucleotide is stimulated by monochromatic light from a laser, and the energy is transferred to a quencher fluorophore at the other end. In this kind of probe, light is either not reemitted by the quencher or is reemitted at a wavelength different from that of the donor fluorophore. During the primer *extension* step of PCR (see Fig. 9.3, second frame) the probe remains intact, and the light remains quenched, until degraded into single nucleotides by the 5′-exonuclease activity of *Taq* as it passes through during polymerization of the target DNA (see Fig. 9.3, third frame). Once cleaved, the emitter nucleotide drifts too far away for its emissions to be quenched, and its fluorescence is recorded in each PCR cycle by a sensitive photodetector and entered into the system's computer. With each PCR cycle, the number of amplicons and the quantity of light double as more probes are cleaved.

Dual-probe real-time PCR (LightCycler® type or *hybridization probes*; see Fig. 9.3, last two frames) uses donor/emitter and acceptor probes, designed to anneal side by side on the same target DNA. Fluorescent energy is transferred from the fluorophore of the 5′-donor probe (on the left) to the 3′-acceptor probe (on the right) during the *annealing* step of PCR. The acceptor immediately reemits at a wavelength unique to its fluorophore, and its signal strength doubles as the number of amplicons doubles with each PCR cycle. The probes separate, and FRET is terminated as the temperature is increased for the denaturation step.

Whether in single- or double-probe real-time PCR, the unique emission spectra of all fluorophores are captured and analyzed throughout each cycle by the real-time instrument software to generate curves such as those in Fig. 9.4. Cycle number is plotted versus fluorescence, and each of the colored curves represents a different specimen. Flat lines indicate no target DNA was in the specimen, and curves that begin rising at a low cycle number represent specimens that had more initial target DNA than those curves that rise later. Note that an exponential increase in fluorescence (and thus amplicon number) occurs only after numerous cycles, when PCR becomes its most efficient. To compare specimen target DNA copy number, a single horizontal *threshold* line can be drawn through all curves at this exponential phase and then a vertical line dropped from each intersection point to the *X* (cycle number)-axis. This point on the *X*-axis is the *cycle threshold* number of the specimen, and it *increases* as the original DNA copy number in a specimen *decreases*. Real-time PCR can be quantitative: a series of controls with a range of known DNA copy numbers are run to generate a standard curve of cycle threshold versus copy number. The cycle threshold values of specimens run simultaneously with the controls can then be converted into original DNA copy number by extrapolating from the standard curve.

Real-time instruments are able to detect fluorescence over a broad range of amplification: dynamic ranges of 7–8 logs can be obtained[61] (e.g., specimen DNA copy numbers from 10^1 to 10^8 or from 10^2 to 10^9 can be detected on the same run). In addition, probes are able to detect even a single base pair change in target DNA, such as in allelic discrimination.[67,68]

In the design of real-time PCR, the target should be short, preferably less than 150 bp, to maximize amplification efficiency. In some real-time instrument programs the annealing and extension steps are combined into one and run at a temperature of ~60°C. Probes should be designed before the primers and can bind the sense or antisense strand. They

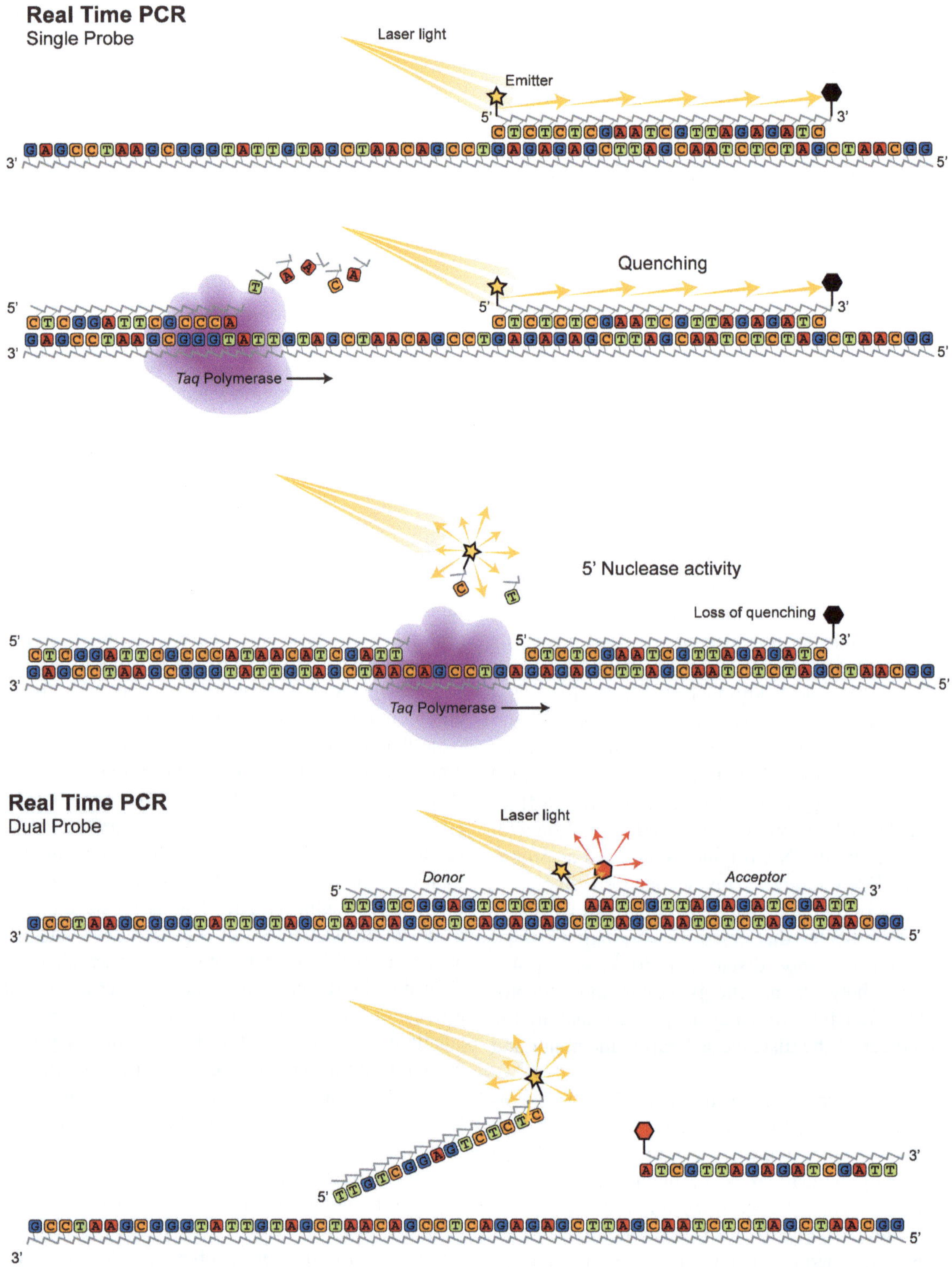

FIG. 9.3. Real-time polymerase chain reaction (PCR). The *top* three frames demonstrate single-probe real-time PCR. Quenching ceases once the nucleotide bound to the emitter (*yellow*) is cleaved by *Taq*. In the third frame, the forward portion of *Taq* cleaves the probe (releasing nucleotides), while the back part of *Taq* extends the forward primer (consuming other nucleotides). The *bottom* two frames show a dual-probe assay in which probes attach to the target DNA and transfer energy from the *yellow* to the *red* fluorophore during the annealing step. They float away during the extension or denaturation step, terminating fluorescence energy transfer, and will not be cleaved by *Taq*.

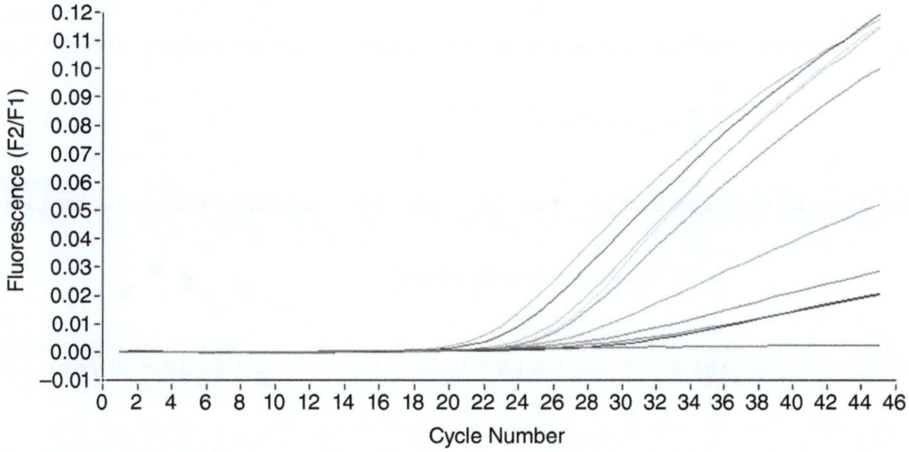

FIG. 9.4. Amplification curves using a dual-probe real-time polymerase chain reaction (PCR). The PCR cycle number is plotted against fluorescence for seven samples of human genomic DNA being tested for a blood coagulation factor. Note that the specimen represented by the rising curve on the far left contains the most target DNA, as it shows exponential amplification at an earlier cycle. (Courtesy of Mai Le, M, ASCP).

should be ≤35 bp long, have a GC content of 30–80%, should not have runs of four or more of the same nucleotide, should not partially anneal to each other or to either primer, and should have the 3′-end blocked by phosphorylation to prevent extension by *Taq*. TaqMan® probes should have a T_m of 68–70°C (about 8–10°C higher than their associated primers, which should have a T_m of 58–60°C), and the 5′-fluorophore should not be bound to a G. LightCycler® probes should optimally be 1–3 bp apart, have a T_m between the extension and annealing step temperatures, and be 5–10°C higher than the primer T_m; the T_m of the two probes should be within 2°C of each other unless mutation detection is desired. Mutation detection is facilitated by looking for differential melting (denaturation and consequently loss of FRET) of probes from wild-type versus mutant target DNA sequences (*melting curve*).

Reverse Transcription

All normal cells in a human's body, with few exceptions, have the same chromosomal DNA sequence, that is, the same genetic code. Thus, genomic information obtained from the DNA of easily obtained normal white blood cells would be applicable to the genetic makeup of normal lung, brain, or colon cells. This idea does not apply to malignant cells, which can have a genetic composition that is profoundly different from that of normal cells.

Function and structure of various cell types differ because of the mRNA that they transcribe and ultimately the proteins that are translated. In other words, it is the protein expression profile of cells that differentiates them. Presently, the most practical way to study the specific genes expressed in a particular cell type is to analyze the mRNA the cells make. Because RNA is unstable and therefore difficult to work with in the laboratory, it can be converted into the complementary DNA (cDNA) by a process known as *reverse transcription.*[69–71] The resultant cDNA is much more stable than the mRNA. Reverse transcription is so named because RNA is used as the template to direct the production of DNA: the reverse of normal cellular transcription, where DNA is used by RNA polymerase to direct the production of mRNA.

A reverse transcriptase enzyme is an RNA-directed DNA polymerase made and used by some RNA viruses to complete their life cycle within a host. Viral reverse transcriptases have been characterized and/or cloned, and the enzymes are commercially available for use in research and clinical molecular laboratories.[72–76]

Reverse transcriptase, similar to DNA polymerase, requires a DNA primer (Fig. 9.5) to initiate its function. Because mRNA has a poly(A) tail at its 3′-end, an ideal primer for reverse transcription of mRNA species would be a poly(T) oligonucleotide (oligo dT).[77,78] A replete collection of short DNA primers with random sequences can also be used; these primers are recommended if reverse transcription of ribosomal RNA (rRNA) is also desired along with mRNA. The enzyme starts transcription at the 3′-end of template RNA [the 5′-end of the nascent (new) cDNA strand] and proceeds in a 5′- → 3′-direction on the nascent strand ("first strand synthesis"). In this fashion, all the mRNA (or total RNA) present in a cell can be transcribed into complementary DNA. Those mRNA sequences that are present at a high copy number in the cell will be reverse transcribed to a high cDNA copy number compared with those mRNA sequences which are rare in a cell.

A typical reverse transcription protocol is given in Table 9.3. The two most commonly used reverse transcriptases are from bird and mouse viruses: avian myeloblastosis virus

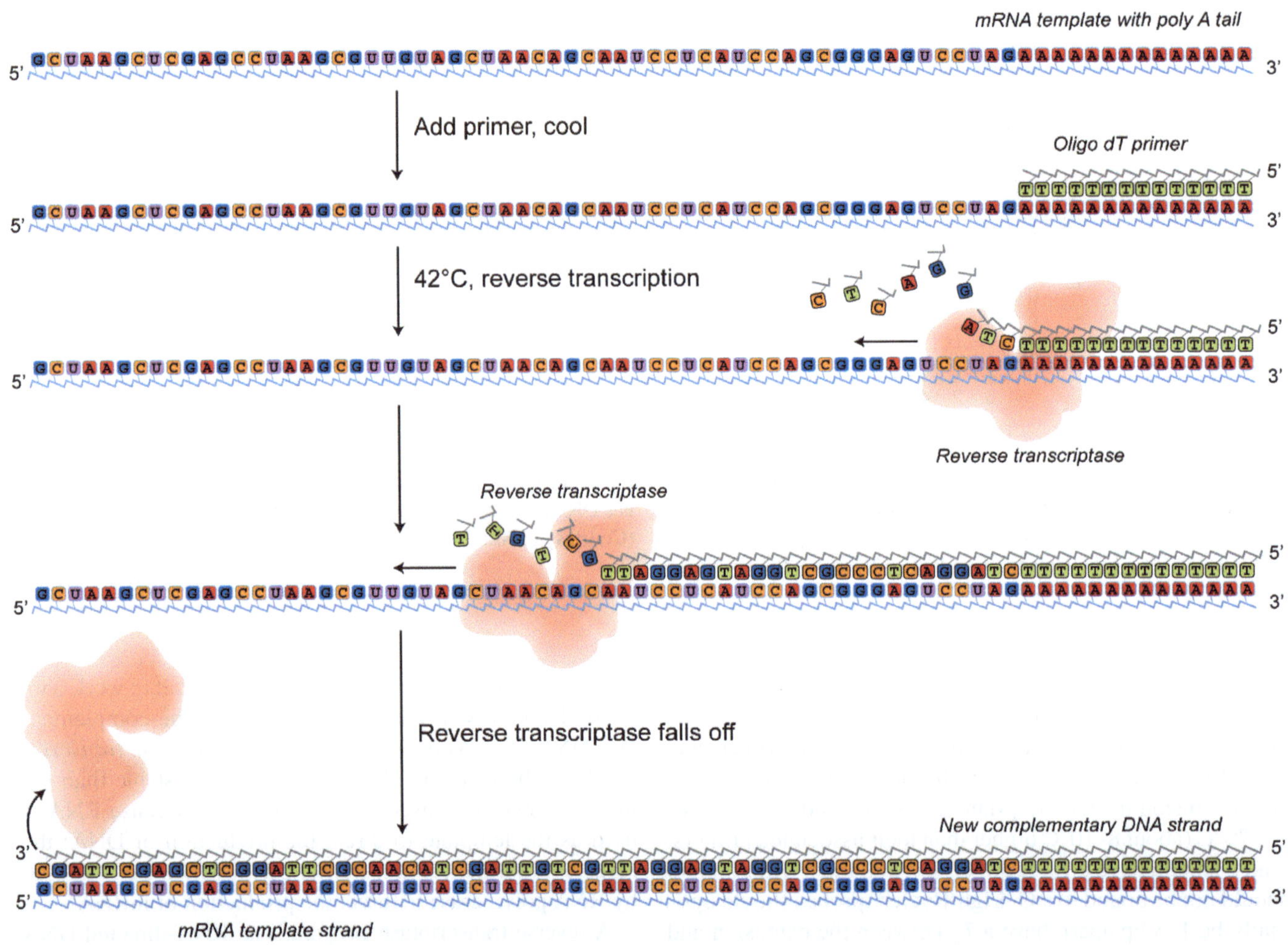

Fig. 9.5. Reverse transcription. Reverse transcriptase uses oligo dT as the primer on the target mRNA and polymerizes in the 5′- → 3′-direction on the new DNA strand. The original mRNA strand is then cleaved by an RNAse domain within the reverse transcriptase (not shown), thus allowing polymerization of the single-stranded DNA into double-stranded DNA during polymerase chain reaction.

Table 9.3. Typical reverse transcription reaction protocol.

RNA 1–2 μg	1.0 μl
DEPC-treated H_2O	8.5 μl
Oligo dT or random primers at 40 μM	2.0 μl
70°C ↓ 5 min	
4°C ↓ 5 min	
Quickly add 20.0 μl of prepared RT master mix	
DEPC-treated H_2O: add to total final volume	20.0 μl
MMLV (10×) or AMV (5×) buffer	2.0–4.0 μl
dNTP at 10 mmole	3.0 μl
RNAse inhibitor 10–40 units	1.0 μl
MMLV reverse transcriptase 200 units[a]	1.0 μl
AMV reverse transcriptase 30 units[a]	
37 or 42°C ↓ 60 min	
90–95°C ↓ 5 min	
Use ~1–3 μl in the PCR reaction	

AMV avian myeloblastosis virus; *DEPC* diethylpyrocarbonate; *dNTP* deoxynucleotide triphosphate; *MMLV* Moloney murine leukemia virus; *PCR* polymerase chain reaction; *RT* reverse transcription

[a]One unit of reverse transcriptase is defined as the amount of enzyme that will incorporate 1 nmole of deoxythymidine triphosphate into acid-insoluble material in 10 min at 37°C using poly(rA), oligo(dT) as template primer

(AMV) and Moloney murine leukemia virus (MMLV).[79] Their recommended buffers should not be interchanged. RNAse inhibitors and diethylpyrocarbonate-treated water are needed to preserve the unstable RNA. The initial 70°C heating is to remove secondary structures from the RNA; the 42°C (AMV and some MMLV products) or 37°C (some MMLV products) incubations are the working temperature of the enzymes. The 90–95°C step is needed to inactivate the enzymes.

The cDNA made by reverse transcription of mRNA (and/or rRNA) can then be used as a template for PCR if the appropriate primers for the target DNA are present. During the first cycle of the PCR, only one (the forward) primer is needed because only one strand is polymerized, but this new strand will serve as the template for the opposite primer during the second PCR cycle, and polymerization of both strands will continue with each cycle. [Note that some bacteria such as *Thermus thermophilus* have an enzyme (*Tth*) that can both reverse transcribe RNA and polymerize DNA, allowing reverse transcription and PCR to proceed simultaneously in a single tube.]

Reverse transcription-PCR (RT-PCR) is thus an important tool that allows the investigator to study the genes expressed or not expressed in specific cells after isolation of the mRNA.[80–83] Additional (post-PCR) techniques such as gel electrophoresis, single-stranded conformation polymorphism gels, restriction fragment length polymorphism analysis, DNA sequencing, microarrays, and so forth, can be applied to also determine if the genes expressed have mutations.

Under- or overexpression of a particular gene in neoplastic or reactive cells can be investigated by comparing their expression levels in normal cells, which could be done by comparing band strengths on Northern (RNA) blots. However, in these methods one must control for the number of tumor/reactive cells being the same as the number of normal cells. Analysis is much easier if done by real-time PCR, in which the ratio of the expression level of the gene of interest is compared with the expression level of a constitutively expressed housekeeping gene such as β-actin, 18S rRNA, cyclophilin, glyceraldehyde-3-phosphate dehydrogenase, and β_2-microglobulin. This ratio is calculated in both the normal and neoplastic/reactive cells, and then the ratios are compared to see if there is relative up- or downregulation of the gene of interest.

References

1. Bell J. The polymerase chain reaction. *Immunol Today*. 1989;10:351–355.
2. Saiki RK, Scharf S, Faloona F, et al. Enzymatic amplification of beta-globin genomic sequences and restriction site analysis for diagnosis of sickle cell anemia. *Science*. 1985;230:1350–1354.
3. Mullis K, Faloona F, Scharf S, et al. Specific enzymatic amplification of DNA in vitro: the polymerase chain reaction. *Cold Spring Harb Symp Quant Biol.* 1986;51(pt 1):263–273.
4. Saiki RK, Gelfand DH, Stoffel S, et al. Primer-directed enzymatic amplification of DNA with a thermostable DNA polymerase. *Science*. 1988;239:487–491.
5. Saboor SA, Johnson NM, McFadden J. Detection of mycobacterial DNA in sarcoidosis and tuberculosis with polymerase chain reaction. *Lancet.* 1992;339:1012–1015.
6. Myerson D, Lingenfelter PA, Gleaves CA, et al. Diagnosis of cytomegalovirus pneumonia by the polymerase chain reaction with archived frozen lung tissue and bronchoalveolar lavage fluid. *Am J Clin Pathol.* 1993;100:407–413.
7. Raad I, Hanna H, Huaringa A, et al. Diagnosis of invasive pulmonary aspergillosis using polymerase chain reaction-based detection of aspergillus in BAL. *Chest.* 2002;121:1171–1176.
8. Sundaresan S, Alevy YG, Steward N, et al. Cytokine gene transcripts for tumor necrosis factor-alpha, interleukin-2, and interferon-gamma in human pulmonary allografts. *J Heart Lung Transplant*. 1995;14:512–518.
9. Lordan JL, Bucchieri F, Richter A, et al. Cooperative effects of Th2 cytokines and allergen on normal and asthmatic bronchial epithelial cells. *J Immunol*. 2002;169:407–414.
10. Nogee LM, Dunbar AE III, Wert SE, et al. A mutation in the surfactant protein C gene associated with familial interstitial lung disease. *N Engl J Med.* 2001;344:573–579.
11. Pan Q, Pao W, Ladanyi M. Rapid polymerase chain reaction-based detection of epidermal growth factor receptor gene mutations in lung adenocarcinomas. *J Mol Diagn.* 2005;7:396–403.
12. Westra WH, Baas IO, Hruban RH, et al. *K-ras* oncogene activation in atypical alveolar hyperplasias of the human lung. *Cancer Res.* 1996;56:2224–2228.
13. Pulte D, Li E, Crawford BK, et al. Sentinel lymph node mapping and molecular staging in nonsmall cell lung carcinoma. *Cancer.* 2005;104:1453–1461.
14. Bohlmeyer T, Le TN, Shroyer AL, et al. Detection of human papillomavirus in squamous cell carcinomas of the lung by polymerase chain reaction. *Am J Respir Cell Mol Biol.* 1998;18:265–269.
15. Bremnes RM, Sirera R, Camps C. Circulating tumour-derived DNA and RNA markers in blood: a tool for early detection, diagnostics, and follow-up? *Lung Cancer.* 2005;49:1–12.
16. Eisenstein BI. The polymerase chain reaction. A new method of using molecular genetics for medical diagnosis. *N Engl J Med.* 1990;322:178–183.
17. Wenham PR. DNA-based techniques in clinical biochemistry: a beginner's guide to theory and practice. *Ann Clin Biochem.* 1992;29(pt 6):598–624.
18. Remick DG, Kunkel SL, Holbrook EA, Hanson CA. Theory and applications of the polymerase chain reaction. *Am J Clin Pathol.* 1990;93:S49–S54.
19. Chakrabarti R, Schutt CE. The enhancement of PCR amplification by low molecular-weight sulfones. *Gene.* 2001;274:293–298.
20. Laksanalamai P, Pavlov AR, Slesarev AI, Robb FT. Stabilization of *Taq* DNA polymerase at high temperature by protein folding pathways from a hyperthermophilic archaeon, Pyrococcus furiosus. *Biotechnol Bioeng.* 2006;93:1–5.
21. Breslauer KJ, Frank R, Blocker H, Marky LA. Predicting DNA duplex stability from the base sequence. *Proc Natl Acad Sci U S A.* 1986;83:3746–3750.
22. SantaLucia J Jr, Allawi HT, Seneviratne PA. Improved nearest-neighbor parameters for predicting DNA duplex stability. *Biochemistry.* 1996;35:3555–3562.
23. Wallace RB, Shaffer J, Murphy RF, et al. Hybridization of synthetic oligodeoxyribonucleotides to phi chi 174 DNA: the effect of single base pair mismatch. *Nucleic Acids Res.* 1979;6:3543–3557.
24. Chien A, Edgar DB, Trela JM. Deoxyribonucleic acid polymerase from the extreme thermophile *Thermus aquaticus. J Bacteriol.* 1976;127:1550–1557.
25. Takagi M, Nishioka M, Kakihara H, et al. Characterization of DNA polymerase from *Pyrococcus* sp. strain KOD1 and its application to PCR. *Appl Environ Microbiol.* 1997;63:4504–4510.
26. Davidson JF, Fox R, Harris DD, et al. Insertion of the T3 DNA polymerase thioredoxin binding domain enhances the processivity and fidelity of *Taq* DNA polymerase. *Nucleic Acids Res.* 2003;31:4702–4709.
27. Keohavong P, Thilly WG. Fidelity of DNA polymerases in DNA amplification. *Proc Natl Acad Sci U S A* 1989;86:9253–9257.
28. Wilson IG. Inhibition and facilitation of nucleic acid amplification. *Appl Environ Microbiol.* 1997;63:3741–2751.
29. Khan G, Kangro HO, Coates PJ, Heath RB. Inhibitory effects of urine on the polymerase chain reaction for cytomegalovirus DNA. *J Clin Pathol.* 1991;44:360–365.

30. Pavlov AR, Pavlova NV, Kozyavkin SA, Slesarev AI. Recent developments in the optimization of thermostable DNA polymerases for efficient applications. *Trends Biotechnol.* 2004;22:253–260.
31. Longo MC, Berninger MS, Hartley JL. Use of uracil DNA glycosylase to control carry-over contamination in polymerase chain reactions. *Gene.* 1990;93:125–128.
32. Eckert KA, Kunkel TA. High fidelity DNA synthesis by the *Thermus aquaticus* DNA polymerase. *Nucleic Acids Res.* 1990;18:3739–3744.
33. McPherson MJ, Møller SG. PCR. Oxford, UK: BIOS Scientific; 2000.
34. Southern EM. Detection of specific sequences among DNA fragments separated by gel electrophoresis. *J Mol Biol.* 1975;98: 503–517.
35. Bevan IS, Rapley R, Walker MR. Sequencing of PCR-amplified DNA. *PCR Methods Appl.* 1992;1:222–228.
36. Moretti T, Koons B, Budowle B. Enhancement of PCR amplification yield and specificity using AmpliTaq Gold DNA polymerase. *Biotechniques.* 1998;25:716–722.
37. Brandwein M, Zeitlin J, Nuovo GJ, et al. HPV detection using "hot start" polymerase chain reaction in patients with oral cancer: a clinicopathological study of 64 patients. *Mod Pathol.* 1994;7: 720–727.
38. Nuovo GJ, Gallery F, MacConnell P. Detection of amplified HPV 6 and 11 DNA in vulvar lesions by hot start PCR in situ hybridization. *Mod Pathol.* 1992;5:444–448.
39. Chou Q, Russell M, Birch DE, et al. Prevention of pre-PCR mis-priming and primer dimerization improves low-copy-number amplifications. *Nucleic Acids Res.* 1992;20: 1717–1723.
40. Tilston P, Corbitt G. A single tube nested PCR for the detection of hepatitis C virus RNA. *J Virol Methods.* 1995;53:121–129.
41. Smit VT, Boot AJ, Smits AM, et al. KRAS codon 12 mutations occur very frequently in pancreatic adenocarcinomas. *Nucleic Acids Res.* 1988;16:7773–7782.
42. Herman JG, Graff JR, Myohanen S, et al. Methylation-specific PCR: a novel PCR assay for methylation status of CpG islands. *Proc Natl Acad Sci U S A.* 1996;93:9821–9826.
43. Curtis CD, Goggins M. DNA methylation analysis in human cancer. *Methods Mol Med.* 2005;103:123–136.
44. Li LC, Dahiya R. MethPrimer: designing primers for methylation PCRs. *Bioinformatics.* 2002;18:1427–1431.
45. Bird A. The essentials of DNA methylation. *Cell.* 1992;70:5–8.
46. Robertson KD, Jones PA. DNA methylation: past, present and future directions. *Carcinogenesis.* 2000;21:461–467.
47. Rhee I, Bachman KE, Park BH, et al. DNMT1 and DNMT3b cooperate to silence genes in human cancer cells. *Nature.* 2002;416: 552–556.
48. Dammann R, Takahashi T, Pfeifer GP. The CpG island of the novel tumor suppressor gene RASSF1A is intensely methylated in primary small cell lung carcinomas. *Oncogene.* 2001;20:3563–3567.
49. Zochbauer-Muller S, Fong KM, Virmani AK, et al. Aberrant promoter methylation of multiple genes in non-small cell lung cancers. *Cancer Res.* 2001;61:249–255.
50. Clark SJ, Harrison J, Paul CL, Frommer M. High sensitivity mapping of methylated cytosines. *Nucleic Acids Res.* 1994;22: 2990–2997.
51. Elnifro EM, Ashshi AM, Cooper RJ, Klapper PE. Multiplex PCR: optimization and application in diagnostic virology. *Clin Microbiol Rev.* 2000;13:559–570.
52. Richards B, Skoletsky J, Shuber AP, et al. Multiplex PCR amplification from the CFTR gene using DNA prepared from buccal brushes/swabs. *Hum Mol Genet.* 1993;2:159–163.
53. Scurto P, Hsu Rocha M, Kane JR, et al. A multiplex RT-PCR assay for the detection of chimeric transcripts encoded by the risk-stratifying translocations of pediatric acute lymphoblastic leukemia. *Leukemia.* 1998;12:1994–2005.
54. Pallisgaard N, Hokland P, Riishoj DC, et al. Multiplex reverse transcription-polymerase chain reaction for simultaneous screening of 29 translocations and chromosomal aberrations in acute leukemia. *Blood.* 1998;92:574–588.
55. Newton CR, Graham A, Heptinstall LE, et al. Analysis of any point mutation in DNA. The amplification refractory mutation system (ARMS). *Nucleic Acids Res.* 1989;17:2503–2516.
56. Saiki RK, Bugawan TL, Horn GT, et al. Analysis of enzymatically amplified beta-globin and HLA-DQ alpha DNA with allele-specific oligonucleotide probes. *Nature.* 1986;324:163–166.
57. Schaefer BC. Revolutions in rapid amplification of cDNA ends: new strategies for polymerase chain reaction cloning of full-length cDNA ends. *Anal Biochem.* 1995;227:255–273.
58. Komminoth P, Long AA. In-situ polymerase chain reaction. An overview of methods, applications and limitations of a new molecular technique. *Virchows Arch B Cell Pathol Incl Mol Pathol.* 1993;64:67–73.
59. Livak KJ, Flood SJ, Marmaro J, et al. Oligonucleotides with fluorescent dyes at opposite ends provide a quenched probe system useful for detecting PCR product and nucleic acid hybridization. *PCR Methods Appl.* 1995;4:357–362.
60. Heid CA, Stevens J, Livak KJ, Williams PM. Real time quantitative PCR. *Genome Res.* 1996;6:986–994.
61. Lie YS, Petropoulos CJ. Advances in quantitative PCR technology: 5′ nuclease assays. *Curr Opin Biotechnol.* 1998;9:43–48.
62. Holland PM, Abramson RD, Watson R, Gelfand DH. Detection of specific polymerase chain reaction product by utilizing the 5′-3′ exonuclease activity of *Thermus aquaticus* DNA polymerase. *Proc Natl Acad Sci U S A.* 1991;88:7276–7280.
63. Longley MJ, Bennett SE, Mosbaugh DW. Characterization of the 5′ to 3′ exonuclease associated with *Thermus aquaticus* DNA polymerase. *Nucleic Acids Res.* 1990;18:7317–7322.
64. Förster T. Zwischemolekulare energiewanderung und fluoreszenz. *Ann Physik.* 1948;2:55–67.
65. Stryer L, Haugland RP. Energy transfer: a spectroscopic ruler. *Proc Natl Acad Sci U S A.* 1967;58:719–726.
66. Grinvald A, Haas E, Steinberg IZ. Evaluation of the distribution of distances between energy donors and acceptors by fluorescence decay. *Proc Natl Acad Sci U S A.* 1972;69:2273–2277.
67. Oliver DH, Thompson RE, Griffin CA, Eshleman JR. Use of single nucleotide polymorphisms (SNP) and real-time polymerase chain reaction for bone marrow engraftment analysis. *J Mol Diagn.* 2000;2:202–208.
68. Lay MJ, Wittwer CT. Real-time fluorescence genotyping of factor V Leiden during rapid-cycle PCR. *Clin Chem.* 1997;43: 2262–2267.
69. Temin HM, Mizutani S. RNA-dependent DNA polymerase in virions of Rous sarcoma virus. *Nature.* 1970;226:1211–1213.
70. Baltimore D. RNA-dependent DNA polymerase in virions of RNA tumour viruses. *Nature.* 1970;226:1209–1211.

71. Spiegelman S, Burny A, Das MR, et al. Characterization of the products of DNA-directed DNA polymerases in oncogenic RNA viruses. *Nature*. 1970;227:563–567.
72. Shinnick TM, Lerner RA, Sutcliffe JG. Nucleotide sequence of Moloney murine leukaemia virus. *Nature*. 1981;293:543–548.
73. Reddy EP, Smith MJ, Aaronson SA. Complete nucleotide sequence and organization of the Moloney murine sarcoma virus genome. *Science*. 1981;214:445–450.
74. Kotewicz ML, D'Alessio JM, Driftmier KM, et al. Cloning and overexpression of Moloney murine leukemia virus reverse transcriptase in Escherichia coli. *Gene*. 1985;35:249–258.
75. Verma IM, Baltimore D. Purification of the RNA-directed DNA polymerase from avian myeloblastosis virus and its assay with polynucleotide templates. *Methods Enzymol*. 1974;29:125–130.
76. Houts GE, Miyagi M, Ellis C, et al. Reverse transcriptase from avian myeloblastosis virus. *J Virol*. 1979;29:517–522.
77. Verma IM. Studies on reverse transcriptase of RNA tumor viruses III. Properties of purified Moloney murine leukemia virus DNA polymerase and associated RNase H. *J Virol*. 1975;15:843–854.
78. Marcus SL, Modak MJ. Observations on template-specific conditions for DNA synthesis by avian myeloblastosis virus DNA polymerase. *Nucleic Acids Res*. 1976;3:1473–1486.
79. Sambrook J, Russell DW. Molecular Cloning: A Laboratory Manual, 3rd ed. Cold Spring Harbor, NY: Cold Spring Harbor Laboratory Press, 2001:A4.24.
80. Broackes-Carter FC, Mouchel N, Gill D, et al. Temporal regulation of CFTR expression during ovine lung development: implications for CF gene therapy. *Hum Mol Genet*. 2002;11:125–131.
81. Dagnon K, Pacary E, Commo F, et al. Expression of erythropoietin and erythropoietin receptor in non-small cell lung carcinomas. *Clin Cancer Res*. 2005;11:993–999.
82. Singhal S, Wiewrodt R, Malden LD, et al. Gene expression profiling of malignant mesothelioma. *Clin Cancer Res*. 2003;9:3080–3097.
83. Lam KM, Oldenburg N, Khan MA, et al. Significance of reverse transcription polymerase chain reaction in the detection of human cytomegalovirus gene transcripts in thoracic organ transplant recipients. *J Heart Lung Transplant*. 1998;17:555–565.

10
Array Comparative Genomic Hybridization in Pathology

Reinhard Ullmann

Principle of Comparative Genomic Hybridization

Comparative genomic hybridization (CGH) is a molecular cytogenetic method for the detection and mapping of chromosomal gains and losses.[1] It is based on the cohybridization of differentially labeled test and reference DNA onto metaphase spreads, which usually have been prepared from peripheral blood lymphocytes of a healthy donor. The signal intensity ratios of the two labels along the chromosomes then reflect DNA copy number changes in the test genome relative to the reference genome. Although CGH has tremendously contributed to our knowledge of chromosomal aberrations, its resolution, unfortunately, is limited to about 3–10 Mb.[2] Resolution of CGH has significantly improved when samples were no longer hybridized to metaphase spreads but to DNA targets that have been arrayed on a glass substrate. This modification to the original technique has been named array CGH[3] or matrix CGH,[4] respectively. In theory, resolution of array CGH is only limited by the number and quality of DNA targets arrayed on the slide. The principle of array CGH is illustrated in Fig. 10.1.

Array CGH Platforms

Arrays Based on Clone Inserts

cDNA Arrays

The first genome-wide application of array CGH was based on cDNA arrays, with each spot representing one reversely transcribed mRNA.[5] Apart from the fact that these arrays were readily available, the main advantage of cDNA arrays was that they facilitated a direct comparison of DNA copy number changes with gene expression data derived from the same tumor.[6,7] However, cDNA arrays are extremely gene focused and thus only a small percentage of the genome is actually represented. DNA copy number changes concerning introns or intergenic sequences are missed. Another problem arises from paralogous genes or shared sequence motifs. Together with the low signal-to-noise ratio usually obtained with this kind of arrays, these disadvantages have limited the use of cDNAs for the detection of DNA copy number changes.

Large-Insert Clone Arrays

Array CGH platforms based on large-insert clones are typically made up of bacterial artificial chromosome (BAC) and P1 artificial chromosome (PAC) clones and, to a lesser extent, also cosmids (Fig. 10.2). The first large-insert clone sets for whole genome analysis provided a resolution of about 1 Mb.[8,9] Subsequently, a more comprehensive clone set has been assembled,[10,11] and in 2004, the first sub-megabase-resolution whole genome tiling path array CGH study has been published, which was based on an array comprising 32,433 large-insert clones covering the whole genome in an overlapping fashion[12] (http://bacpac.chori.org/). Because of the overlap of clones, tiling path arrays can provide a theoretical resolution that is below the average insert size of a BAC of 150 kb. Figure 10.3 shows the array CGH analysis of a squamous cell lung carcinoma using a sub-megabase-resolution whole genome tiling path with more than 36,000 clones.

Despite their robustness and widespread application, BAC arrays have some limitations. The low copy number of BACs and PACs within the propagating bacteria requires special isolation methods that preserve the integrity of the clone insert while at the same time eliminate as much bacterial genomic DNA as possible. Most laboratories are not spotting this isolated DNA directly but instead amplify the material to generate a renewable stock of amplicons that can be printed several times (see below for discussion on amplification

P.T. Cagle and T.C. Allen (eds.), *Basic Concepts of Molecular Pathology*, Molecular Pathology Library 2,
DOI: 10.1007/978-0-387-89626-7_10, © Springer Science+Business Media, LLC 2009

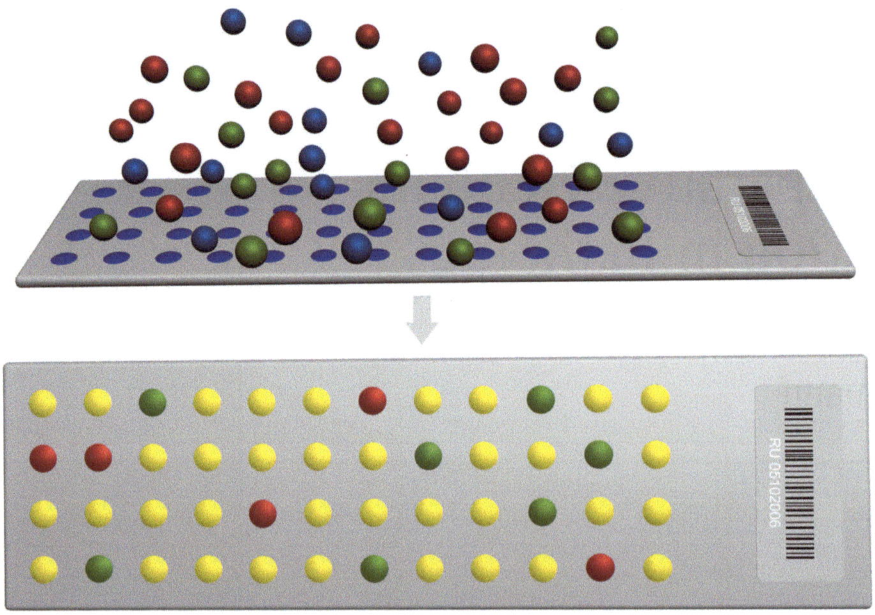

FIG. 10.1. Principle of array comparative genomic hybridization (CGH). Differentially labeled test and reference DNA (*green and red spheres*, respectively) are cohybridized onto an array of DNA spots printed on a glass slide. In case of a deletion in the test DNA, fewer tests DNA will bind to the corresponding spots and the red label of the reference DNA will prevail; gains in the test genome can be identified by a dominance of the green label of the test DNA. Spots representing sequences with the same copy number in the test genome relative to the reference genome appear *yellow*. For BAC arrays, an excess of repetitive Cot DNA (*blue spheres*) has to be added to suppress otherwise unspecifically binding repetitive sequences.

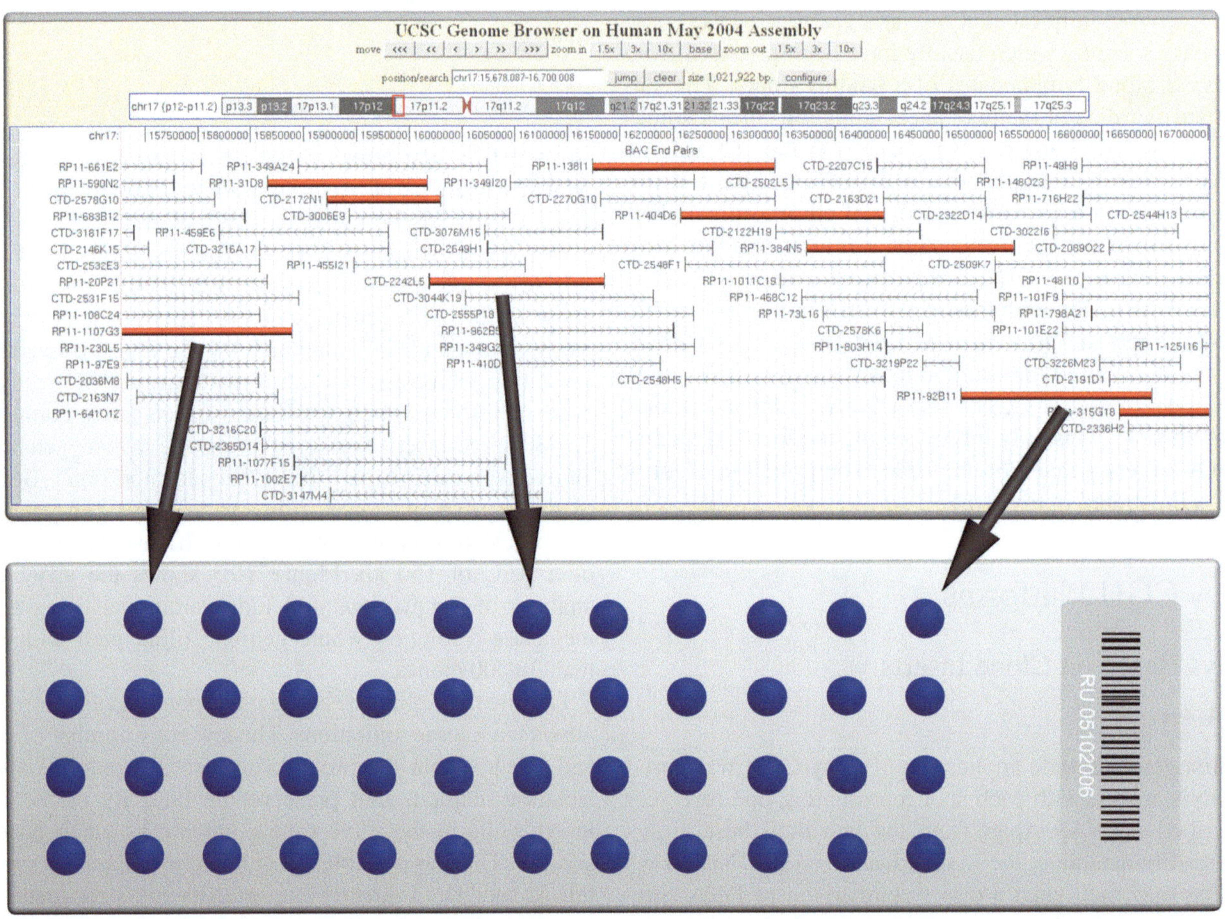

FIG. 10.2. Bacterial artificial chromosome (BAC) array. Each spot on a BAC array represents the very specific part of the genome that is contained in the BAC. A whole genome tiling path BAC array comprises as many BAC clones as necessary to cover the whole genome in an overlapping manner (~32,400 for the human genome). The *upper part* of the image is based on a screenshot of the UCSC Human Genome Browser. The *red lines* illustrate the selection of overlapping clones from a comprehensive BAC library.

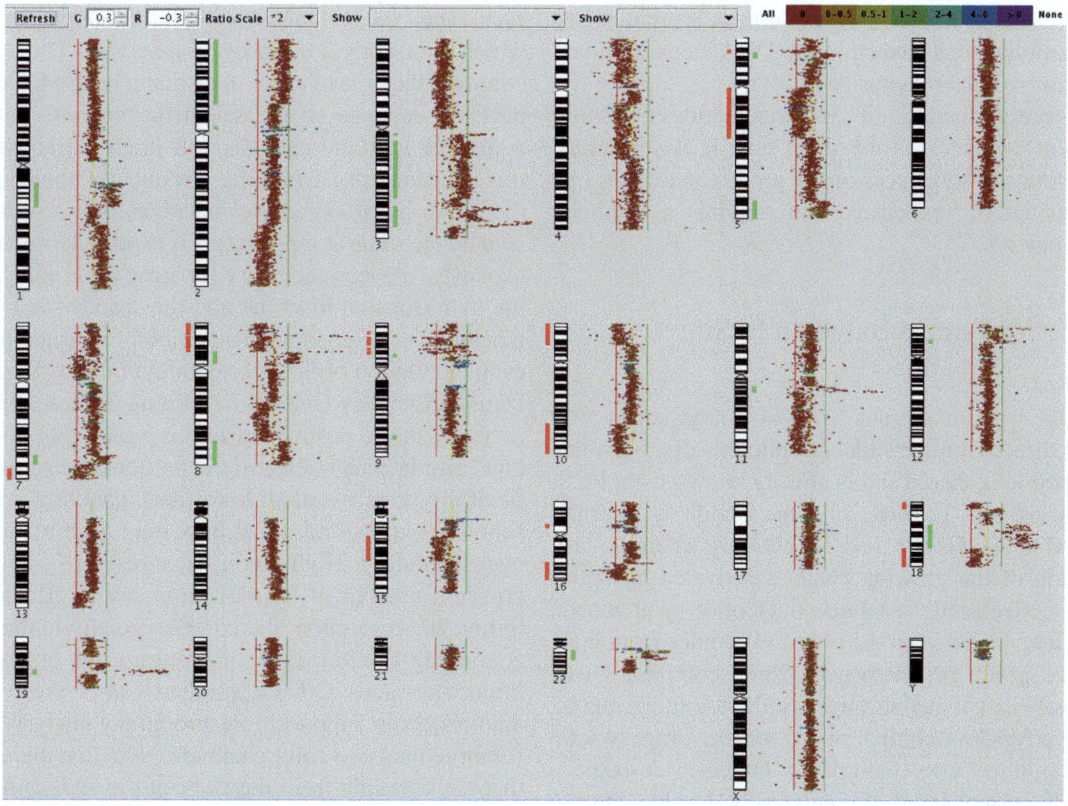

FIG. 10.3. Array CGH analysis of a squamous cell lung carcinoma using a sub-megabase-resolution whole genome tiling path BAC array comprising more than 36,000 spots. Cy3:Cy5 intensity ratios of each clone are plotted in a size-dependent manner along the chromosome ideograms. The *red and green lines* indicate the log 2 ratio thresholds −0.3 (loss) and 0.3 (gain), respectively. Note the very small high copy amplicons that would have been missed by low-resolution methods.

methods). A protocol for the high-throughput isolation and purification of BAC/PAC clone inserts can be downloaded from our website (http://www.molgen.mpg.de/~abt_rop/molecular_cytogenetics/Protocols.html). Whatever protocol is followed, the setup of a comprehensive BAC array platform remains time consuming and costly. Other shortcomings of BAC arrays are directly related to specific features of the respective genomic sequence. Otherwise unspecific binding repetitive sequences must be blocked using a considerable excess of (expensive) Cot-DNA. Low copy repeats, that is, stretches of DNA that are longer than 1 kb and have a sequence similarity of more than 90% to other locations in the genome, can lead to ambiguous results. This problem especially applies to tiling path arrays because low-resolution arrays usually avoid low copy repeats.[13] Finally, with the coming of tiling path arrays, BAC arrays have met the limits of resolution, which are simply given by clone insert size.

Repeat-Free and Nonredundant Sequence Arrays

In the light of the problems connected to the presence of repetitive sequences in the genome, researches have set out to generate genomic arrays that are depleted for repetitive and redundant sequences. This depletion, for example, has been accomplished by means of selective amplification using sequence-specific primers. Mantipragada et al have used this approach to create arrays focusing on the DiGeorge region (22q11 deletion syndrome).[14] Another array, based on sequence-specific polymerase chain reaction (PCR) products of 162 exons of five genes, has been generated to test a spectrum of inherited human disorders.[15] However, workload and high costs associated with this approach, as well as the upcoming of commercial oligonucleotide arrays, have hampered the widespread use of this approach.

Oligonucleotide Arrays Using Presynthesized Oligonucleotides

In contrast to the oligonucleotide platforms described below, these arrays are either generated by printing prefabricated, commercially available sets of oligonucleotides on glass slides or by coupling oligonucleotides to beads that are assembled on the slide afterward. One example for the use of printed arrays is reported by Carvalho et al, who have used a set of 18,861 oligos to identify DNA copy number changes in several tumor cell lines.[16] Presynthesized oligos coupled to beads are used for a platform developed by Illumina (http://www.illumina.com). The company provides several designs that are dedicated to either gene expression, linkage, or DNA copy number analysis. Some of their array designs enable the simultaneous detection of DNA copy number changes and loss of heterozygosity (LOH).[17]

Although prefabrication of oligonucleotides enables highly efficient synthesis, at the same time it also reduces flexibility in terms of sequences on the array. Custom design becomes considerably expensive and requires a minimal batch size to pay off.

Oligonucleotide Arrays Based on In Situ Synthesis

Meanwhile, there are numerous ways to synthesize an oligonucleotide directly on the slide. Despite this diversity, the common principle is shared and is already known from PCR primer synthesis: the growing oligonucleotide is alternatively exposed to As, Gs, Cs, and Ts, but only when the last oligonucleotide of the growing chain is activated by splitting off a protective group can a new nucleotide be attached. The main difference between the platforms comes from how this protective group is inactivated. Some companies use light, selectively distributed through fixed photolithographic masks (www.affymetrix.com) or micro-mirrors (http://www.nimblegen.com/http://www.febit.de). Others de-protect by means of a current-induced change of pH value (http://www.combimatrix.com/) or control synthesis by specifically addressing each spot separately with high-resolution printers (http://www.home.agilent.com). Usually, oligonucleotides on such arrays are designed to be both isothermal (i.e., they share the same melting temperature, T_m) and single copy sequences; this sometimes results in an uneven distribution of oligonucleotides, leading to considerable variability in terms of resolution across the genome. Nevertheless, given the current developments, it can be expected that oligonucleotide arrays will replace all other platforms in the near future. Shortcomings with respect to hybridization kinetics (see below) and coverage will be compensated by the incredible increase of features on the array. Oligonucleotide arrays with more than 500,000 features are readily available. Those platforms that are not dependent on fixed photolithographic masks especially can offer extreme flexibility, which is limited only by the setup fees charged by some companies. Single nucleotide polymorphism (SNP) arrays, consisting of short oligonucleotides in the range of 16–20 mers and originally dedicated to linkage analysis, have been successfully used for the simultaneous detection of LOH (loss of heterozygosity) and DNA copy number changes.[18]

General Platform Considerations

Before setting up an array CGH facility, several decisions have to be made. The first one refers to the expected number of array CGH experiments. In many cases, it will be much cheaper to cooperate with other groups that already have established the technique or send the samples to a company which is offering a hybridization service. The next decision concerns the type of array that should be used. Certainly, this decision depends on the scientific problem that should be addressed with the analysis, but often the consequences of this decision are far ranging. Frequently, the choice implies the purchase of expensive machines only useful for arrays sold by the same company; this is especially true for the most expensive devices necessary for array CGH analysis, namely, the hybridization machine and the scanner. A hybridization machine is designed to provide controlled temperature and even circulation of the hybridization mix to promote hybridization efficiency (see below). Some of these machines also accomplish the posthybridization washing of slides. Important criteria when selecting a particular machine could be flexibility in terms of slide formats, handling, maintenance/follow-up costs, and, most important, performance in one's own laboratory. High-quality scanners are essential for the errorless readout of hybridization results. Reliability, flexibility, and resolution, the latter especially in the light of the continuing minimization of feature sizes on the array, are important issues. Other arguments can be the availability of autoloaders to support high-throughput analysis or the need for more than two color channels. Note that there are devices that are scanning from the back of the slide and need transparent substrates.

DNA Preparation and Hybridization

DNA Isolation and Quality

DNA quality has a great influence on the outcome of an array CGH experiment. In general, oligonucleotide platforms, especially when used for one-color experiments, are more sensitive to compromised DNA quality than are BAC arrays. One source of trouble can come from the DNA isolation procedure, for example, detergents that could hamper the subsequent labeling process and introduce noise in the data. Although such problems can be avoided quite easily by improving or changing the isolation protocol, the situation is much more difficult when it comes to retrospective studies using formalin-fixed, paraffin-embedded (FFPE) material. This kind of fixation very frequently results in damage of DNA.[19] To cope with such damage, several adaptations of the DNA isolation and labeling protocols are required. Modifications can include the prolongation of tissue digestion time, with adding fresh proteinase each day, or the switch from enzymatic to chemical labeling systems. For other applications, some labs have even tried to repair fixation-induced DNA strand breaks by protocols following the principle of nick translation assay, or employed sodium thiocyanate to revert DNA-protein cross-linking. Unfortunately, it is not always possible to reliably predict the performance of FFPE DNA in an array CGH experiment, but verifying the average fragment size of the single-stranded DNA by gel electrophoresis and testing DNA performance in PCR reactions with differently sized amplicons can give a good estimate.

In the light of the increasing impact of molecular (cytogenetic) techniques on research and routine diagnostics, optimization of fixation protocols gains increasing importance. Short-term measurements include the correct buffering of formalin and regular control of the pH value, but also reducing fixation time to the minimum necessary to ensure proper histological evaluation, and storage needs, should be considered.

In the long run, it may be beneficial to overcome old traditions and switch to alternative fixation protocols.[19] In this context, it is also worth noting that many institutes have started to establish tissue banks where, in addition to the usual formalin fixation, tissues are especially preserved for various applications.

Microdissection and DNA Amplification

In many instances, for example, when trying to avoid normal cell contamination in compact growing tumors, it is sufficient to scratch out regions of interest from the section manually using a needle. Unfortunately, this low-budget solution may not be applicable to other problems in pathology. For single cells or small cell clusters interspersed in the tissue, such as suspected tumor stem cells, micrometastasis, or preneoplasias, laser microdissection must be employed.[20]Figure 10.4 demonstrates the microdissection of a bronchiolar columnar cell dyplasia.[21]

The amount of DNA isolated from laser-microdissected samples usually is not sufficient for array CGH analysis. Thus, uniform amplification of the whole genome is required to make such samples amenable to array CGH. Fortunately, in the meantime, numerous protocols for whole genome amplification (WGA) exist and many companies are providing advanced WGA kits. WGA methods can be roughly divided into PCR-based and non-PCR-based approaches. Probably the best known PCR-based technique is degenerate oligonucleotide primer (DOP) PCR.[22] The name of this technique refers to the specific primer, which contains a central cassette of six degenerated bases (at each of these six positions, every one of four bases can be present). In the first cycles of PCR with very low annealing temperature, DOP promotes the priming of DNA synthesis from multiple evenly dispersed sites within the genome, resulting in amplicons that are flanked by the primer sequence, which then can serve as the annealing site in the subsequent amplification step using high stringency conditions (Fig. 10.5a). The initial priming is supposed to occur at approximately 10^6 sites in

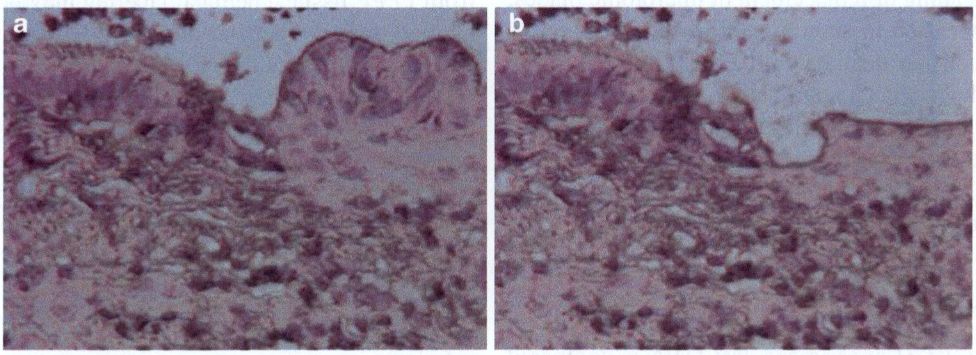

FIG. 10.4. Laser microdissection. The selective isolation of bronchiolar columnar cell dysplasia is shown (**a**) before and (**b**) after laser microdissection.

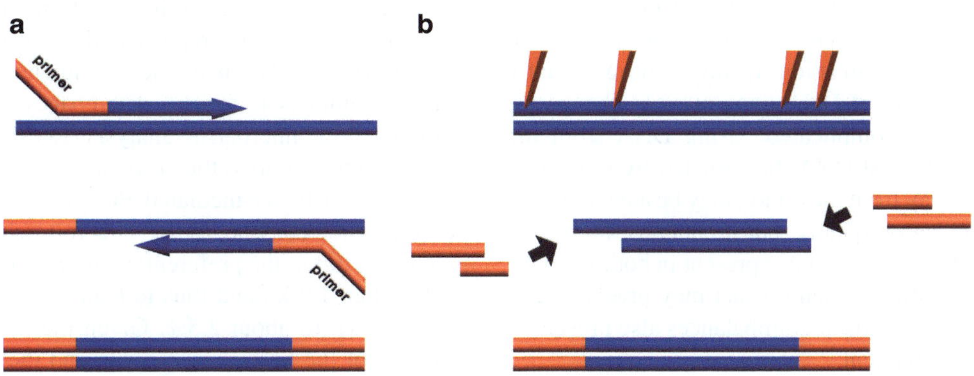

FIG. 10.5. Whole genome amplification. Two examples of PCR-based whole genome amplification techniques are shown: (**a**) degenerate oligonucleotides primer (DOP)-PCR; (**b**) linker-mediated PCR. Primer and oligo-complexes, respectively, are depicted in *red*; template DNA is drawn in *blue*. See text for explanation.

the human genome,[22] resulting in a genomic representation sufficient for chromosomal CGH. However, with the coming of high-resolution array-based CGH techniques, many laboratories have shifted to alternative methods, which promise better coverage of the genomic complexity. Among the PCR-based methods, this capability is mainly ascribed to ligation-mediated PCR techniques, which have already been successfully used for the analysis of single cells by CGH.[23] The principle of ligation-mediated PCR is based on either enzymatic,[23–25] hydrodynamic,[26] or chemical fragmentation (http://www.rubicongenomics.com) of DNA followed by the ligation of an adapter complex, which serves as a universal priming site for uniform amplification (Fig. 10.5b). Certainly, PCR-based approaches imply the risk of amplification bias and experimental errors, and therefore some researchers prefer non-PCR-based approaches such as strand displacement amplification using such enzymes as phi29 DNA polymerase. A more comprehensive review and discussion about the pros and cons of diverse WGA is given by Hughes et al.[27]

Unfortunately, WGA cannot distinguish DNA that should be amplified from incidentally introduced DNA. Therefore, precautions have to be taken to avoid any contamination with DNAs from other sources. This caution starts with the cleaning of the carving board before sectioning the resected tumor, regular cleaning or replacement of microtome blades, exchange of water used to stretch paraffin sections, and so forth. Especially when dealing with FFPE, one has to be aware that nondegraded, contaminating DNA is preferentially amplified. Amplicons generated in previous PCR reactions, distributed through aerosols, are an extremely good template. Therefore, strict compliance to the general rules of PCR setup is mandatory to produce reliable results.

Reference DNA

In a typical array CGH experiment, test DNA is compared to DNA of a healthy donor or a DNA pool of healthy individuals. Many laboratories hybridize in a sex-matched manner, that is, test and reference DNA have the same sex, while others prefer sex-mismatched hybridizations, where the ratio changes at the sex chromosomes can serve as an internal control of hybridization quality, but eventually render the interpretation of DNA copy number changes at the sex chromosomes complicated. Using DNA from the same individual as the test DNA, but isolated from normal-looking tissue far away from the tumor, may be advantageous in terms of hybridization quality, but it contains the risk of missing DNA copy number changes present in both DNAs[28]; for example, constitutional changes that may predispose to tumor formation or chromosomal imbalances also present in the tumor microenvironment.[29]

For chromosomal CGH, it has already been observed that matching the quality of test and reference DNA can significantly improve the results of an experiment. Thus, it has been realized that combining amplified test DNA with nonamplified reference DNA can introduce hybridization artifacts, which can be avoided when a PCR-amplified reference DNA is used.[30] A related observation has been made in our laboratory when working with FFPE DNA, where the use of reference DNA also isolated from FFPE tissue has significantly improved results. In one-color array CGH experiments using short oligonucleotides arrays, the reference DNA is not tested within the same experiment, but data are compared to a reference dataset, either generated in the same laboratory by analyzing a cohort of several normal individuals or provided by the array-selling company. This in silico comparison can hardly compensate for intraexperimental variation and thus is more prone to noise.

DNA Labeling

Problems in labeling DNA can result in artifacts and failure of the array CGH experiment. Despite a great variety of direct and indirect labeling protocols, at the moment the prevailing method for DNA labeling for noncommercial slides is direct incorporation of fluorochromes into the DNA by means of a random priming assay, which, as a nice side effect, also results in a net gain of DNA. In some instances, however, for example when dealing with highly degraded DNA isolated from FFPE tissue, chemical labeling may be superior as this method does not depend on long DNA fragments and DNA synthesis.[31] As it is true for DNA quality, in our experience it is essential that test and reference DNA do also match in the way they have been labeled, so that a possible bias introduced by problems in labeling of specific sequences can be compensated. Commercial arrays frequently require different labeling protocols and the use of specialized kits, usually sold by the same company.

Hybridization

Disadvantageous hybridization kinetics is the greatest problem of array CGH based on oligonucleotide arrays. Even long oligonucleotides represent an extreme low complex target: a diversity of 3 billion bases in the hybridization mix versus ~60 bases represented by each specific spot (Fig. 10.6). Circulation of the hybridization mix can only partially compensate for this discrepancy. Representational oligonucleotide microarray analysis (ROMA) is one technical approach to address the problem of complexity.[32] ROMA is based on a linker-mediated PCR (see Fig. 10.5b) using the restriction enzyme *BglI*I. The following PCR is optimized to result in the preferential amplification of fragments smaller than 1.2 kb and thus to lead in a reduction of complexity down to about 2.5%. Given the known recognition sequence of the restriction enzyme and the maximum spacing between two sites, the authors expected the amplification of about 200,000 sites across the genome and designed a corresponding array specific for the expected sites. It is clear

FIG. 10.6. Comparison of hybridization efficiency of oligonucleotide and BAC arrays. With more than 3 billion different bases, the human genome is very complex (indicated by the *differentially colored spheres*). Therefore, the relative concentration for a given oligonucleotide that is complementary to the various oligonucleotides on the array (exemplarily shown as four single-color disks) is extremely low. In contrast, an average BAC clone represents 150 kb (exemplarily illustrated by the colored disks), which increases the chance of binding and thus signal intensity.

that this elegant approach is technically and computationally demanding and requires optimal settings and standardization to avoid the introduction of artifacts.

Data Analysis

A typical array CGH experiment yields several thousands of data points, which have to be displayed in a comprehensive and illustrative way. Hardly ever are raw data presented in a manuscript. Instead, data have already passed several steps of manipulation. For many readers, the computational analysis appears like a black box, and the brief descriptions of the procedures in many manuscripts are not always able to shed light into this "box." However, availability and understanding of information about data analysis are not only essential for appreciating the quality of a microarray experiment and the functional (real) resolution of an experiment but also to judge whether a comparison of two different studies is feasible. It is beyond the scope of this chapter to provide in-depth insights into data analysis. Instead, the following is meant as a rough overview that should enable a basic understanding of the data analysis workflow and highlight key points of data interpretation of an array CGH experiment. For a more comprehensive review on computational aspects of array CGH, see Chari et al.[33]

Image Processing

The wet lab part of a typical two-color array CGH experiment ends with putting the hybridized array into the scanner, where the signal intensities of the two different fluorochromes are recorded as two gray-scale images, either simultaneously or one by one. Usually the scanner output is two 16-bit TIFF (tagged image file format) images, which provide 2^{16} (65,536) different gray scales. These images are then imported into specialized image processing software, frequently sold as a package together with the laser scanner. There, DNA spots are defined by superimposing a grid that reflects the architecture of the array in terms of rows and columns and links each spot with information on clone/sequence identity and chromosomal location.

Background Subtraction

Global background subtraction is based on the averaged signal intensities of all pixels outside those areas that have been identified as DNA spots. This method does not take into account the possible uneven distribution of background signal intensities across the array. Therefore, most people favor local subtraction methods based on background intensity values determined in the vicinity of each spot to cope with spatial bias. In our laboratory, we do not subtract background at all.

At this stage, data already can be exported to array CGH-specific analysis tools.

Normalization

The absolute signal intensities measured in each channel can be subject to systematic, spatial, or intensity-dependent bias, which can influence array CGH results[34] and has to be removed before calculating the signal intensity ratios. Systematic bias of signal intensities can arise from differences in input of test and reference DNA, fluorochromes and labeling efficiency, laser settings (laser power as well as amplification through photomultiplier tubes), etc. The simplest way to eliminate systematic bias is to normalize by equalizing the median intensities of the two channels.

However, calculating the global median intensity does not take into account spatial effects, that is, areas that appear to show trends of higher and lower intensity, respectively. For this reason, many researchers apply this normalization method for small subgrids separately. Meanwhile, more sophisticated programs are also available that can detect and remove spatial bias.[35] Sometimes evaluation of intensity scatter-plots reveals an intensity-dependent trend; for example, the ratios of low-intensity spots always show the same tendency. Normalization algorithms such as LOWESS or subgrid LOWESS, which are based on regression models, can be employed to tackle this problem.

Whatever algorithm is applied to normalize the data, it relies on the assumption that for many, if not most, of the data points of DNA copy number are the same in test and reference DNA. This assumption may be a problem in tumors with complex aberrations or customized arrays focusing on very small regions of interest. Therefore, for some projects it may be advisable to skip normalization completely to avoid loss of dynamic range.

Identifying an Aberration

Despite the availability of several sophisticated tools dedicated to the analysis and visualization of array CGH data,[36] defining the presence or absence of an aberration remains challenging. The simplest way to identify an aberration would be to determine fixed thresholds, and every clone/oligo whose signal ratio exceeds these thresholds would be considered as an indicator of aberrant copy number of the respective genomic region. Alternatively to the use of fixed thresholds, which does not take into account variable data quality, thresholds can be defined based on standard deviation. However, given the considerable number of outliers that can occur in one experiment, false positives would pose a problem. Smoothing data by means of sliding windows (moving average) has been frequently applied to be less sensitive to outliers (Fig. 10.7). Although this method dramatically reduces the false-positive rate, it goes along with loss of resolution, which depends on the size of the window, i.e., the number of clones/oligos that are averaged. To avoid this disadvantage, advanced algorithms such as hidden Markow models (HMM[37]), BioHMM,[38] circular binary segementation (CBS[39]), and wavelet-based approaches[40] have been designed to determine copy number changes in an objective and automatic fashion.

Especially in tumor cytogenetics, biology introduces another layer of complexity that renders data interpretation difficult. Just imagine the following scenario: a tetraploid cell looses one chromosome. In a tetraploid cell with four copies of each chromosome, this means a relative loss of 25%. There may also be some polyclonality in the tumor specimen, with some cells having this aberration and others not. Some cells may be even euploid, especially those normal cells that are contaminating the tumor sample. To make a long story short, DNA copy number changes are not necessarily changing as integers, and it is up to the scientist to determine those thresholds that reflect the best comprise between sensitivity and specificity. This decision can dramatically influence the outcome of an array CGH study and can hamper the comparability of different array CGH studies. In this context, it is of great interest that authors of array CGH papers are increasingly following the so-called MIAME (minimal information about a microarray experiment) criteria,[41] which also include the deposition of primary data to a public database such as Gene Expression Omnibus (http://www.ncbi.nlm.nih.gov/geo/; GEO) or Array Express (http://www.ebi.ac.uk/arrayexpress/). Every scientist interested in a specific published dataset can download the respective files and perform an in silico reanalysis of the data sets with different analysis parameters or different focus. Meanwhile, a number of array CGH analysis and visualization tools are available for free ([42]; for review[36]) that can assist in the exploitation of this valuable data source.

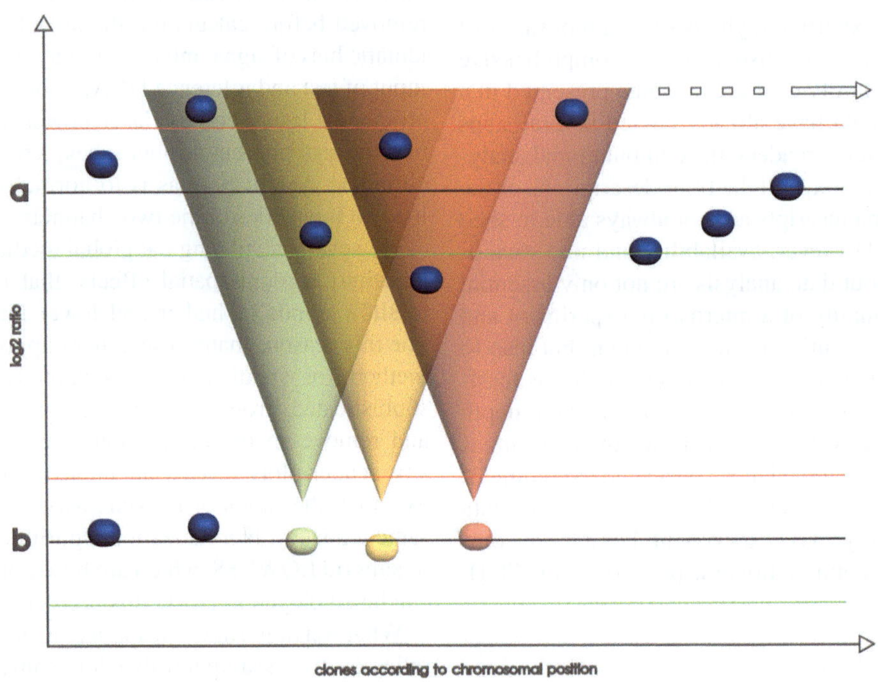

FIG. 10.7. Data smoothening by sliding window. In this schematic illustration, a sliding window of three clones/oligonucleotides is shown. Instead of the original ratio (*row a;* upper part of the triangles), the averaged ratio of three adjacent clones/oligonucleotides is plotted (*row b;* tip of each triangle). The differentially colored triangles symbolize the moving of the windows (*blue triangle* is the latest window). Note that the size of the window determines the quality of data smoothening and also the functional resolution of the array.

However, not only technical issues complicate the identification of causative aberrations. Because of the high resolution of array CGH, a number of DNA copy number variants have been found to occur in the normal population (a comprehensive database can be found under the following link: http://projects.tcag.ca/variation/). Although these variants can be several hundred kilobases in size, many of them are supposed to be phenotypically neutral, but they also may predispose to disease.

Conclusion

The accumulation of chromosomal imbalances is a typical feature of tumor development. The specific composition of chromosomal changes creates a unique genomic setting that influences the biological behavior of the tumor. Currently, array CGH using high-resolution platforms is the technique most often applied to detect DNA copy number changes. Because of the introduction of commercial platforms, the technique is no longer restricted to highly specialized laboratories, and therefore molecular cytogenetics profiling by means of array CGH will gain increasing importance in routine pathology also. Array CGH data can be employed to identify chromosomal signatures useful for differential diagnosis and therapeutic decisions. Some aberrations may serve as discriminators in differential diagnosis, especially for tumors that are hardly distinguishable by classical pathological approaches, while the finding of other chromosomal changes may be useful to anticipate the risk of metastasis or the likelihood of recurrence. The detection of genes frequently amplified in tumors will help to identify new therapy targets, and customized therapy arrays will prove useful to predict response to such therapies.

References

1. Kallioniemi A, Kallioniemi OP, Sudar D, et al. Comparative genomic hybridization for molecular cytogenetic analysis of solid tumors. *Science.* 1992;258:818–821.
2. Kirchhoff M, Gerdes T, Maahr J, et al. Deletions below 10 megabase pairs are detected in comparative genomic hybridization by standard reference intervals. *Genes Chromosomes Cancer.* 1999;25:410–413.
3. Pinkel D, Segraves R, Sudar D, et al. High resolution analysis of DNA copy number variation using comparative genomic hybridization to microarrays. *Nat Genet.* 1998;20:207–211.
4. Solinas-Toldo S, Lampel S, Stilgenbauer S, et al. Matrix-based comparative genomic hybridization: biochips to screen for genomic imbalances. *Genes Chromosomes Cancer.* 1997;20:399–407.
5. Pollack JR, Perou CM, Alizadeh AA, et al. Genome-wide analysis of DNA copy-number changes using cDNA microarrays. *Nat Genet.* 1999;23:41–46.
6. Pollack JR, Sorlie T, Perou CM, et al. Microarray analysis reveals a major direct role of DNA copy number alteration in the transcriptional program of human breast tumors. *Proc Natl Acad Sci USA.* 2002;99:12963–12968.
7. Monni O, Barlund M, Mousses S, et al. Comprehensive copy number and gene expression profiling of the 17q23 amplicon in human breast cancer. *Proc Natl Acad Sci USA.* 2001;98:5711–5716.
8. Fiegler H, Carr P, Douglas EJ, et al. DNA microarrays for comparative genomic hybridization based on DOP-PCR amplification of BAC and PAC clones. *Genes Chromosomes Cancer.* 2003;36:361–374.
9. Snijders AM, Nowak N, Segraves R, et al. Assembly of microarrays for genome-wide measurement of DNA copy number. *Nat Genet.* 2001;29:263–264.
10. Krzywinski M, Bosdet I, Smailus D, et al. A set of BAC clones spanning the human genome. *Nucleic Acids Res.* 2004;32:3651–3660.
11. Osoegawa K, Mammoser AG, Wu C, et al. A bacterial artificial chromosome library for sequencing the complete human genome. *Genome Res.* 2001;11:483–496.
12. Ishkanian AS, Malloff CA, Watson SK, et al. A tiling resolution DNA microarray with complete coverage of the human genome. Methods for high throughput validation of amplified fragment pools of BAC DNA for constructing high resolution CGH arrays. *Nat Genet.* 2004;36:299–303. [Epub 2004 Feb 2015.]
13. Garnis C, Lockwood WW, Vucic E, et al. High resolution analysis of non-small cell lung cancer cell lines by whole genome tiling path array CGH. *Int J Cancer.* 2006;118:1556–1564.
14. Mantripragada KK, Tapia-Paez I, Blennow E, Nilsson P, Wedell A, Dumanski JP. DNA copy-number analysis of the 22q11 deletion-syndrome region using array-CGH with genomic and PCR-based targets. *Int J Mol Med.* 2004;13:273–279.
15. Dhami P, Coffey AJ, Abbs S, et al. Exon array CGH: detection of copy-number changes at the resolution of individual exons in the human genome. *Am J Hum Genet.* 2005;76:750–762.
16. Carvalho B, Ouwerkerk E, Meijer GA, Ylstra B. High resolution microarray comparative genomic hybridisation analysis using spotted oligonucleotides. *J Clin Pathol.* 2004;57:644–646.
17. Peiffer DA, Le JM, Steemers FJ, et al. High-resolution genomic profiling of chromosomal aberrations using Infinium whole-genome genotyping. *Genome Res.* 2006;16:1136–1148.
18. Bignell GR, Huang J, Greshock J, et al. High-resolution analysis of DNA copy number using oligonucleotide microarrays. *Genome Res.* 2004;14:287–295.
19. Srinivasan M, Sedmak D, Jewell S. Effect of fixatives and tissue processing on the content and integrity of nucleic acids. *Am J Pathol.* 2002;161:1961–1971.
20. Hernandez S, Lloreta J. Manual versus laser micro-dissection in molecular biology. *Ultrastruct Pathol.* 2006;30:221–228.
21. Ullmann R, Bongiovanni M, Halbwedl I, et al. Bronchiolar columnar cell dysplasia: genetic analysis of a novel preneoplastic lesion of peripheral lung. *Virchows Arch.* 2003;442:429–436.
22. Telenius H, Pelmear AH, Tunnacliffe A, et al. Cytogenetic analysis by chromosome painting using DOP-PCR amplified flow-sorted chromosomes. *Genes Chromosomes Cancer.* 1992;4:257–263.
23. Klein CA, Schmidt-Kittler O, Schardt JA, Pantel K, Speicher MR, Riethmuller G. Comparative genomic hybridization, loss of heterozygosity, and DNA sequence analysis of single cells. *Proc Natl Acad Sci USA.* 1999;96:4494–4499.
24. Ludecke HJ, Senger G, Claussen U, Horsthemke B. Cloning defined regions of the human genome by microdissection of banded chromosomes and enzymatic amplification. *Nature.* 1989;338:348–350.

25. Saunders RD, Glover DM, Ashburner M, et al. PCR amplification of DNA microdissected from a single polytene chromosome band: a comparison with conventional microcloning. *Nucleic Acids Res.* 1989;17:9027–9037.
26. Tanabe C, Aoyagi K, Sakiyama T, et al. Evaluation of a whole-genome amplification method based on adaptor-ligation PCR of randomly sheared genomic DNA. *Genes Chromosomes Cancer.* 2003;38:168–176.
27. Hughes S, Arneson N, Done S, Squire J. The use of whole genome amplification in the study of human disease. *Prog Biophys Mol Biol.* 2005;88:173–189.
28. Petzmann S, Ullmann R, Halbwedl I, Popper HH. Analysis of chromosome-11 aberrations in pulmonary and gastrointestinal carcinoids: an array comparative genomic hybridization-based study. *Virchows Arch.* 2004;445:151–159.
29. Moinfar F, Man YG, Arnould L, Bratthauer GL, Ratschek M, Tavassoli FA. Concurrent and independent genetic alterations in the stromal and epithelial cells of mammary carcinoma: implications for tumorigenesis. *Cancer Res.* 2000;60:2562–2566.
30. Huang Q, Schantz SP, Rao PH, Mo J, McCormick SA, Chaganti RS. Improving degenerate oligonucleotide primed PCR-comparative genomic hybridization for analysis of DNA copy number changes in tumors. *Genes Chromosomes Cancer.* 2000;28:395–403.
31. van Gijlswijk RP, Talman EG, Janssen PJ, et al. Universal Linkage System: versatile nucleic acid labeling technique. *Expert Rev Mol Diagn.* 2001;1:81–91.
32. Lucito R, Healy J, Alexander J, et al. Representational oligonucleotide microarray analysis: a high-resolution method to detect genome copy number variation. *Genome Res.* 2003;13:2291–2305.
33. Chari R, Lockwood WW, Lam WL. Computational methods in array CGH. *Cancer Inform.* 2006;2:48–58.
34. Khojasteh M, Lam WL, Ward RK, MacAulay C. A stepwise framework for the normalization of array CGH data. *BMC Bioinformatics.* 2005;6:274.
35. Neuvial P, Hupe P, Brito I, et al. Spatial normalization of array-CGH data. *BMC Bioinformatics.* 2006;7:264.
36. Chari R, Lockwood WW, Lam WL. Computational methods for the analysis of array comparative genomic hybridization. *Cancer Inform.* 2006;2:48–58.
37. Fridlyand J, Snijders AM, Pinkel D, Albertson DG, Jain AN. Application of Hidden Markov models to the analysis of the array CGH data. *Special Genomic Issue J Multivariate Anal.* 2004;90:132–153.
38. Marioni JC, Thorne NP, Tavare S. BioHMM: a heterogeneous hidden Markov model for segmenting array CGH data, *Bioinformatics.* 2006;22:1144–1146.
39. Olshen AB, Venkatraman ES, Lucito R, Wigler M. Circular binary segmentation for the analysis of array-based DNA copy number data. *Biostatistics.* 2004;5:557–572.
40. Hsu L, Self SG, Grove D, et al. Denoising array-based comparative genomic hybridization data using wavelets. *Biostatistics.* 2005;6:211–226.
41. Brazma A, Hingamp P, Quackenbush J, et al. Minimum information about a microarray experiment (MIAME)-toward standards for microarray data. *Nat Genet.* 2001;29: 365–371.
42. Chen W, Erdogan F, Ropers HH, Lenzner S, Ullmann R. CGH-PRO: a comprehensive data analysis tool for array CGH. *BMC Bioinformatics.* 2005;6:85.

11
Loss of Heterozygosity

Belinda J. Wagner and Sharon C. Presnell

Introduction

The most common molecular alteration observed in human cancers,[1] loss of heterozygosity (LOH), is a significant mecha-nism by which critical genes involved in growth regulation and homeostasis become inactivated, or silenced, during disease evolution. This chapter provides a review of LOH and its implications in various cancers as well as a review of LOH in nonmalignant diseases. Only 0.08% of those base pairs within the entire human genome (3 billion base pairs) vary between any two humans, and only 0.02% of those variations actually result in an expressed protein with a different amino acid as a result of the change.[2] Even more remarkable, 90% of those variations are changes that are common in the population and lead to normal variation in traits among individuals; eye color, for example.

In a normal individual's DNA, every genetic locus is composed of two alleles, one inherited from each parent. Genetic loci can be homozygous, meaning that both copies of the gene are exactly the same, or heterozygous, meaning that the two copies of the chromosome are different. The most common differences between two chromosomes are single nucleotide polymorphisms (SNPs), where the DNA sequence differs by one nucleotide. SNPs can be found on average every 1,200 base pairs.[1] Most (90%) of this variance is normal within the population, but approximately 10% of the heterozygous sites in any individual's DNA are the result of deleterious genetic variations that have the potential to cause disease.

Loss of one allele can occur at any genetic locus, homozygous, or heterozygous (Fig. 11.1). Allelic loss at a heterozygous site, called *loss of heterozygosity* (LOH), leaves the cell with a single version of a gene. If the remaining copy is aberrant, LOH has the potential to disrupt cellular homeostasis. LOH can occur when a whole or part of a chromosome is lost (Fig. 11.1b) or if part of a chromosome is replaced by translocation (Fig. 11.1c). During mitosis, recombination between chromosomes can also cause localized LOH (Fig. 11.1d).

The importance of LOH in cancer was recognized by Knudsen in what has become a classic model of tumor suppressor gene inactivation.[3] The *two–hit model* recognized that one copy of an important regulatory or homeostatic gene could become defective by mutation or gene silencing (e.g., via promoter hypermethylation or small intragenic deletion) and the remaining copy could be lost through an LOH event. In hereditary cancers, individuals inherit one defective copy and thus are predisposed to disease because LOH can occur subsequently by several mechanisms (Fig. 11.1). The two-hit model is now a well-documented mechanism by which tumor suppressor genes are lost during the formation and progression of many types of cancer, including lung, breast, bladder, esophagus, and liver.[1,4,5]

Detecting LOH

Methods to detect LOH rely on the ability to assess copy number and determine that one copy is missing or reduced compared to the other copy. Technical methods such as comparative genomic hybridization (CGH) allow detection of regions of chromosomal gain or loss but generally do not have the resolution to detect LOH at specific gene loci.[6] Classical methods for detection of LOH involved using heterozygous restriction fragment length polymorphisms (RFLP) and Southern blotting. Two drawbacks to this classical approach are low throughput and the relatively sparse coverage of the genome available for study with RFLP. Much like geopositioning satellite (GPS) technology, which has brought about more accurate pinpointing of our physical position on the globe, higher-density markers that leverage simple sequence length polymorphisms (SSLPs, or microsatellites) and single nucleotide polymorphisms (SNPs) have generated more comprehensive polymerase chain reaction (PCR)-based "allelotyping" studies of LOH.[7] LOH can now be surveyed in larger sample sets, and studies can identify clusters or linkages of multiple loci altered by LOH in specific disease

P.T. Cagle and T.C. Allen (eds.), *Basic Concepts of Molecular Pathology*, Molecular Pathology Library 2,
DOI: 10.1007/978-0-387-89626-7_11, © Springer Science+Business Media, LLC 2009

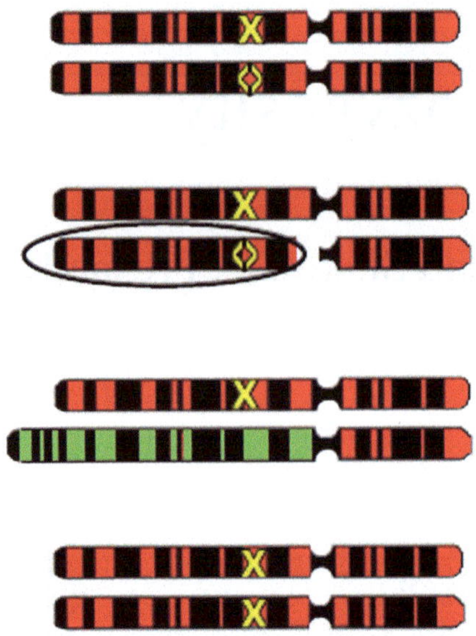

FIG. 11.1. Mechanisms of loss of heterozygosity (LOH). (**a**) Normal heterozygous locus. (**b**) LOH by whole or partial chromosome deletion. (**c**) LOH by translocation (replacement of part of a chromosome with a portion of another chromosome). (**d**) LOH by mitotic recombination.

states. The availability of high-density polymorphic probes, gene arrays, and analysis algorithms is expanding the scope and increasing the throughput of genome-wide LOH assessment. *Minimally deleted regions* (MDR) have been identified for many different cancers, pointing researchers to the most likely locations for tumor suppressor genes whose inactivation contributes to the progression of primary tumor development or metastases.[8–15]

Although the classic LOH detection methods established the importance of LOH in a wide variety of cancers, SNP-based screening technologies are becoming the dominant technology. SNPs are high-density markers. When multiple adjacent SNPs show a consistent genetic alteration, researchers' confidence in concluding that an LOH event has occurred is higher. Affymetrix, in particular, has invested in a series of SNP gene array chips at various densities, providing the research community with a commercially available resource that allows high-throughput testing of target populations and a common database on which the results of different groups can be compared.[1]

Regardless of the methods used, studies of LOH typically provide several key measurements helpful in understanding the potential role of LOH in various cancers. In addition to defining hotspots for allelic loss, or minimally deleted regions, likely to contain key genes whose alteration plays a role in the development and/or progression of the cancer, LOH analysis defines the extent of allelic loss in a given sample. *Fractional allelic loss* (FAL) is defined as the number of LOH events in a sample divided by the total number of informative heterozygous markers in the corresponding normal DNA.[16] Breakpoints can also be determined, and are defined as the junction between a specific marker displaying LOH and an adjacent marker that retains heterozygosity (LOH:HET).[16]

Conducting LOH analysis requires gaining access to tumor cells. In the case of surgical resection or postmortem analysis of specimens, cells can be obtained directly from tissue via macro- or microdissection. Cells have also been harvested successfully from biopsies, bronchial brushings, and urine.[17,18] The purity of the harvested cell population must be assessed because contaminating stroma, blood vessels, lymphocytes, and other normal cells are inevitably present in clinical specimens. LOH sensitivity to nontumor DNA depends on the method and type of probe.

Classical LOH detection methods that rely on copy number analyses (e.g., CGH, RFLP) lose sensitivity if the proportion of DNA contributed by tumor cells is not ≥80% of the total sample DNA.[1,19,20] Flow cytometry has been used to increase the percentage of tumor cells in the sample DNA.[1,21] Alternatively, LOH has been detected in tissues with fluorescent in situ hybridization (FISH) in clinical situations where pure samples cannot be easily obtained and the loci of LOH are well known.[17,22] However, SNP arrays are significantly less sensitive to nontumor DNA contamination because they do not rely on copy number analysis.[21] Studies using genome-wide SNP probes have revealed new LOH loci that would have gone undetected by copy number analysis techniques.[1,21,23] Another caveat of using copy number analysis techniques to detect LOH is the phenomenon by which one allele becomes genetically amplified. When one allele is present in significantly more copies than the other, copy number analysis data can sometimes indicate LOH when in reality the nonamplified allele remains intact.[20,23] Yet another source of false negatives using classical LOH detection methods with <100% pure samples is when both copies of a locus are missing in tumor cells and the heterozygous signal arises solely from contaminating nontumor DNA.[17]

SNP-based technologies have expanded the scope and utility of LOH detection. SNPs have helped pinpoint minimally deleted regions and focus researchers' efforts to identify new tumor suppressor genes. SNPs have helped dissect the temporal order of mutations that affect prognosis and the potential for metastatic spread of tumors. The high throughput available with SNP-based LOH detection will likely continue to spur new applications and new understanding of the genetic basis of tumor initiation, disruption of cellular homeostasis, and metastases.

Environmental Factors Influencing LOH

Mechanisms by which LOH occurs are varied (see Fig. 11.1), and all have been associated with at least one type of cancer. Environmental factors can contribute to LOH and have been linked to certain cancers. Carcinogenic compounds in the air we breathe, the food we eat, and even some prescription drugs can induce mutation either directly or through metabolic by-products.[24] Pathogens have also been shown to play a contributing role in cancer progression.[25,26]

LOH is a downstream event that happens some time after exposure to a carcinogen. The initial effect is usually a small change at the single nucleotide level. Carcinogens cause mutations through highly reactive intermediate compounds produced by the metabolic or elimination processes of the body – either in the digestive tract or by immune system surveillance. These compounds can chemically react with DNA, modifying its structure. These so-called DNA adducts occur in multiple forms, depending on the chemical and molecular structure of the carcinogen.[27,28] Some DNA adducts cause DNA polymerases to insert an incorrect nucleotide during DNA replication. Other DNA adducts link DNA strands to proteins or other DNA molecules, causing a physical barrier to replication.[29] Other carcinogens cause DNA strand breaks.[28]

The body has mechanisms that monitor and repair DNA or initiate the death of cells that harbor deleterious mutations. However, chronic or high-level exposures to a carcinogen can overwhelm the body's intrinsic repair mechanisms. Some individuals inherit DNA repair-associated alleles that make them more[30] or less[31] susceptible to disease from many sources of mutation, including environmental carcinogens.[32] In most cases, disease emerges months or years after an exposure event. Therefore, establishing linkages between environmental factors and cancers can take many years of research with large study populations.

Tobacco

Exposures to tobacco smoke from cigarette use and tobacco in smokeless products have both been linked to increased risk of cancer.[33] Lung cancer is the most well known association, but tobacco smoking is a major cause of bladder cancer as well. The evidence linking exposure to tobacco smoke and lung cancer is overwhelming, with smoking named as the greatest etiological factor contributing to an individual's risk of developing lung cancer.[34–37] The mechanisms by which tobacco carcinogens contribute to the development and progression of lung cancer are multivariate and are only partially understood. Tobacco smoke produces reactive chemical species, including benzo[*a*]pyrenes and polycyclic aromatic hydrocarbons (PAH). Smokeless tobacco is a major source of nitrosamines and is linked to increased risk of oral, esophageal, and pancreatic cancers.[33]

When adolescents smoke, it is believed that normal lung epithelium is "preconditioned" by exposure to tobacco carcinogens at a time of critical lung development, resulting in somatic mutations in an entire "field" of epithelium rather than in a single cell or cluster of cells.[37] These genetically altered cells are replicated during normal adolescent lung growth, leading to the presence of a large number of cells that are "primed" for subsequent events (such as allelic loss), thus rendering the individual highly susceptible to the development of lung cancer. Smoking during adolescence may increase the risk that if lung cancer develops it will develop as many clonal cancers that progress concomitantly and aggressively, a concept first described as "field cancerization."[37,38]

Asbestos

Asbestos is a well-known pulmonary carcinogen, linked in particular to the formation of mesotheliomas, lung tumors that usually develop in individuals with a history of asbestos exposure, typically with a long period of latency between exposure and clinical presentation.[39] The specific mechanisms by which asbestos contributes to LOH are not known. Mutagenicity studies carried out using a lymphocyte model demonstrated that the LOH rate upon exposure to asbestos fibers was greater than the spontaneous LOH rate of the assay,[40] suggesting a direct connection between the fibers and LOH. In patients exposed to asbestos, allelic deletion of the FHIT gene on chromosome 3p14[41] and LOH on regions of chromosome 6p[42] have been reported, showing a correlation between asbestos exposure and LOH in human disease.

Carcinogens in Food

Food is another source of carcinogens that lead to LOH and disease. Aflatoxin, a metabolite of certain fungi, can be introduced into food sources by grains such as corn and livestock feed. Aflatoxins often occur in crops in the field before harvest. Postharvest contamination can occur if crop drying is delayed and during storage of the crop if water is allowed to exceed critical values for mold growth. Insect or rodent infestations facilitate mold invasion of some stored commodities.

Whether sodium nitrate and sodium nitrate, used to preserve meats, are carcinogens has been a matter of debate.

Chemically, the possibility exists for these compounds to be converted into nitrosoamines, which have been demonstrated as participating in the carcinogenesis resulting from tobacco smoke exposure. However, regulations on maximal levels imposed by the FDA and USDA and changes in industry processes have lowered the risk of cancer from dietary intake of preserved meats.[43] Despite the lower risk of exposure from preserved meats, nitrosamines can be generated from the digestion of meats that have been charred during cooking and are ingested with smokeless tobacco products, as noted above.

Alcohol

Alcoholic beverage consumption was first added as a known human carcinogen to the U.S. Department of Health and Human Services biannual *Report on Carcinogens* in 2000 because it was demonstrated to increase the risk of upper gastrointestinal tract cancers (e.g., mouth, pharynx, larynx, and esophagus)[44] and because evidence of a linkage to pancreatic cancer is accumulating.[45] The formation of acetaldehyde seems to be the most important mechanism by which alcoholic beverage consumption causes human cancer.[27] Although the N^2-ethyl-2′-deoxyguanosine DNA adduct is increased in white blood cells obtained from human alcohol abusers, a more likely cause of the mutagenic effects of alcohol is a different DNA adduct, 1,N^2-propano-2′-deoxyguanosine (PdG), that can form when acetaldehyde is in the presence of histones and other basic molecules. PdG has been shown to be responsible for the genotoxic and mutagenic effects of crotonaldehyde through the formation of secondary lesions, including DNA-protein cross-links and DNA interstrand cross-links. Individuals who smoke tobacco and are heavy consumers of alcoholic beverages are at even greater risk of upper gastrointestinal cancers than individuals who are exposed to either substance alone.

Pathogens

Inflammation associated with chronic infection can be a contributing risk factor for carcinogenesis. Examples include infection of the gastric wall with *Helicobacter pylori* as a predisposing factor for gastric cancer[26]; chronic infections of hepatitis virus leading to hepatocellular carcinoma (HCC); and the linkage of lung cancer with chronic *Chlamydia pneumoniae* and *Mycobacterium tuberculosis* (M-TB) infections.[25,46] Studies indicate that the mechanism of DNA damage in relation to infections is indirect and comes from the host's activated inflammatory cells sent to fight the infection. Cumulative evidence suggests that free radicals, such as NO, produced by activated inflammatory cells can contribute to cancer, as they are known to be able to cause direct damage to DNA.[47]

Patterns of LOH in Cancers

LOH has been detected on all chromosomes. Table 11.1 summarizes some associations between LOH and cancers, and Table 11.2 lists individual genes that are targets for allele loss in cancers. The following sections outline in more detail selected research on LOH in disease, with particular focus on known or putative tumor suppressor genes and their role in disease initiation or progression.

LOH at 1p

LOH at 1p has been associated several cancers, including lung,[48–51] colorectal,[52] and gastrointestinal stromal tumors[53] and hepatocellular carcinoma.[54] LOH at 1p36 was found to

Table 11.1. Associations between loss of heterozygosity (LOH) and some cancers.

Chromosome	Cancers	References
1p	Oligodendroglial tumors	9
	Hepatocellular carcinoma	54
2p	Breast carcinoma	136
3p	Eye, central nervous system, kidney, pancreas, and other tissues	137
	Esophageal adenocarcinoma	68
	Lung cancer	41
5q	Pulmonary large cell neuroendocrine carcinoma	14
6p	Cervical carcinoma	8
	Acinic cell carcinoma of the parotid gland	10
8q	Hepatocellular carcinoma	54
9	Bladder cancer	138
10q	Primary and secondary glioblastoma	139
	Small cell lung cancer	11,116
11p	Epithelial ovarian carcinoma (invasive)	12
16q	Ductal breast carcinoma	140
18q	Esophageal carcinoma	68
22q	Secondary glioblastoma	13

Table 11.2. Individual genes associated with loss of heterozygosity (LOH) in disease.

Gene	Disease	References
GPX1 (selenium-containing antioxidant enzyme)	Head and neck, lung, breast, and colon cancers	141
Von Hippel-Lindau (VHL) (inherited cancer predisposition)	Multiple organs can develop cancer in VHL patients	137
BRCA1, -2	Breast carcinoma	142·143
HRPT2 gene	Sporadic renal tumor	144
CDH13	Ductal breast carcinoma	140
FHIT gene	Lung, gastric, and cervical cancer	41,145,146

be a prognostic indicator for decreased overall survival in hepatocellular carcinoma patients.[54] 1p36 encompasses several candidate tumor suppressor genes, including p73 and tumor necrosis factor receptor-2 (TNFR2).[48] P73 is a member of the p53 family and is capable of mimicking some of the effector functions of p53, including induction of permanent growth arrest and promotion of apoptosis.[55] The TNFR2 binds to and mediates signals from lymphotoxin-α (LTα), lymphotoxin-β (LTβ), and TNF, three cytokines associated with receptor-mediated induction of cell death.[56]

In oligodendroglial tumors, LOH at 1p is associated with better prognosis.[57] Further characterization of these tumors revealed that in cells with LOH at 1p, the expression of GLUT-1, a component of the glucose transport machinery, was reduced compared to cells that retained heterozygosity at 1p. Stockhammer et al[57] speculated that the disappearance of GLUT-1 in cells might affect the tumor's sensitivity to chemotherapy drugs. Further study is needed to elucidate the details, but this research highlights how LOH detection can inform treatment planning.

LOH at 3p

The best described and best documented gene targeted for LOH in nearly all lung cancers is the fragile histidine triad (FHIT), on 3p14. LOH at 3p14 is strongly correlated with exposure to tobacco smoke and occurs early in carcinogen-exposed lung epithelium.[20,58–63] The linkage between tobacco carcinogen exposure and LOH on chromosome 3p is particularly strong in individuals who initiated smoking at an early age.[37] In two studies, LOH at a locus near the hMLH1 gene on chromosome 3p21 was correlated strongly with early age of smoking initiation and with the level of DNA adducts.[35,64]

The FHIT gene encodes a small protein with diadenosine triphosphate hydrolase activity, but its tumor suppressor activity is presumed to be independent from this enzymatic activity.[62] The microregion of the FHIT gene is frequently targeted for LOH in tumors; however, this association alone does not confirm tumor suppressor activity. Evidence for FHIT as a tumor suppressor gene includes the observations that FHIT gene-deficient mice are more susceptible to carcinogen-induced tumor formation,[65] and that expression of FHIT suppresses the growth of cancer cells through promotion of apoptosis and growth inhibition.[66]

FHIT genetic and protein losses have been shown in tumors from other cancers in addition to lung: esophageal,[67–69] breast,[70] and kidney.[71,72] In addition, other tumor suppressor gene candidates targeted by LOH are located in the same region (3p12–3p22) as the FHIT gene. These other genes include transforming growth factor receptor beta-2 (TGFβR2), MLH/HNPCC2, deleted in lung cancer-1 (DLC1), RASSF1A, retinoic acid receptor-beta (RARβ), and BRCA1-associated protein-1 (BAP1).[20,48,58,59,73,74] LOH at 3p25, which includes the von Hippel-Lindau (VHL) locus, has also been reported as a frequent event in several cancers.[20,75]

LOH at 4q

LOH at 4q has been associated with breast,[76] cervical,[77–79] mouth,[80] throat,[81–83] bladder,[84–86] colon,[87,88] lung,[48,89,90] and liver cancers[91–94] as well as acute lymphoblastic leukemia.[95] Jiang et al[87] further analyzed 4q LOH patterns in colorectal carcinoma tumors from 83 patients to evaluate relationships between LOH patterns and clinicopathological features. LOH in 4q13.2 was significantly higher in tumors larger than 5 cm in diameter. LOH in 4q21.23 was significantly associated with metastasis. Suspected tumor suppressor genes are located in each region, but cause-and-effect relationships between specific genes and tumor diameter or metastasis have not yet been established.

LOH at 5q

SCLC is characterized by a high frequency of LOH at 5q32-ter, and although no candidate tumor suppressor genes have been identified, the SPARC (secreted protein acidic and rich in cysteine) gene maps to 5q32 and has been associated with LOH in idiopathic pulmonary fibrosis (IPF).[96] SPARC, a protein involved in the regulation of cell adhesion and growth, is methylated aberrantly in lung cancers.[97] Although nonsmall cell lung cancer (NSCLC) has not been associated with LOH at 5q32, there is often LOH at 5q21.3–31, a region that contains several key tumor suppressor genes including MCC (mutated in colorectal cancer), APC (adenomatous polyposis coli), and IRF1.[48] LOH at 5q21 has been documented in preneoplastic cells, suggesting it may be an early genetic change, and it is more prevalent in squamous cell carcinomas compared to adenocarcinomas.[98,99] However, as in other LOH events, studies have failed to provide strong correlation between these events and patient prognosis. In lung cancer, it appears that LOH at the APC locus is relatively frequent but that the mechanism by which the remaining allele is silenced is usually promoter methylation, rather than mutation.[100]

LOH at 8p

LOH on chromosome 8p21–23 is a frequent event in many cancers and is often associated with more aggressive disease.[101,102] Intense research has focused on 8p21 to identify tumor suppressor genes and look for connections between LOH and clinical characteristics. Shi et al[103] investigated DBC2 (deleted in breast cancer 2) expression in bladder cancer. In these tumors, LOH was not frequent, but they discovered that promoter methylation caused downregulation of DBC2 expression, mimicking the effect of LOH at DBC2 observed in other cancers.

In head and neck squamous cell cancer (SCC), 8p21–22 was the most frequent site of LOH.[104] Ye et al[104] investigated expression of another candidate tumor suppressor gene identified in the region, MTUS1 (mitochondrial tumor suppressor gene 1), and found downregulation of transcription and mutations in exon-coding sequences. Coon et al[102] found that LOH in a region of 8p that included 8p21 was strongly associated with shorter survival in head and neck SCC patients. Ye et al. discovered that LOH was observed mainly in invasive and metastatic SCC

LOH at 8p21 is a frequent and early occurrence in NSCLC, believed to occur after LOH events at 3p and 9p[105,106] and hepatocellular carcinoma.[54] Despite consistent observations of LOH in this region in lung cancer, specific tumor suppressor genes have not been mapped to 8p21–23 and implicated in the development and progression of lung cancer. LOH at 8p21–23 is also a frequent event in hepatocellular carcinoma (HCC), and a gene encoding a growth inhibitory protein, HCRP1 (hepatocellular carcinoma related protein-1) has been mapped to that area and implicated in HCC.[107]

LOH at 9p

Chromosome 9p is frequently altered in NSCLC, not only by LOH but also by homozygous deletion and gene silencing via promoter hypermethylation.[48,108,109] The redundancy in mechanisms aimed at silencing genes in this region and the prevalence of these alterations in smoke-exposed preneoplastic epithelium point to the importance of 9p in the development of lung cancer. The gene most frequently identified in LOH studies of 9p (9p21) in NSCLC is $P16^{INK4A}$, a gene that encodes a cell-cycle protein which inhibits CDK4, CDK6, and cyclin-dependent phosphorylation of the Rb gene product.[110–112] Loss of the $P16^{INK4A}$ gene effectively removes key negative regulation of the cell cycle at the G_1- → S-phase transition.[113] Interestingly, experimental evidence suggests that the silencing of the remaining $P16^{INK4A}$ allele after allelic loss is predominantly caused by epigenetic methylation rather than mutation.[114]

LOH at 10q

LOH at 10q22–23 is a frequent observation in SCLC. Although no tumor suppressor genes in this region have been definitively associated with lung cancer, multiple studies have identified LOH targeted to the PTEN/MMAC locus at 10q 23.[115–117] The encoded PTEN protein is a lipid phosphatase that negatively regulates the phosphatidylinositol 3-kinase/AKT (PI3K/AKT) pathway. Loss of PTEN function results in reduced apoptosis and stimulation of cellular proliferation and migration.[117]

LOH at 13q

Chromosome 13q12–14 is a prevalent hotspot for LOH in many cancers. The retinoblastoma (RB1) gene is a well-characterized tumor suppressor gene, at 13q12, the product of which is a key regulator of entry into S phase of the cell cycle.[118] The critical pathway regulated by RB, $p16^{INK\$A}$, cyclin-D1, and cyclin-dependent kinases (CDKs) is disturbed, usually via multiple components, in nearly every case of lung cancer, thus rende-ring cells insensitive to the signaling involved in regulating mitosis.[118] LOH on 13q12.1–13.1 can be identified in cells obtained from bronchial washing specimens from SCLC and NSCLC patients and is proposed as one of a set of markers that could be used for early detection of lung cancer.[119] Another gene targeted to 13q12–14 is BRCA2, a gene associated with cancer susceptibility and with possible linkages to p53 in the context of DNA damage repair.[120]

LOH at 17p

The TP53 gene at 17p13 is the most frequently altered tumor suppressor gene in human cancers. TP53 is targeted by mutation, methylation, and homozygous deletion, in addition to LOH. The observations of redundant mechanisms of TP53 inactivation and frequency of LOH at TP53 gene point to its critical role in the development of malignant disease. TP53 modulates a broad network of cellular responses, including cell-cycle arrest, apoptosis, DNA repair, cellular senescence, and inflammation, and therefore plays a central role in homeostasis.[121] TP53 mutations can be found in precancerous lesions, but to date studies of LOH at 17p indicate that allelic loss of TP53 occurs during disease progression, after LOH events on chromosome 3p.[122]

LOH at 19p

LOH at 19p13.3 is a very frequent event in NSCLC but is not targeted in SCLC.[48] One tumor suppressor gene candidate residing at this locus is STK11/LKB1, a gene implicated in Peutz-Jeghers syndrome.[123,124] LOH at this locus has been documented in breast cancer and in brain metastases from a variety of human cancers.[124,125]

Summary

Abundant research has demonstrated that certain patterns of LOH are seen in multiple cancers. Other LOH loci are seen in only certain cancers or occur at a particular stage of disease progression. The availability of high-throughput assays and high-resolution LOH probes is increasing the breadth and depth of studies to elucidate the contribution of LOH to disease, prognosis, or effectiveness of treatment regimens. Currently, there are few loci for which a cause-and-effect relationship to cancer has been established. However, researchers are now armed with an arsenal of tools with which to discover the temporal sequence of LOH events in tumor initiation and progression and contribution of specific tumor suppression genes to cellular and tissue growth regulation.

LOH in Benign Diseases

There is evidence to suggest that benign conditions, such as chronic obstructive pulmonary disease (COPD) and asthma, have a genetic predisposition and probably arise via the interaction between multiple gene products.[126] Studies of microsatellite instability and LOH have been employed to study these diseases with the goals of identifying causative genes and/or develop genetic screening tools for use in epidemiology. LOH in several prevalent noncancer diseases is discussed next.

LOH in Idiopathic Pulmonary Fibrosis

Idiopathic pulmonary fibrosis (IPF) is a serious disease believed to be the result of immune response to tissue damage in the lung.[127] Although IPF is a benign disease, it progresses to bronchogenic carcinoma in about 10% of IPF patients.[126] The observation that the incidence of lung cancer in IPF patients is much higher than the incidence in the general population lends support to the hypothesis that IPF contains precancerous lesions that may progress to peripheral-type lung tumors through the inactivation of critical tumor suppressor genes.[128] Indeed, LOH at several common loci (3p21, 5q32, 9p21, and 17p13) has been documented in IPF.[96]

LOH in Sarcoidosis

Sarcoidosis is a multisystem disease characterized by the formation of noncaseating granulomatous lesions in affected organs, especially the lungs. These lesions can progress to cause fibrosis and, similar to IPF, lead to a higher incidence of lung cancer.[126] Studies of LOH have identified targeted loci on 9p, 9q, and 17q.[129,130] The genes targeted by LOH were within or proximal to DNA mismatch repair (MMR) genes and genes associated with lymphocyte activation. This pattern is somewhat distinct from lung cancer or IPF and may reflect the absence of linkage with exposure to tobacco smoke.[129]

LOH in Chronic Obstructive Pulmonary Disease and Asthma

LOH has been detected in chronic obstructive pulmonary disease (COPD) and asthma, although the number of studies is limited and these have focused on analysis of chromosomal regions that contain genes suspected in the diseases.[79] There are genes implicated in the development and progression of asthma throughout the genome, although the "hotspot" for alterations via LOH is chromosome 14q, which contains several target genes that have been implicated in asthma, including prostaglandin E receptor 2 (PTGER2), arginase II (ARG2), and alpha-1-antichymotrypsin precursor (AACT).[131,132] Within asthmatic patients, those with the greater number of genetic alterations have higher mean immunoglobulin E and blood eosinophils,[133] both of which are indicators of inflamma-tion and bronchial hyperresponsiveness. Analyses of COPD samples demonstrate that LOH occurs most frequently at the thyroid hormone receptor alpha-1 (THRA1) locus on chromosome 17q21.[134] It is believed that exposure to tobacco smoke is a risk factor for the genetic changes associated with COPD, and individuals with COPD (similar to those with IPF) carry a greater risk for the development of lung cancer.[135]

Summary

Clearly, LOH is a well-documented occurrence in benign, premalignant, and malignant disease. Although some genes targeted by LOH represent key early events in pathogenesis, many occur at later stages of disease progression, thus making their specific contributions difficult to discern. Comprehensive genome-wide studies of LOH in diseases using SNPs and bioinformatics analysis to analyze patterns of multiple genetic alterations have given clinicians more precise tools for diagnosis and treatment of cancers. Of particular interest is the importance of LOH research for discerning the incremental effects of accumulating mutations and how specific LOH events or clusters of LOH events can influence a patient's response to treatment or inform prognosis. The availability of scalable, high-throughput LOH screening techniques promises to give researchers and clinicians the information they need to design more effective and customized treatment plans for patients with cancer or at higher risk for cancer.

References

1. Zheng HT, Peng ZH, Li S, He L. Loss of heterozygosity analyzed by single nucleotide polymorphism array in cancer. *World J Gastroenterol.* 2005;11:6740–6744.
2. Cecil RL, Goldman L, Ausiello DA. *Cecil Textbook of Medicine.* Philadelphia: Saunders Elsevier; 2007.
3. Knudson AG Jr. Nakahara memorial lecture. Hereditary cancer, oncogenes, and anti-oncogenes. *Princess Takamatsu Symp.* 1989; 20:15–29.
4. Fong KM, Sekido Y, Minna JD. Molecular pathogenesis of lung cancer. *J Thorac Cardiovasc Surg.* 1999;118:1136–1152.
5. Sekido Y, Fong KM, Minna JD. Progress in understanding the molecular pathogenesis of human lung cancer. *Biochim Biophys Acta.* 1998;1378:F21–F59.

6. Pinkel D, Segraves R, Sudar D, et al. High resolution analysis of DNA copy number variation using comparative genomic hybridization to microarrays. *Nat Genet.* 1998;20:207–211.
7. Mao X, Young BD, Lu YJ. The application of single nucleotide polymorphism microarrays in cancer research. *Curr Genomics.* 2007;8:219–228.
8. Chatterjee A, Pulido HA, Koul S, et al. Mapping the sites of putative tumor suppressor genes at 6p25 and 6p21.3 in cervical carcinoma: occurrence of allelic deletions in precancerous lesions. *Cancer Res.* 2001;61:2119–2123.
9. Dong Z, Pang JS, Ng MH, Poon WS, Zhou L, Ng HK. Identification of two contiguous minimally deleted regions on chromosome 1p36.31-p36.32 in oligodendroglial tumours. *Br J Cancer.* 2004;91:1105–1111.
10. el-Naggar AK, Abdul-Karim FW, Hurr K, Callender D, Luna MA, Batsakis JG. Genetic alterations in acinic cell carcinoma of the parotid gland determined by microsatellite analysis. *Cancer Genet Cytogenet.* 1998;102:19–24.
11. Kim SK, Ro JY, Kemp BL, et al. Identification of two distinct tumor-suppressor loci on the long arm of chromosome 10 in small cell lung cancer. *Oncogene.* 1998;17:1749–1753.
12. Lu KH, Weitzel JN, Kodali S, Welch WR, Berkowitz RS, Mok SC. A novel 4-cM minimally deleted region on chromosome 11p15.1 associated with high grade nonmucinous epithelial ovarian carcinomas. *Cancer Res.* 1997;57:387–390.
13. Nakamura M, Ishida E, Shimada K, et al. Frequent LOH on 22q12.3 and TIMP-3 inactivation occur in the progression to secondary glioblastomas. *Lab Invest.* 2005;85:165–175.
14. Shin JH, Kang SM, Kim YS, Shin DH, Chang J, Kim SK. Identification of tumor suppressor loci on the long arm of chromosome 5 in pulmonary large cell neuroendocrine carcinoma. *Chest.* 2005;128:2999–3003.
15. Simoneau AR, Spruck CH III, Gonzalez-Zulueta M, et al. Evidence for two tumor suppressor loci associated with proximal chromosome 9p to q and distal chromosome 9q in bladder cancer and the initial screening for GAS1 and PTC mutations. *Cancer Res.* 1996;56:5039–5043.
16. Vogelstein B, Fearon ER, Kern SE, et al. Allelotype of colorectal carcinomas. *Science.* 1989;244:207–211.
17. Frigerio S, Padberg BC, Strebel RT, et al. Improved detection of bladder carcinoma cells in voided urine by standardized microsatellite analysis. *Int J Cancer.* 2007;121:329–338.
18. Powell CA, Klares S, O'Connor G, Brody JS. Loss of heterozygosity in epithelial cells obtained by bronchial brushing: clinical utility in lung cancer. *Clin Cancer Res.* 1999;5:2025–2034.
19. Lindblad-Toh K, Tanenbaum DM, Daly MJ, et al. Loss-of-heterozygosity analysis of small-cell lung carcinomas using single-nucleotide polymorphism arrays. *Nat Biotechnol.* 2000;18:1001–1005.
20. Zabarovsky ER, Lerman MI, Minna JD. Tumor suppressor genes on chromosome 3p involved in the pathogenesis of lung and other cancers. *Oncogene.* 2002;21:6915–6935.
21. Mei R, Galipeau PC, Prass C, et al. Genome-wide detection of allelic imbalance using human SNPs and high-density DNA arrays. *Genome Res.* 2000;10:1126–1137.
22. Haltrich I, Kost-Alimova M, Kovacs G, et al. Multipoint interphase FISH in childhood T-acute lymphoblastic leukemia detects subpopulations that carry different chromosome 3 aberrations. *Cancer Genet Cytogenet.* 2007;172:54–60.
23. Carr J, Bown NP, Case MC, Hall AG, Lunec J, Tweddle DA. High-resolution analysis of allelic imbalance in neuroblastoma cell lines by single nucleotide polymorphism arrays. *Cancer Genet Cytogenet.* 2007;172:127–138.
24. Wogan GN. Molecular epidemiology in cancer risk assessment and prevention: recent progress and avenues for future research. *Environ Health Perspect.* 1992;98:167–178.
25. Song L, Yan W, Zhao T, et al. Mycobacterium tuberculosis infection and FHIT gene alterations in lung cancer. *Cancer Lett.* 2005;219:155–162.
26. Wu MS, Shun CT, Wang HP, et al. Genetic alterations in gastric cancer: relation to histological subtypes, tumor stage, and Helicobacter pylori infection. *Gastroenterology* 1997;112:1457–1465.
27. Brooks PJ, Theruvathu JA. DNA adducts from acetaldehyde: implications for alcohol-related carcinogenesis. *Alcohol.* 2005;35:187–193.
28. Ray G, Husain SA. Oxidants, antioxidants and carcinogenesis. *Indian J Exp Biol.* 2002;40:1213–1232.
29. Broyde S, Wang L, Zhang L, Rechkoblit O, Geacintov NE, Patel DJ. DNA adduct structure-function relationships: comparing solution with polymerase structures. *Chem Res Toxicol.* 2008;21:45–52.
30. Roos PH, Bolt HM. Cytochrome P450 interactions in human cancers: new aspects considering CYP1B1. *Expert Opin Drug Metab Toxicol.* 2005;1:187–202.
31. D'Agostino J, Zhang X, Wu H, et al. Characterization of CYP2A13*2, a variant cytochrome P450 allele previously found to be associated with decreased incidences of lung adenocarcinoma in smokers. *Drug Metab Dispos.* 2008;36(11):2316–2323.
32. Vineis P. Individual susceptibility to carcinogens. *Oncogene.* 2004;23:6477–6483.
33. Boffetta P, Hecht S, Gray N, Gupta P, Straif K. Smokeless tobacco and cancer. *Lancet Oncol.* 2008;9:667–675.
34. Halpern MT, Gillespie BW, Warner KE. Patterns of absolute risk of lung cancer mortality in former smokers. *J Natl Cancer Inst.* 1993;85:457–464.
35. Hirao T, Nelson HH, Ashok TD, et al. Tobacco smoke-induced DNA damage and an early age of smoking initiation induce chromosome loss at 3p21 in lung cancer. *Cancer Res.* 2001;61:612–615.
36. Knoke JD, Shanks TG, Vaughn JW, Thun MJ, Burns DM. Lung cancer mortality is related to age in addition to duration and intensity of cigarette smoking: an analysis of CPS-I data. *Cancer Epidemiol Biomarkers Prev.* 2004;13:949–957.
37. Wiencke JK, Kelsey KT. Teen smoking, field cancerization, and a "critical period" hypothesis for lung cancer susceptibility. *Environ Health Perspect.* 2002;110:555–558.
38. Slaughter DP, Southwick HW, Smejkal W. Field cancerization in oral stratified squamous epithelium; clinical implications of multicentric origin. *Cancer.* 1953;6:963–968.
39. Craighead JE, Mossman BT. The pathogenesis of asbestos-associated diseases. *N Engl J Med.* 1982;306:1446–1455.
40. Both K, Henderson DW, Turner DR. Asbestos and erionite fibres can induce mutations in human lymphocytes that result in loss of heterozygosity. *Int J Cancer.* 1994;59:538–542.
41. Pylkkanen L, Wolff H, Stjernvall T, et al. Reduced Fhit protein expression and loss of heterozygosity at FHIT gene in tumours from smoking and asbestos-exposed lung cancer patients. *Int J Oncol.* 2002;20:285–290.
42. Tug E, Tug T, Elyas H, Coskunsel M, Emri S. Tumor suppressor gene alterations in patients with malignant mesothelioma due to environmental asbestos exposure in Turkey. *J Carcinog.* 2006;5:23.

43. Safety of sodium nitrite in cured meats. http://www.medem.com/MedLB/article_detaillb.cfm?article_ID=ZZZ80XEN0IC&sub_cat = 380; 2008 Accessed 12.09.08.
44. U.S. Department of Health and Human Services. *Report on Carcinogens*. 11th ed. Research Triangle Park, NC; 2005.
45. Go VL, Gukovskaya A, Pandol SJ. Alcohol and pancreatic cancer. *Alcohol.* 2005;35:205–211.
46. Song L, Yan W, Deng M, Song S, Zhang J, Zhao T. Aberrations in the fragile histidine triad (FHIT) gene may be involved in lung carcinogenesis in patients with chronic pulmonary tuberculosis. *Tumour Biol.* 2004;25:270–275.
47. Brenner C, Bieganowski P, Pace HC, Huebner K. The histidine triad superfamily of nucleotide-binding proteins. *J Cell Physiol.* 1999;181:179–187.
48. Girard L, Zochbauer-Muller S, Virmani AK, Gazdar AF, Minna JD. Genome-wide allelotyping of lung cancer identifies new regions of allelic loss, differences between small cell lung cancer and non-small cell lung cancer, and loci clustering. *Cancer Res.* 2000;60:4894–4906.
49. Dacic S, Ionescu DN, Finkelstein S, Yousem SA. Patterns of allelic loss of synchronous adenocarcinomas of the lung. *Am J Surg Pathol.* 2005;29:897–902.
50. Sasatomi E, Johnson LR, Aldeeb DN, et al. Genetic profile of cumulative mutational damage associated with early pulmonary adenocarcinoma: bronchioloalveolar carcinoma vs. stage I invasive adenocarcinoma. *Am J Surg Pathol.* 2004;28:1280–1288.
51. Ragnarsson G, Eiriksdottir G, Johannsdottir JT, Jonasson JG, Egilsson V, Ingvarsson S. Loss of heterozygosity at chromosome 1p in different solid human tumours: association with survival. *Br J Cancer.* 1999;79:1468–1474.
52. Kurashina K, Yamashita Y, Ueno T, et al. Chromosome copy number analysis in screening for prognosis-related genomic regions in colorectal carcinoma. *Cancer Sci.* 2008;99(9): 1835–1840.
53. Yang J, Du X, Lazar AJ, et al. Genetic aberrations of gastrointestinal stromal tumors. *Cancer.* 2008;113:1532–1543.
54. Pang JZ, Qin LX, Ren N, et al. Loss of heterozygosity at D8S298 is a predictor for long-term survival of patients with tumor-node-metastasis stage I of hepatocellular carcinoma. *Clin Cancer Res.* 2007;13:7363–7369.
55. Fang L, Lee SW, Aaronson SA. Comparative analysis of p73 and p53 regulation and effector functions. *J Cell Biol.* 1999;147: 823–830.
56. Gupta S. Molecular steps of tumor necrosis factor receptor-mediated apoptosis. *Curr Mol Med.* 2001;1:317–324.
57. Stockhammer F, von Deimling A, Synowitz M, Blechschmidt C, van Landeghem FK. Expression of glucose transporter 1 is associated with loss of heterozygosity of chromosome 1p in oligodendroglial tumors WHO grade II. *J Mol Histol.* 2008; 39(5):553–560.
58. Wistuba II, Behrens C, Virmani AK, et al. High resolution chromosome 3p allelotyping of human lung cancer and preneoplastic/preinvasive bronchial epithelium reveals multiple, discontinuous sites of 3p allele loss and three regions of frequent breakpoints. *Cancer Res.* 2000;60:1949–1960.
59. Yokota J, Wada M, Shimosato Y, Terada M, Sugimura T. Loss of heterozygosity on chromosomes 3, 13, and 17 in small-cell carcinoma and on chromosome 3 in adenocarcinoma of the lung. *Proc Natl Acad Sci USA.* 1987;84:9252–9256.
60. Toledo G, Sola JJ, Lozano MD, Soria E, Pardo J. Loss of FHIT protein expression is related to high proliferation, low apoptosis and worse prognosis in non-small-cell lung cancer. *Mod Pathol.* 2004;17:440–448.
61. Ohta M, Inoue H, Cotticelli MG, et al. The FHIT gene, spanning the chromosome 3p14.2 fragile site and renal carcinoma-associated t(3;8) breakpoint, is abnormal in digestive tract cancers. *Cell.* 1996;84:587–97.
62. Fong KM, Biesterveld EJ, Virmani A, et al. FHIT and FRA3B 3p14.2 allele loss are common in lung cancer and preneoplastic bronchial lesions and are associated with cancer-related FHIT cDNA splicing aberrations. *Cancer Res.* 1997;57:2256–2267.
63. Woenckhaus M, Grepmeier U, Wild PJ, et al. Multitarget FISH and LOH analyses at chromosome 3p in non-small cell lung cancer and adjacent bronchial epithelium. *Am J Clin Pathol.* 2005;123: 752–761.
64. Zienolddiny S, Ryberg D, Arab MO, Skaug V, Haugen A. Loss of heterozygosity is related to p53 mutations and smoking in lung cancer. *Br J Cancer.* 2001;84:226–231.
65. Zanesi N, Fidanza V, Fong LY, et al. The tumor spectrum in FHIT-deficient mice. *Proc Natl Acad Sci USA.* 2001;98: 10250–10255.
66. Ji L, Fang B, Yen N, Fong K, Minna JD, Roth JA. Induction of apoptosis and inhibition of tumorigenicity and tumor growth by adenovirus vector-mediated fragile histidine triad (FHIT) gene overexpression. *Cancer Res.* 1999;59:3333–3339.
67. Kuroki T, Trapasso F, Yendamuri S, et al. Allele loss and promoter hypermethylation of VHL, RAR-beta, RASSF1A, and FHIT tumor suppressor genes on chromosome 3p in esophageal squamous cell carcinoma. *Cancer Res.* 2003;63:3724–3728.
68. Nancarrow DJ, Handoko HY, Smithers BM, et al. Genome-wide copy number analysis in esophageal adenocarcinoma using high-density single-nucleotide polymorphism arrays. *Cancer Res.* 2008;68:4163–4172.
69. Wiech T, Nikolopoulos E, Weis R, et al. Genome-wide analysis of genetic alterations in Barrett's adenocarcinoma using single nucleotide polymorphism arrays. Lab Invest 2009;89:385–397.
70. Diab SG, Clark GM, Osborne CK, Libby A, Allred DC, Elledge RM. Tumor characteristics and clinical outcome of tubular and mucinous breast carcinomas. *J Clin Oncol.* 1999;17: 1442–1448.
71. Sukosd F, Kuroda N, Beothe T, Kaur AP, Kovacs G. Deletion of chromosome 3p14.2-p25 involving the VHL and FHIT genes in conventional renal cell carcinoma. *Cancer Res.* 2003;63:455–457.
72. Toma MI, Grosser M, Herr A, et al. Loss of heterozygosity and copy number abnormality in clear cell renal cell carcinoma discovered by high-density affymetrix 10 K single nucleotide polymorphism mapping array. *Neoplasia.* 2008;10:634–642.
73. Marsit CJ, Hasegawa M, Hirao T, et al. Loss of heterozygosity of chromosome 3p21 is associated with mutant TP53 and better patient survival in non-small-cell lung cancer. *Cancer Res.* 2004;64:8702–8707.
74. Chmara M, Wozniak A, Ochman K, et al. Loss of heterozygosity at chromosomes 3p and 17p in primary non-small cell lung cancer. *Anticancer Res.* 2004;24:4259–4263.
75. Ho WL, Chang JW, Tseng RC, et al. Loss of heterozygosity at loci of candidate tumor suppressor genes in microdissected primary non-small cell lung cancer. *Cancer Detect Prev.* 2002;26:343–349.
76. Schwendel A, Richard F, Langreck H, et al. Chromosome alterations in breast carcinomas: frequent involvement of DNA losses including chromosomes 4q and 21q. *Br J Cancer.* 1998;78:806–811.

77. Backsch C, Rudolph B, Kuhne-Heid R, et al. A region on human chromosome 4 (q35.-->qter) induces senescence in cell hybrids and is involved in cervical carcinogenesis. *Genes Chromosomes Cancer.* 2005;43:260–272.
78. Mitra AB, Murty VV, Li RG, Pratap M, Luthra UK, Chaganti RS. Allelotype analysis of cervical carcinoma. *Cancer Res.* 1994;54:4481–4487.
79. Sherwood JB, Shivapurkar N, Lin WM, et al. Chromosome 4 deletions are frequent in invasive cervical cancer and differ between histologic variants. *Gynecol Oncol.* 2000;79:90–96.
80. Wang XL, Uzawa K, Imai FL, Tanzawa H. Localization of a novel tumor suppressor gene associated with human oral cancer on chromosome 4q25. *Oncogene.* 1999;18:823–825.
81. Hammoud ZT, Kaleem Z, Cooper JD, Sundaresan RS, Patterson GA, Goodfellow PJ. Allelotype analysis of esophageal adenocarcinomas: evidence for the involvement of sequences on the long arm of chromosome 4. *Cancer Res.* 1996;56:4499–4502.
82. Rumpel CA, Powell SM, Moskaluk CA. Mapping of genetic deletions on the long arm of chromosome 4 in human esophageal adenocarcinomas. *Am J Pathol.* 1999;154:1329–1334.
83. Sterian A, Kan T, Berki AT, et al. Mutational and LOH analyses of the chromosome 4q region in esophageal adenocarcinoma. *Oncology.* 2006;70:168–172.
84. Hurst CD, Fiegler H, Carr P, Williams S, Carter NP, Knowles MA. High-resolution analysis of genomic copy number alterations in bladder cancer by microarray-based comparative genomic hybridization. *Oncogene.* 2004;23:2250–2263.
85. Polascik TJ, Cairns P, Chang WY, Schoenberg MP, Sidransky D. Distinct regions of allelic loss on chromosome 4 in human primary bladder carcinoma. *Cancer Res.* 1995;55:5396–5399.
86. Rosin MP, Cairns P, Epstein JI, Schoenberg MP, Sidransky D. Partial allelotype of carcinoma in situ of the human bladder. *Cancer Res.* 1995;55:5213–5216.
87. Jiang LX, Xu J, Wang ZW, et al. Tumor suppress genes screening analysis on 4q in sporadic colorectal carcinoma. *World J Gastroenterol.* 2008;14:5606–5611.
88. Kurashina K, Yamashita Y, Ueno T, et al. Chromosome copy number analysis in screening for prognosis-related genomic regions in colorectal carcinoma. *Cancer Sci.* 2008;99: 1835–1840.
89. Cho ES, Chang J, Chung KY, Shin DH, Kim YS, Kim SK. Identification of tumor suppressor loci on the long arm of chromosome 4 in primary small cell lung cancers. *Yonsei Med J.* 2002; 43:145–151.
90. Shivapurkar N, Virmani AK, Wistuba , II, et al. Deletions of chromosome 4 at multiple sites are frequent in malignant mesothelioma and small cell lung carcinoma. *Clin Cancer Res.* 1999;5:17–23.
91. Bluteau O, Beaudoin JC, Pasturaud P, et al. Specific association between alcohol intake, high grade of differentiation and 4q34–q35 deletions in hepatocellular carcinomas identified by high resolution allelotyping. *Oncogene.* 2002;21:1225–1232.
92. Chang J, Kim NG, Piao Z, et al. Assessment of chromosomal losses and gains in hepatocellular carcinoma. *Cancer Lett.* 2002;182:193–202.
93. Chou YH, Chung KC, Jeng LB, Chen TC, Liaw YF. Frequent allelic loss on chromosomes 4q and 16q associated with human hepatocellular carcinoma in Taiwan. *Cancer Lett.* 1998;123:1–6.
94. Rashid A, Wang JS, Qian GS, Lu BX, Hamilton SR, Groopman JD. Genetic alterations in hepatocellular carcinomas: association between loss of chromosome 4q and p53 gene mutations. *Br J Cancer.* 1999;80:59–66.
95. Takeuchi S, Seriu T, van Dongen JJ, et al. Allelotype analysis in relapsed childhood acute lymphoblastic leukemia. *Oncogene.* 2003;22:6970–6976.
96. Demopoulos K, Arvanitis DA, Vassilakis DA, Siafakas NM, Spandidos DA. MYCL1, FHIT, SPARC, p16(INK4) and TP53 genes associated to lung cancer in idiopathic pulmonary fibrosis. *J Cell Mol Med.* 2002;6:215–222.
97. Suzuki M, Hao C, Takahashi T, et al. Aberrant methylation of SPARC in human lung cancers. *Br J Cancer.* 2005;92: 942–948.
98. Sanz-Ortega J, Bryant B, Sanz-Esponera J, et al. LOH at the APC/MCC gene (5Q21) is frequent in early stages of non-small cell lung cancer. *Pathol Res Pract.* 1999;195:677–680.
99. Yoshino I, Osoegawa A, Yohena T, et al. Loss of heterozygosity (LOH) in non-small cell lung cancer: difference between adenocarcinoma and squamous cell carcinoma. *Respir Med.* 2005;99: 308–312.
100. Brabender J, Usadel H, Danenberg KD, et al. Adenomatous polyposis coli gene promoter hypermethylation in non-small cell lung cancer is associated with survival. *Oncogene.* 2001;20:3 528–3532.
101. Adams J, Cuthbert-Heavens D, Bass S, Knowles MA. Infrequent mutation of TRAIL receptor 2 (TRAIL-R2/DR5) in transitional cell carcinoma of the bladder with 8p21 loss of heterozygosity. *Cancer Lett.* 2005;220:137–144.
102. Coon SW, Savera AT, Zarbo RJ, et al. Prognostic implications of loss of heterozygosity at 8p21 and 9p21 in head and neck squamous cell carcinoma. *Int J Cancer.* 2004;111:206–212.
103. Shi Y, Chen JY, Yang J, Li B, Chen ZH, Xiao CG. DBC2 gene is silenced by promoter methylation in bladder cancer. *Urol Oncol.* 2008;26:465–469.
104. Ye H, Pungpravat N, Huang BL, et al. Genomic assessments of the frequent loss of heterozygosity region on 8p21.3–p22 in head and neck squamous cell carcinoma. *Cancer Genet Cytogenet.* 2007;176:100–106.
105. Kurimoto F, Gemma A, Hosoya Y, et al. Unchanged frequency of loss of heterozygosity and size of the deleted region at 8p21–23 during metastasis of lung cancer. *Int J Mol Med.* 2001;8:89–93.
106. Wistuba II, Behrens C, Virmani AK, et al. Allelic losses at chromosome 8p21–23 are early and frequent events in the pathogenesis of lung cancer. *Cancer Res.* 1999;59: 1973–1979.
107. Xu Z, Liang L, Wang H, Li T, Zhao M. HCRP1, a novel gene that is downregulated in hepatocellular carcinoma, encodes a growth-inhibitory protein. *Biochem Biophys Res Commun.* 2003; 311:1057–1066.
108. Marsit CJ, Wiencke JK, Nelson HH, et al. Alterations of 9p in squamous cell carcinoma and adenocarcinoma of the lung: association with smoking, TP53, and survival. *Cancer Genet Cytogenet.* 2005;162:115–121.
109. Virmani AK, Fong KM, Kodagoda D, et al. Allelotyping demonstrates common and distinct patterns of chromosomal loss in human lung cancer types. *Genes Chromosomes Cancer.* 1998;21: 308–319.

110. Kratzke RA, Greatens TM, Rubins JB, et al. Rb and p16INK4a expression in resected non-small cell lung tumors. *Cancer Res.* 1996;56:3415–3420.
111. Ohtani N, Yamakoshi K, Takahashi A, Hara E. The p16INK4a-RB pathway: molecular link between cellular senescence and tumor suppression. *J Med Invest.* 2004;51:146–153.
112. Sumitomo K, Shimizu E, Shinohara A, Yokota J, Sone S. Activation of RB tumor suppressor protein and growth suppression of small cell lung carcinoma cells by reintroduction of p16INK4A gene. *Int J Oncol.* 1999;14:1075–1080.
113. Quesnel B, Preudhomme C, Fenaux P. p16ink4a gene and hematological malignancies. *Leuk Lymphoma.* 1996;22: 11–24.
114. Awaya H, Takeshima Y, Amatya VJ, et al. Inactivation of the p16 gene by hypermethylation and loss of heterozygosity in adenocarcinoma of the lung. *Pathol Int.* 2004;54:486–489.
115. Hosoya Y, Gemma A, Seike M, et al. Alteration of the PTEN/MMAC1 gene locus in primary lung cancer with distant metastasis. *Lung Cancer.* 1999;25:87–93.
116. Kim SK, Su LK, Oh Y, Kemp BL, Hong WK, Mao L. Alterations of PTEN/MMAC1, a candidate tumor suppressor gene, and its homologue, PTH2, in small cell lung cancer cell lines. *Oncogene.* 1998;16:89–93.
117. Marsit CJ, Zheng S, Aldape K, et al. PTEN expression in non-small-cell lung cancer: evaluating its relation to tumor characteristics, allelic loss, and epigenetic alteration. *Hum Pathol.* 2005;36:768–776.
118. Wikman H, Kettunen E. Regulation of the G1/S phase of the cell cycle and alterations in the RB pathway in human lung cancer. *Expert Rev Anticancer Ther.* 2006;6:515–530.
119. Arvanitis DA, Papadakis E, Zafiropoulos A, Spandidos DA. Fractional allele loss is a valuable marker for human lung cancer detection in sputum. *Lung Cancer.* 2003;40:55–66.
120. Gorgoulis VG, Zacharatos P, Kotsinas A, et al. Alterations of the p16-pRb pathway and the chromosome locus 9p21–22 in non-small-cell lung carcinomas: relationship with p53 and MDM2 protein expression. *Am J Pathol.* 1998;153:1749–1765.
121. Hussain SP, Harris CC. p53 biological network: at the crossroads of the cellular-stress response pathway and molecular carcinogenesis. *J Nippon Med Sch.* 2006;73:54–64.
122. Pan H, Califano J, Ponte JF, et al. Loss of heterozygosity patterns provide fingerprints for genetic heterogeneity in multistep cancer progression of tobacco smoke-induced non-small cell lung cancer. *Cancer Res.* 2005;65:1664–1669.
123. von Herbay A, Arens N, Friedl W, et al. Bronchioloalveolar carcinoma: a new cancer in Peutz-Jeghers syndrome. *Lung Cancer.* 2005;47:283–288.
124. Yang TL, Su YR, Huang CS, et al. High-resolution 19p13.2–13.3 allelotyping of breast carcinomas demonstrates frequent loss of heterozygosity. *Genes Chromosomes Cancer.* 2004; 41:250–2546.
125. Sobottka SB, Haase M, Fitze G, Hahn M, Schackert HK, Schackert G. Frequent loss of heterozygosity at the 19p13.3 locus without LKB1/STK11 mutations in human carcinoma metastases to the brain. *J Neuro-Oncol.* 2000;49:187–195.
126. Samara K, Zervou M, Siafakas NM, Tzortzaki EG. Microsatellite DNA instability in benign lung diseases. *Respir Med.* 2006;100: 202–211.
127. King T. Idiopathic pulmonary fibrosis. In: Schwartz M, King T, eds. *Interstitial Lung Disease.* Ontario, Canada: Decker BC; 1998:597–644.
128. Uematsu K, Yoshimura A, Gemma A, et al. Aberrations in the fragile histidine triad (FHIT) gene in idiopathic pulmonary fibrosis. *Cancer Res.* 2001;61:8527–8533.
129. Demopoulos K, Arvanitis DA, Vassilakis DA, Siafakas NM, Spandidos DA. Genomic instability on hMSH2, hMLH1, CD48 and IRF4 loci in pulmonary sarcoidosis. *Int J Biol Markers.* 2002;17:224–230.
130. Vassilakis DA, Sourvinos G, Pantelidis P, Spandidos DA, Siafakas NM, Bouros D. Extended genetic alterations in a patient with pulmonary sarcoidosis, a benign disease. *Sarcoidosis Vasc Diffuse Lung Dis.* 2001;18:307–310.
131. Sandford AJ, Pare PD. The genetics of asthma. The important questions. *Am J Respir Crit Care Med.* 2000;161:S202–S206.
132. Xu J, Meyers DA, Ober C, et al. Genome-wide screen and identification of gene-gene interactions for asthma-susceptibility loci in three U.S. populations: collaborative study on the genetics of asthma. *Am J Hum Genet.* 2001;68:1437–1446.
133. Paraskakis E, Sourvinos G, Passam F, et al. Microsatellite DNA instability and loss of heterozygosity in bronchial asthma. *Eur Respir J.* 2003;22:951–955.
134. Siafakas NM, Tzortzaki EG, Sourvinos G, et al. Microsatellite DNA instability in COPD. *Chest.* 1999;116:47–51.
135. Anderson GP, Bozinovski S. Acquired somatic mutations in the molecular pathogenesis of COPD. *Trends Pharmacol Sci.* 2003;24:71–76.
136. Argos M, Kibriya MG, Jasmine F, et al. Genomewide scan for loss of heterozygosity and chromosomal amplification in breast carcinoma using single-nucleotide polymorphism arrays. *Cancer Genet Cytogenet.* 2008;182:69–74.
137. Chan CC, Collins AB, Chew EY. Molecular pathology of eyes with von Hippel-Lindau (VHL) disease: a review. *Retina.* 2007;27:1–7.
138. Simoneau AR, Spruck CH, 3rd, Gonzalez-Zulueta M, et al. Evidence for two tumor suppressor loci associated with proximal chromosome 9p to q and distal chromosome 9q in bladder cancer and the initial screening for GAS1 and PTC mutations. *Cancer Res.* 1996;56:5039–5043.
139. Ohgaki H, Kleihues P. Genetic pathways to primary and secondary glioblastoma. *Am J Pathol.* 2007;170:1445–1453.
140. Riener MO, Nikolopoulos E, Herr A, et al Microarray comparative genomic hybridization analysis of tubular breast carcinoma shows recurrent loss of the CDH13 locus on 16q. *Hum Pathol.* 2008;39:1621–1629.
141. Hu Y, Benya RV, Carroll RE, Diamond AM. Allelic loss of the gene for the GPX1 selenium-containing protein is a common event in cancer. *J Nutr.* 2005;135:3021S–3024S.
142. Margolin S, Lindblom A. Familial breast cancer, underlying genes, and clinical implications: a review. *Crit Rev Oncog.* 2006;12:75–113.
143. Palacios J, Robles-Frias MJ, Castilla MA, Lopez-Garcia MA, Benitez J. The molecular pathology of hereditary breast cancer. *Pathobiology.* 2008;75:85–94.
144. Zhao J, Yart A, Frigerio S, et al. Sporadic human renal tumors display frequent allelic imbalances and novel mutations of the HRPT2 gene. *Oncogene.* 2007;26:3440–3449.
145. Choi CH, Lee KM, Choi JJ, et al. Hypermethylation and loss of heterozygosity of tumor suppressor genes on chromosome 3p in cervical cancer. *Cancer Lett.* 2007;255:26–33.
146. Xiao YP, Wu DY, Xu L, Xin Y. Loss of heterozygosity and microsatellite instabilities of fragile histidine triad gene in gastric carcinoma. *World J Gastroenterol.* 2006;12:3766–3769.

12
In Situ Hybridization: Principles and Applications

Kevin C. Halling and Amy J. Wendel

Introduction

In situ hybridization (ISH) has become an extremely useful tool for the clinical pathology laboratory to aid oncologists, geneticists, and infectious disease specialists in the diagnosis and treatment of their patients. ISH utilizes nucleic acid (DNA or RNA) probes to assess intact cells for various types of genetic alterations. Common applications of in situ hybridization include its use to detect cancer cells in cytological specimens, chromosomal alterations in resected tumor specimens that predict prognosis, and response to therapy of certain cancer types and microorganisms in various specimen types. ISH is typically performed either with fluorescently labeled probes or with probes that are subsequently visualized with a chromogen such as diaminobenzidine. If performed with fluorescently labeled probes, the technique is referred to as Fluorescence In Situ Hybridization, or FISH. Alternatively, if the technique is performed with a probe that requires subsequent visualization with a chemical reaction that produces a colored chemical at the site of the probe (a chromogen), the technique is referred to as Chromogenic In Situ Hybridization, or CISH.

In contrast to other techniques, such as polymerase chain reaction (PCR), which is also used to assess cells for genetic alterations, ISH has the advantage of keeping cells intact, allowing one to determine which specific cells have the abnormality. An additional advantage that ISH has over some genetic techniques is that it can sensitively detect alterations that occur in only a small subset of cells analyzed. This point is quite important, for example, in oncology because tumors tend to be quite heterogeneous, and the alteration that is being assessed for may be present in only a small percentage of the cells. Other molecular techniques in which the tissue is homogenized before analysis often cannot detect these alterations because of dilution by cells not carrying the alteration (e.g., by benign stromal or inflammatory cells that are invariably present in tumors).

K.C. Halling
Molecular Cytology and Imaging Laboratory, Department of Laboratory Medicine and Pathology, Mayo Clinic, Rochester, MN, USA

A.J. Wendel
Department of Laboratory Medicine and Pathology, Mayo Clinic, Rochester, MN, USA

In Situ Hybridization: General Principles

To understand how in situ hybridization works, one should be familiar with (1) DNA and RNA composition and structure, (2) principles of base pairing, (3) denaturation and hybridization, and (4) factors that influence denaturation and hybridization. It is not possible to cover these topics extensively; however, the following paragraphs briefly touch on the most important aspects of these topics.

DNA Composition and Structure

DNA and RNA are each composed of four nucleotides that form long chains. The nucleotides that comprise DNA contain one of four bases: adenine, cytosine, guanine, or thymine (Fig. 12.1). RNA is composed of nucleotides that also contain the same nucleotides except that uracil substitutes for thymine. Another important difference between DNA and RNA is that DNA is double stranded and RNA is single stranded.

Principles of Base Pairing

To comprehend hybridization, it is essential to understand how base pairing occurs in double-stranded DNA or DNA-RNA hybrids. In double-stranded DNA, nucleotides with an adenine (A) or thymine (T) pair with one another (referred to as "base pairing") through two hydrogen bonds whereas guanine (G) and cytosine (C) pair with one another through three hydrogen bonds (see Fig. 12.1). The only difference with DNA-RNA hybrids is that uracil (U) rather than thymine

P.T. Cagle and T.C. Allen (eds.), *Basic Concepts of Molecular Pathology,* Molecular Pathology Library 2,
DOI: 10.1007/978-0-387-89626-7_12, © Springer Science+Business Media, LLC 2009

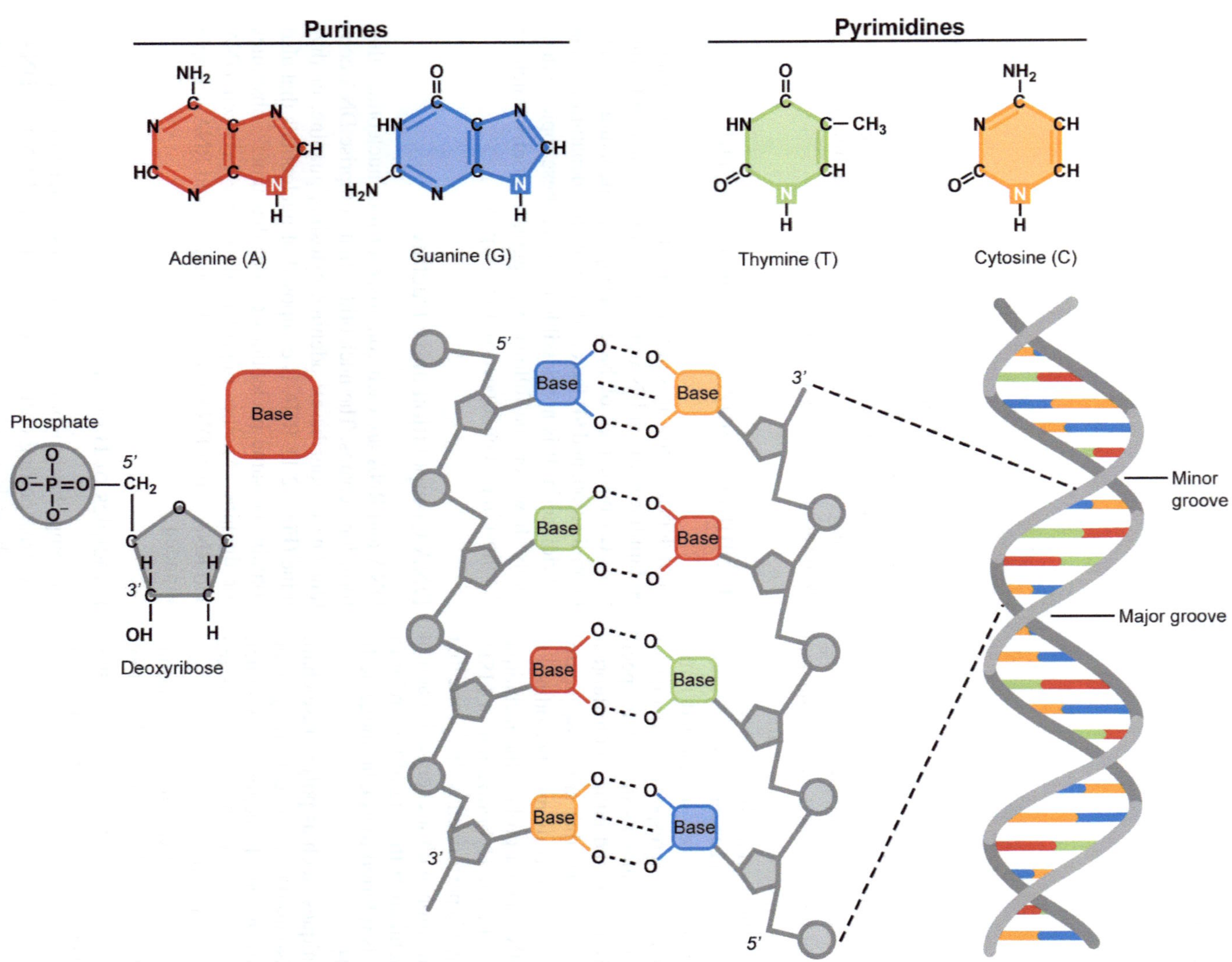

FIG. 12.1. DNA structure. Nucleotides are the building blocks of both DNA and RNA. Nucelotides are composed of the sugar deoxyribose (DNA) or ribose (RNA), a phosphate group, and a base. The four bases found in DNA are the purines guanine and adenine, and the pyrimidines cytosine and thymine. Uracil substitutes for thymine in RNA. Single-stranded DNA is a polymer of nucleotides. The 5′-end of a single-stranded DNA molecule is the end that has a free phosphate group and the 3′-end is the end that has a free 3′-OH group. Double-stranded DNA is composed of two single-stranded DNA molecules that run in opposite directions. In double-stranded DNA, guanine pairs with cytosine through three hydrogen bonds and adenine pairs with thymine through two hydrogen bonds.

base pairs with adenine. It is the pairing of A's to T's (or U's in the case of RNA) and of C's to G's that keeps the two strands of DNA or the two strands of a DNA-RNA hybrid paired to one another. Because GC base pairs have three hydrogen bonds, they are stronger and require more energy to break than AT base pairs. If two strands of DNA have bases that pair perfectly, that is, every A, C, G, or T on one strand is matched with a T, G, C, or A, respectively, on the opposite strand, then the two strands are said to be 100% complementary.

Denaturation, Renaturation, and Influencing Factors

Denaturation refers to the process of making double-stranded DNA (or DNA-RNA hybrids) single stranded. Denaturation is also sometimes referred to as "melting" the DNA. Denaturation is brought about by breaking the hydrogen bonds that hold the two strands of DNA together; this is most commonly achieved by applying heat (i.e., raising the temperature of the sample). In addition, certain chemicals such as formamide can be used to promote denaturation. Because G/C base pairs are stronger than A/T or A/U base pairs, double-stranded DNA fragments (or DNA-RNA hybrids) with a higher percentage of G's and C's require more heat (i.e., higher temperatures) to denature than will double-stranded DNA fragments (or DNA-RNA hybrids) with a higher percentage of A's and T's. Renaturation refers to the process of bringing two complementary strands of DNA (or complementary DNA and RNA) back together.

Hybridization occurs when two strands of complementary single-stranded DNA or RNA molecules "stick to" one another. The greater the complementarity of the two single-stranded molecules, the more likely that two strands will hybridize to one another. In addition, due to stronger base pairing, DNA-DNA or DNA-RNA hybrids that are 100% complementary will be more stable and require more energy to break apart than those which are less complementary. Two strands that are less complementary (e.g., only 95% complementary) can hybridize to one another under certain conditions, which are referred to as "less stringent" conditions. The tendency for a DNA or RNA probe to stick to sequences that are not complementary is referred to as nonspecific hybridization. Nonspecific hybridization can be prevented by maintaining conditions, referred to as "stringent conditions," that favor hybridization of the most complementary sequences but prevent the hybridization of less-complementary sequences. Stringent conditions are generally achieved by keeping the temperature at the highest possible temperature that allows the most complementary single-stranded DNA or RNA probe to anneal to its completely complementary target without allowing it to anneal to targets that are less complementary. Depending on how similar the noncomplementary sequence is to the complementary sequence, the difference between stringent and nonstringent conditions may be just a degree or two. For example, 37°C may favor the annealing of a probe that is 100% complementary to its target sequence while preventing a probe that is only 95% complementary from binding to its target sequence. However, dropping the hybridization temperature just a couple of degrees, to 35°C, may favor the binding of both probes to the target sequence.

The melting temperature of double-stranded DNA in solution is approximated by the following formula:

$$Tm = 81.5 + 16.6 \log (Na+) + 0.41(\%\, GC) - 0.63 (\%\, formamide) - (300 + 2000(Na+)/N$$

where T_m is the melting temperature in degrees centigrade, Na^+ is the molar concentration of sodium ions, %GC is the percentage of GC base pairs in the hybridized molecules, %formamide is the percentage of formamide (volume/volume), and N is the length in bases of the hybrid.[1] This formula may appear somewhat daunting for those of us who are less mathematically inclined, but conceptually, the important points to note are that increasing the salt concentration inhibits melting and increasing formamide promotes melting; this is because high salt concentrations favor hydrogen bonding of the two strands of a DNA-DNA or DNA-RNA hybrid while formamide helps break those hydrogen bonds. Similarly, it is important to note that under identical conditions, shorter fragments of DNA melt at lower temperatures than longer fragments of DNA. A familiarity with this formula can be helpful when trying to troubleshoot problems with nonspecific or incomplete hybridization.

Probes

Types of Probes

All in situ hybridization techniques utilize DNA or RNA probes. These probes are designed to hybridize to specific target sequences of interest such as genes that have been implicated in inherited diseases or cancer as well as to microorganisms of various types. DNA probes that are commonly used for genetic and/or oncologic applications have been categorized as chromosome enumeration probes (CEP), locus-specific indicator (LSI) probes, telomeric probes, and chromosome paints.

CEP Probes

CEP probes hybridize to repetitive DNA sequences found near the centromeres of chromosomes, which are referred to as alpha-satellite DNA. These regions are composed of ~171-bp sequences that are tandemly repeated hundreds to thousands of times and span approximately

250,000–5,000,000 bases.[2] The repeat regions of the different chromosomal centromeres exhibit substantial sequence divergence of approximately 20–40%.[2] Because of this, probes specific for each of the centromeres are available for most, but not all, chromosomes. A few chromosomes, for example, chromosomes 13 and 21, 14 and 22, and 5 and 19, have alpha-satellite repeat sequences that are too similar to allow one to distinguish these chromosomes from one another as a consequence of cross-hybridization.

CEP probes are used to enumerate the number of copies of a given chromosome in a cell. CEP probes are able to enumerate chromosome copy number because if the centromere of a chromosome is lost, the whole chromosome will generally be lost. One advantage of CEP probes is that because they hybridize to sequences that have high copy number, they provide strong (i.e., bright) signals. In addition, because these regions are tightly compacted in the heterochromatic regions of the chromosome, the signals provided with CEP probes are generally tight ("crisp") rather than diffuse.

Locus-Specific Probes

Locus-specific probes hybridize to unique sequences (i.e., nonrepetitive DNA sequences) and are generally used to determine if specific genes are amplified (e.g., *HER2*), deleted (e.g., *p53* or *p16*), or translocated (e.g., *BCR/ABL* translocation). These probes typically hybridize to a region that ranges from 40 to 500 kilobases (kb). Probes that hybridize to regions smaller than 40 kb often produce weak signals that can be difficult to see. Probes larger than this can produce diffuse signals that are difficult to distinguish as single spots.

Telomeric Probes

Telomeric probes are not actually probes to telomeric sequences but to unique DNA sequences found very near the telomeres (so called subtelomeric probes).[3,4] Unique telomeric probes are available for 41 of the 46 chromosomal telomeres. The only telomeres for which there are not probes are for the p arms of the acrocentric chromosomes 13, 14, 15, 21, and 22. However, these regions are composed primarily of ribosomal DNA, and deletions or duplications of these regions are not generally thought to have clinical significance.

Chromosomal Paints

Chromosomal paints are mixtures of probes that hybridize to the entire length of one or more chromosomes. The probes that comprise chromosomal paints are generally prepared by isolating individual chromosomes by flow cytometry and then performing PCR amplification with degenerate oligonucleotide primers. The fluorescently labeled probes generated from this can then be used as a "paint" that highlights the entire chromosome homogeneously along its length. Chromosomal paints can be used to identify specific chromosomes in a metaphase spread. This can be particularly helpful when standard karyotyping is unable to identify a chromosome (e.g., marker chromosomes).

Probe Preparation

Probes are generally prepared from fragments of DNA that have been cloned into bacterial (BAC), P1 (PAC), or yeast (YAC) artificial chromosomes. Probes that are used for FISH can either be directly or indirectly labeled with a fluorophore.[5] A fluorophore is a molecule that fluoresces when excited by light of specific wavelength. There are a wide variety of fluorophores that can be used for FISH.[5] Two of the more commonly used fluorophores are fluorescein isothiocyanate (FITC) and Texas Red, which fluoresce green and red, respectively.

Direct Labeled Probes

Direct labeled probes are prepared by incorporating a fluorophore-labeled nucleotide into the probe, usually by nick translation or random priming. Because the fluorescently labeled probe is bound to its cellular target in a single hybridization step, no further processing is required to visualize the probe. Directly labeled CEP and LSI probes are commercially available singly or as probe mixtures of up to four different probes and are generally labeled with red, green, aqua, or yellow fluorophores.

Indirect Labeled Probes

Indirectly labeled probes contain nucleotides that have covalently attached reporter molecules such as biotin or digoxigenin. These probes require an additional step after hybridization in which the probe is generally visualized by applying fluorophore-labeled avidin or fluorophore-labeled antidigoxigenin in a second step. A significant disadvantage of the use of indirectly labeled probes is that it adds extra steps to the procedure. However, potential advantages of indirectly labeled probes include that it is often possible to obtain stronger signals with indirectly labeled probes because of the ability to achieve greater signal amplification and that indirect labeling allows for greater versatility in probe use as one can use this technique to visualize any probe that has been labeled with biotinylated 2'-deoxyuridine 5'-triphosphate (dUTP) or digoxigenin.

"In-House" Developed Probes

Most diagnostic laboratories utilize commercially available probes. However, a lab may want to prepare their own "in-house" probe if a probe is not commercially available. If a lab prepares its own probes for diagnostic use, it is important to extensively validate the probe before clinical implementation to make sure that it hybridizes to the intended target and provides the expected results on normal and abnormal specimens. Wiktor et al. have recently provided recommendations for the validation of probes for clinical use.[6]

Fluorescence In Situ Hybridization (FISH)

Overview of Steps Involved in FISH

The steps involved in performing FISH are as follows: (1) obtain specimen for FISH, (2) prepare specimen for hybridization ("prehybridization"), (3) hybridize probes to target DNA in specimen, (4) remove nonspecifically bound probe by washing, and (5) assess signals in specimen using fluorescence microscopy (Fig. 12.2). Each of these steps is described in more detail below.

Specimens

FISH can be performed on a wide range of specimens including peripheral blood, paraffin-embedded tissue, or cytology specimens (e.g., urine, sputum, or endoscopic brushings). Frozen tissue will not work well because freezing results in intracellular ice crystals that disrupt the morphology of the cells. The fixative that gives the best FISH results for paraffin-embedded tissue is neutral buffered formalin. However, tissue that has been left in formalin too long (>48 h) before paraffin embedding will not provide

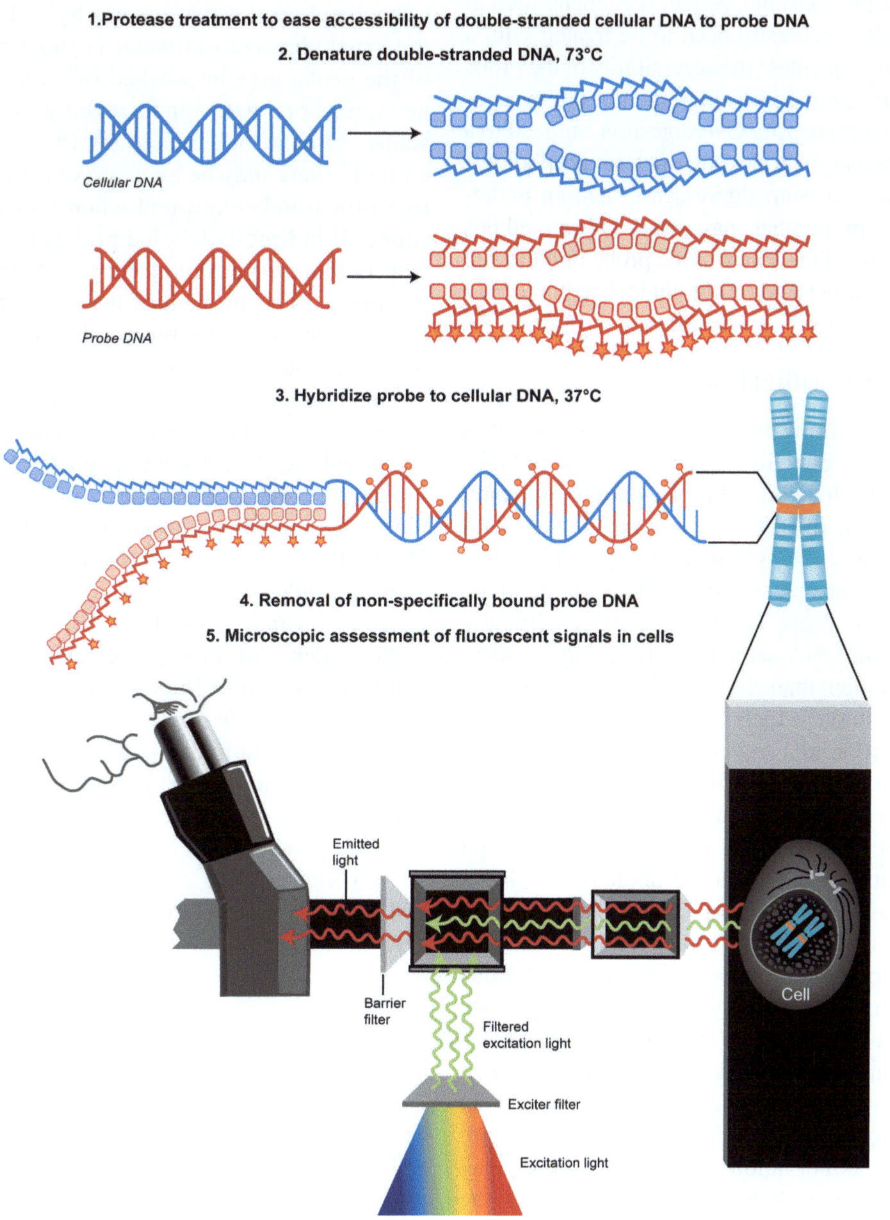

FIG. 12.2. Steps involved in performing fluorescent in situ hybridization (FISH).

good FISH results.[7] FISH on specimens that have been fixed in B5 fixative, which is often used for hematological specimens, may require isolation of nuclei before FISH to obtain good results.[8] Some fixatives such as Prefer work poorly if at all for FISH.[9]

Prehybridization

The goal of this step is to prepare the cells in the specimen so that the probe can efficiently hybridize to its cellular DNA target but do so without significantly disrupting the morphology of the cells. Detailed protocols for prehybridization and hybridization can be found elsewhere.[10,11] The pretreatment required for different specimen types (e.g., fresh cells vs. paraffin-embedded tissue) is similar but frequently has some important differences. For example, certain specimens such as paraffin-embedded tissue generally need to be treated with a protease such as pepsin to increase the accessibility of the probe DNA to its nuclear DNA target. It is important to not over- or underdigest the tissue with pepsin. Overdigestion can lead to a decrease in signal intensity and destroy nuclear morphology. Underdigestion can lead to auto-fluorescence, and an underestimation of signal copy number may result. Other methods that have been used to facilitate entry of the probe into the cell include treatment with dilute acid or non-ionic detergents.[12]

Denaturation and Hybridization

This step involves denaturing the probe and cellular DNA and then allowing the probe to hybridize to its cellular DNA target. The probe hybridization solution contains not only the probe DNA but a type of DNA that is referred to as COT-DNA. COT-DNA is DNA that hybridizes to highly repetitive DNA sequences which are present throughout the genome. Without COT-DNA, the probe DNA would nonspecifically bind to these repetitive DNA sequences, resulting in many nonspecific signals rather than the specific signals that are desired. In one of the more widely utilized FISH hybridization protocols, the probe and target DNA are codenatured at about 73°C for about 3 min in 50% formamide. Formamide is added to lower the temperature at which the probe and cellular DNA melt. This step is important because high temperatures can destroy the cellular morphology that one is attempting to maintain. The temperature is then lowered to about 37°C to allow the probe DNA to hybridize to its specific target. The hybridization temperature is less fastidious for FISH assays than for PCR assays because FISH probes are much longer than "PCR probes." Hybridization is generally allowed to occur over about 4–12 h. Factors that can influence the efficiency and specificity of the hybridization include the probe sequence and hybridization temperature. If the probe sequence is not unique, additional signals may present that represent cross-hybridizing sequences. If the hybridization temperature is too high, the signals may be weak, and if the hybridization temperature is too low, there may be an abundance of nonspecific background signal.

Removing Nonspecifically Bound Probe

Following hybridization, the slide is washed with a specific washing solution. The goal of this step is to remove any probe that is not specifically bound to the desired target without removing the probe which has bound to the desired target. This step is generally done by washing the slide in a solution that is heated to 73°C and contains 0.4× SSC/0.3% NP40 (nonformalin-fixed samples) or 2.0× SSC/0.3% NP40 (formalin-fixed samples). A higher salt concentration (less-stringent wash) is used for paraffin-embedded specimens because the probe-target hybrids are less strong for paraffin-embedded specimens than for non-formalin-fixed samples; this may result from the shorter length of target DNA sequences in paraffin-embedded tissue or the persistence of protein in the hybridized region that causes hybrid destabilization.[10] If the posthybridization wash is too stringent, all or most of the probe may be washed off, and there may be either no signals or weak signals when viewed under the microscope. On the other hand, if the wash is not stringent enough, there may be an excessive amount of nonspecific hybridization background when viewed under the microscope. If indirectly labeled probes are used, an additional step is required at this point to attach the fluorophore to the probe. The procedure used will depend on the type of reporter molecule (biotin, digoxigenin, other) that has been previously chosen.

Finally, approximately 10 μl of a solution that contains compounds known as DAPI and "Antifade" is placed on the slide, and the slide is cover-slipped. DAPI (4′-6-diamidino-2-phenylindole) is a nuclear counterstain that fluoresces a blue-gray hue when viewed with a fluorescent microscope. DAPI allows one to see the nuclei. Without DAPI, the FISH signals would appear to be free floating in a dark background. Antifade is used to inhibit photobleaching, which is the tendency of fluorophores to fade with exposure to light and heat.[13] Because fluorophores are susceptible to photobleaching, it is best to store hybridized slides and probe in a refrigerator and in the dark. In addition, when viewing the slide on the microscope, it is important to remove the slide from the path of light by closing the condenser when not viewing. If a slide has lost its fluorescence, it can frequently be rehybridized with good results because the target DNA is still intact. We have successfully rehybridized slides that are several years old.

Fluorescence Microscopy

Basic Principles

Following hybridization, FISH signals are assessed with a fluorescence microscope (usually an epi-fluorescence microscope) equipped with the appropriate filters necessary to see the probe signals. A detailed discussion of fluorescence microscopy is beyond the scope of this chapter, but there are several excellent reviews on this subject.[14]

Nonetheless, it is important to understand at least a few of the basic aspects of how fluorescence microscopy works. As already noted, a fluorophore is a molecule that absorbs light of a certain wavelength (actually a certain range of wavelengths) and then reemits that energy as light of longer wavelengths. For example, as shown in Fig. 12.3, the fluorophore that is referred to as Spectrum Green absorbs light that ranges from about 470 to 510 nm and then reemits it (i.e., fluoresces) as light of higher wavelengths that range from about 500 to 550 nm. The goal of FISH is to cause the fluorophore-labeled DNA or RNA probe to fluoresce and highlight the target of interest in the specimen. To do this, the fluorescence microscope must have a light source that is capable of exciting the fluorophore. Mercury or xenon arc lamps are typically relied upon as light sources for fluorescence microscopes. These lamps emit light of wavelengths that can be absorbed by the fluorophores. A combination of filters (excitation, beam splitting interference, and barrier) select an appropriate wavelength of light to excite the fluorophore and ensure that the main light that reaches the observer's eye is the desired fluorescence wavelength. These filters are generally incorporated into a single housing device that is referred to as a filter cube. The appropriate combination of filters depends on the fluorophore(s) that is (are) being utilized. Cubes are available for most fluorophores. Most cubes contain filters that allow one to look at only one probe at a time. However, dual-pass and triple-pass filters that allow one to see red and green signals or the red signal, green signals, and DAPI counterstain simultaneously are available and very useful.

It is frequently necessary to take a picture of the cells and their fluorescent signals. This requires the use of imaging systems and software that enable one to capture the different colored probe signals in a single composite image. When a multicolor probe set has been used, it may be particularly difficult to take a picture that shows all the signals when using a normal camera because the signals are often present in different planes. This limitation can be overcome with a device known as a Z-stack camera that allows one to easily capture all the signals in a cell as a single image.

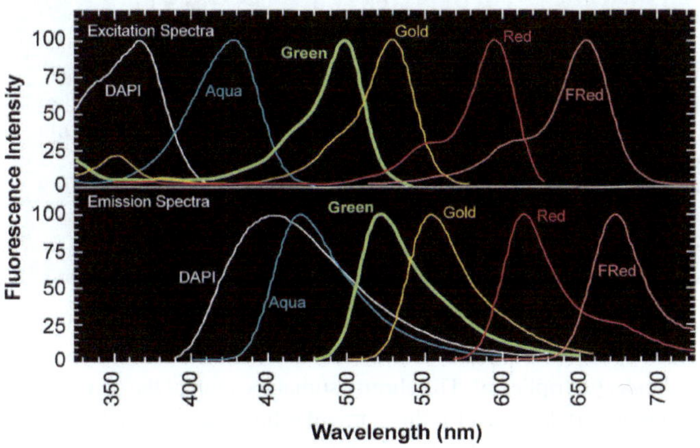

FIG. 12.3. Excitation (absorption) and emission spectra for more commonly used fluorophores.

Chromogenic In Situ Hybridization (CISH)

CISH is another type of in situ hybridization that is very similar to FISH except that the probe is visualized by a chromogenic reaction after it has hybridized to its target rather than being visualized as a fluorescent signal. There are several ways in which one can perform CISH. One method incorporates biotinylated-dUTP into the probe. After hybridization of the probe to its target, horseradish peroxidase-labeled streptavidin is added to the slide. The streptavidin attaches to the biotin. Diaminobenzidine (DAB) is then added, and the horseradish peroxidase converts the DAB to a brown chromogen at the site of the probe. Another CISH method utilizes digoxigenin-labeled probes instead of biotin-labeled probes but also utilizes peroxidase and DAB to visualize the probe.

CISH has several advantages over FISH. A major advantage that CISH has over FISH is that CISH slides can be viewed with a light microscope, and the tissue architecture is much easier to discern with CISH; this makes it easier to correlate molecular alterations with histological or cytological features with CISH than with FISH. In addition, CISH can be performed in conjunction with immunohistochemistry to assess for correlations between chromosomal alterations and protein expression.[15] A final important advantage of CISH is that the permanent slides obtained after CISH analysis do not fade and thus are suitable for long-term archiving.

Disadvantages of CISH relative to FISH include that it can only easily identify a single target at a time, and the signals produced with CISH are less discrete than those produced by FISH, which makes signal enumeration less precise than that which is possible by FISH. The inability of CISH to identify more than one target may or may not be important depending on the application. For example, CISH is frequently used to identify viruses such as Epstein-Barr virus or cytomegalovirus in lung tissue sections (Fig. 12.4). For this particular application, the inability of CISH to identify more than a single target is not a limitation.

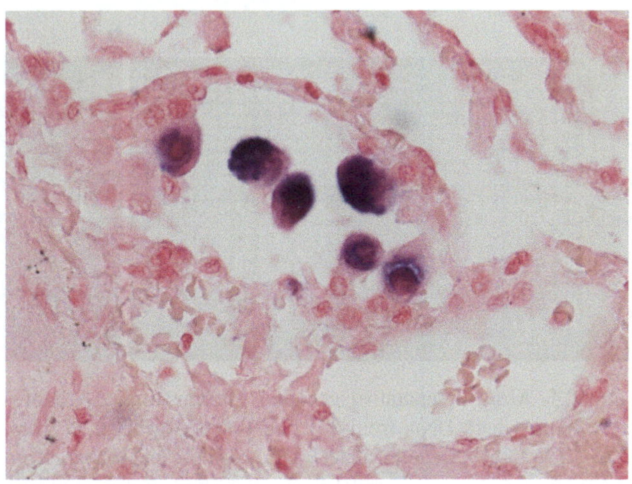

FIG. 12.4. Identification of cytomegalovirus (CMV) in lung tissue section utilizing chromogenic in situ hybridization (CISH). The *purple-stained nuclei* contain CMV.

However, the inability to detect more than one target at a time could be a limitation for some applications. For example, one commercially available FISH assay for HER2 amplification utilizes two probes, a green probe to the chromosome 17 centromere and a red probe to the HER2 gene. With this FISH assay, a HER2/CEP17 signal ratio >2.0 is considered evidence of HER2 amplification and justification for Herceptin therapy. The rationale for including the CEP17 probe in this probe set was that some tumors may exhibit gains (i.e., more than two copies) of the HER2 gene because of polysomy (i.e., extra copies) of chromosome 17 in the tumor but not because of true amplification of the HER2 gene. It is not fully known at this time whether the absolute HER2 copy number by itself is predictive of response to Herceptin or whether knowing the ratio of HER2/CEP17 is important. However, if knowing the ratio is important, that information would not be available with the CISH assay. Other probe sets, for example, a FISH probe set for bladder cancer detection, utilize as many as four different probes in a single cocktail to achieve high sensitivities for bladder tumor detection.[16] Studies have shown that the use of a multiprobe FISH cocktail provides higher sensitivity than that which is achievable with one or two probe sets.

Other Techniques that Utilize In Situ Hybridization

Comparative Genomic Hybridization (CGH)

CGH is a powerful technique that can, similar to conventional karyotyping, be used to analyze the entire genome of cells for chromosomal abnormalities.[17] CGH is particularly useful for assessing solid tumors for chromosomal abnormalities. As first described, CGH is preformed by isolating DNA from tumor tissue and labeling it with a green fluorophore, then isolating DNA from normal tissue and labeling it with a red fluorophore.[18] Equal amounts of the green-labeled tumor DNA and red-labeled normal DNA are then cohybridized to a normal metaphase spread (i.e., a karyotype without any abnormalities). If there are no chromosomal abnormalities in the tumor tissue, then equal amounts of green and red probe will hybridize to the chromosomes in the normal metaphase spread, and the color of all the chromosomes in the metaphase spread will be yellow (because equal amounts of red and green make yellow). However, if there are, for example, three copies of chromosome 1 in the tumor, then there will be more green-labeled chromosome 1 DNA than red-labeled chromosome 1 DNA, and chromosome 1 in the metaphase spread will appear green. Likewise, areas in the metaphase spread that appear green represent chromosomal areas that have been gained in the tumor while areas that appear red represent chromosomal areas which have been deleted in the tumor. In this fashion, CGH can be used to identify many of the chromosomal abnormalities present in a tumor. CGH is generally not performed with metaphase spreads anymore but instead with microarrays of DNA clones that are largely representative of all the loci along the length of the 24 chromosomes.[19] This technique, which is referred to as array CGH, has the potential for higher resolution and greater reproducibility than can be obtained with metaphase spreads. A representative example of a CGH array is shown in Fig. 12.5.

CGH has several advantages over conventional karyotyping, including (1) the fact that the tumor be grown in culture before analysis, (2) the cells do not have to be in metaphase to determine the types of chromosomal abnormalities present, and (3) it has higher resolution, typically down to about 1 megabase, whereas karyotyping has a resolution down to about 5 megabases. The only disadvantage of CGH relative to conventional karyotyping is that it cannot detect balanced chromosomal alterations (e.g., recip-

a b

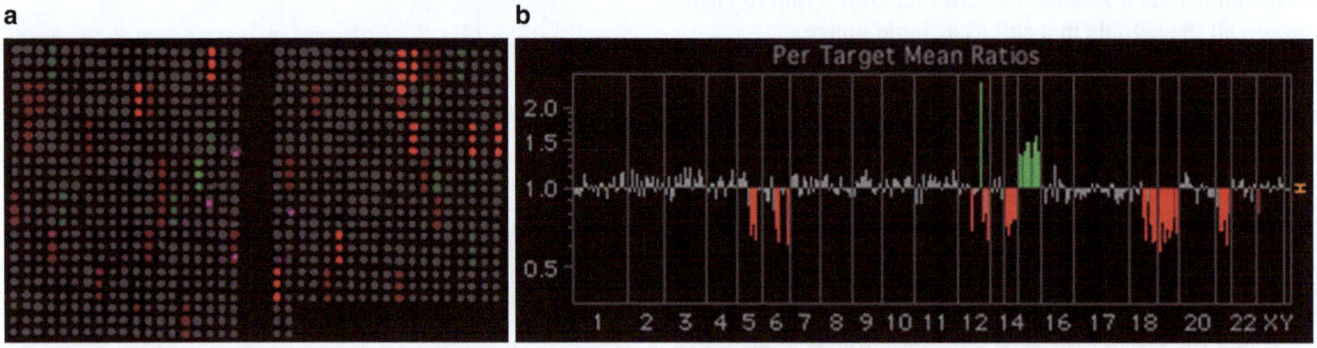

Fig. 12.5. Array comparative genomic hybridization. A genomic array spotted with DNA from 287 targets in the genome is shown in (**a**). The targets include subtelomeric regions, oncogenes, tumor suppressor genes, and microdeletion loci. Each target is spotted in triplicate. *Green-labeled* tumor DNA from a single tumor (a bile duct tumor) and *red-labeled* normal DNA were cohybridized to the DNA spots on the array. Gene amplifications within the tumor appear as *green spots* (in triplicate) and deletions within the tumor appear as *red spots* (in triplicate). The chromosomal location of the gene amplifications and deletions for this particular tumor are illustrated in (**b**). The tumor shows deletions of 5q, 6q, 12q, 14q, 18q, 19, and 21, gain of 15, and amplification of the MDM2 gene at 12q14.3-q15.

rocal translocations). Although CGH is generally not used clinically, it has been a powerful research tool for identifying the most common chromosomal abnormalities that are present in various tumor types, for example, lung cancer.[20–37]

Multicolor Whole Chromosome Painting (M-FISH and Sky)

Spectral karyotyping (SKY) and multicolor FISH (M-FISH) are highly sophisticated techniques that allow one to generate a color-coded karyotype.[38,39] In other words, these techniques produce a karyotype in which each chromosome has its own unique color. The generation of chromosomal paints for each chromosome is accomplished with combinatorial labeling of the chromosome specific probes with just five fluorophores. SKY and M-FISH can be helpful in identifying chromosomes that are difficult to identify, such as marker chromosomes or chromosomes which are involved in complex rearrangements. This aspect can be particularly helpful for solid tumor genetics because solid tumors often have very complex karyotypes that are difficult to elucidate (Figs. 12.6 and 12.7).

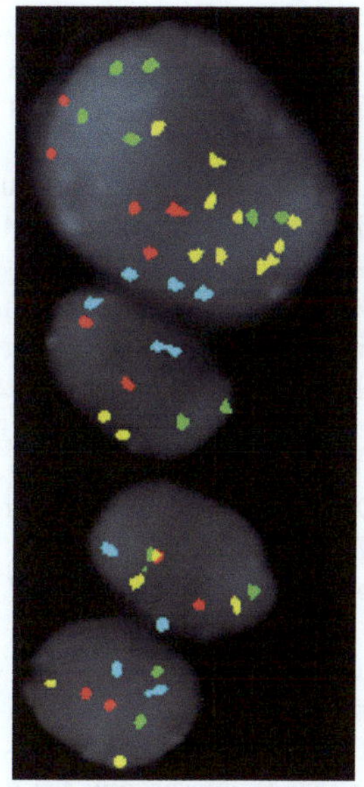

FIG. 12.7. FISH detection of lung cancer in bronchoscopic brushing specimens using the LAVysion probe set: 5p15 (*green*), CEP6 (*aqua*), 7p12 (*red*), and 8q24 (*gold*). Note that the two normal cells on the *left* show only two copies for each of the four probes whereas the malignant cell on the *right* shows extra copies for all four probes.

Summary

ISH is a technique that utilizes nucleic acid (DNA or RNA) probes to assess intact cells for various types of alterations. To understand how in situ hybridization works, one should be familiar with DNA and RNA composition and structure, principles of base pairing, and factors that influence denaturation and hybridization. Techniques that utilize ISH include FISH, CISH, comparative genomic hybridization, spectral karyotyping, and multicolor FISH. These techniques have found many important applications in oncology, genetics, and infectious disease.

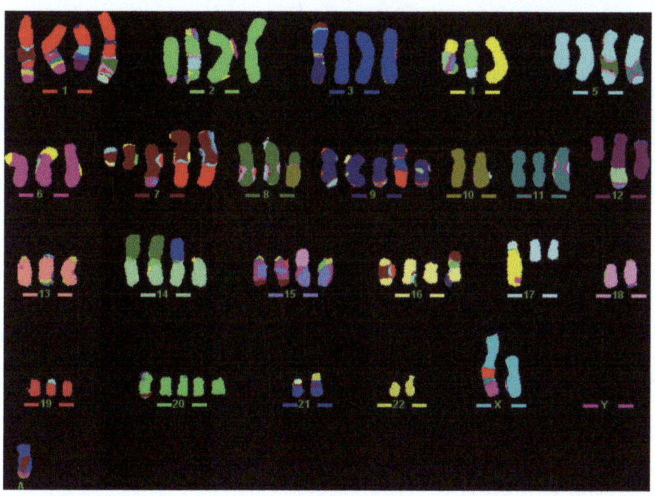

FIG. 12.6. Multicolor FISH (M-FISH) analysis of a karyotype from a pediatric osteosarcoma. M-FISH reveals numerous complex rearrangements in this tumor. For example, the *arrow* points to a complex rearrangement involving the X chromosome (*light aqua*), chromosome 1 (*red*), chromosome 11 (*dark aqua*), and chromosome 6 (*dark pink*).

References

1. Beltz GA, Jacobs KA, Eickbush TH, et al. Isolation of multigene families and determination of homologies by filter hybridization methods. *Methods Enzymol.* 1983;100:266–285.
2. Lee C, Wevrick R, Fisher RB, et al. Human centromeric DNAs. *Hum Genet.* 1997;100:291–304.
3. Knight SJ, Flint J. Perfect endings: a review of subtelomeric probes and their use in clinical diagnosis. *J Med Genet.* 2000;37:401–409.
4. Moog U, Arens YH, van Lent-Albrechts JC, et al. Subtelomeric chromosome aberrations: still a lot to learn. *Clin Genet.* 2005;68:397–407.
5 Morrison LE, Ramakrishnan R, Ruffalo TM, et al. Labeling fluorescence in situ hybridization probes for genomic targets. In: Fan Y-S, ed. *Molecular Cytogenetics: Protocols and Applications.* Vol 204. Totowa: Humana Press; 2002:21–40.
6. Wiktor AE, Van Dyke DL, Stupca PJ, et al. Preclinical validation of fluorescence in situ hybridization assays for clinical practice. *Genet Med.* 2006;8:16–23.
7. Petersen BL, Sorensen MC, Pedersen S, et al. Fluorescence in situ hybridization on formalin-fixed and paraffin-embedded tissue: optimizing the method. *Appl Immunohistochem Mol Morphol.* 2004;12:259–265.

8. Schurter MJ, LeBrun DP, Harrison KJ. Improved technique for fluorescence in situ hybridisation analysis of isolated nuclei from archival, B5 or formalin fixed, paraffin wax embedded tissue. *Mol Pathol.* 2002;55:121–124.
9. Tubbs RR, Hsi ED, Hicks D, et al. Molecular pathology testing of tissues fixed in prefer solution. *Am J Surg Pathol.* 2004;28:417–419.
10. Van Stedum S, King W. Basic FISH techniques and troubles hooting. In: Fan Y-S, ed. *Molecular Cytogenetics: Protocols and Applications*. Vol 204. Totowa: Humana Press; 2002:51–63.
11. Solovei I, Walter J, Cremer C, et al. FISH on three-dimensionally preserved nuclei. In: Beatty B, Mai S, Squire J, eds. *FISH: A Practical Approach.* Oxford: Oxford University Press; 2002:119–158.
12. McNicol AM, Farquharson MA. In situ hybridization and its diagnostic applications in pathology. *J Pathol.* 1997;182: 250–261.
13. Florijn RJ, Bonnet J, Vrolijk H, et al. Effect of chromatic errors in microscopy on the visualization of multi-color fluorescence in situ hybridization. *Cytometry.* 1996;23:8–14.
14. Tanke HJ. Fluorescence microscopy for quantitative fluorescence in situ hybridization analysis. In: Andreeff M, Pinkel D, eds. *Introduction to Fluorescence In Situ Hybridization: Principles and Clinical Applications*. New York: Wiley-Liss; 1999:33–52.
15. Downs-Kelly E, Pettay J, Hicks D, et al. Analytical validation and interobserver reproducibility of EnzMet GenePro: a second-generation bright-field metallography assay for concomitant detection of HER2 gene status and protein expression in invasive carcinoma of the breast. *Am J Surg Pathol.* 2005;29:1505–1511.
16. Sokolova IA, Halling KC, Jenkins RB, et al. The development of a multitarget, multicolor fluorescence in situ hybridization assay for the detection of urothelial carcinoma in urine. *J Mol Diagn.* 2000;2:116–123.
17. Lapierre JM, Tachdjian G. Detection of chromosomal abnormalities by comparative genomic hybridization. *Curr Opin Obstet Gynecol.* 2005;17:171–177.
18. Kallioniemi OP, Kallioniemi A, Sudar D, et al. Comparative genomic hybridization: a rapid new method for detecting and mapping DNA amplification in tumors. *Semin Cancer Biol.* 1993;4:41–46.
19. Vissers LE, Veltman JA, van Kessel AG, et al. Identification of disease genes by whole genome CGH arrays. *Hum Mol Genet.* 2005;14(Spec. No. 2):R215–R223.
20. Kim TM, Yim SH, Lee JS, et al. Genome-wide screening of genomic alterations and their clinicopathologic implications in non-small cell lung cancers. *Clin Cancer Res.* 2005;11: 8235–8242.
21. Peng WX, Shibata T, Katoh H, et al. Array-based comparative genomic hybridization analysis of high-grade neuroendocrine tumors of the lung. *Cancer Sci.* 2005;96:661–667.
22. Shibata T, Uryu S, Kokubu A, et al. Genetic classification of lung adenocarcinoma based on array-based comparative genomic hybridization analysis: its association with clinicopathologic features. *Clin Cancer Res.* 2005;11:6177–6185.
23. Tonon G, Wong KK, Maulik G, et al. High-resolution genomic profiles of human lung cancer. *Proc Natl Acad Sci USA.* 2005; 102:9625–9630.
24. Petersen I. Comparative genomic hybridization of human lung cancer. *Methods Mol Med.* 2003;75:209–237.
25. Ashman JN, Brigham J, Cowen ME, et al. Chromosomal alterations in small cell lung cancer revealed by multicolour fluorescence in situ hybridization. *Int J Cancer.* 2002;102:230–236.
26. Chujo M, Noguchi T, Miura T, et al. Comparative genomic hybridization analysis detected frequent overrepresentation of chromosome 3q in squamous cell carcinoma of the lung. *Lung Cancer.* 2002;38:23–29.
27. Massion PP, Kuo WL, Stokoe D, et al. Genomic copy number analysis of non-small cell lung cancer using array comparative genomic hybridization: implications of the phosphatidylinositol 3-kinase pathway. *Cancer Res.* 2002;62:3636–3640.
28. Lindstrom I, Nordling S, Nissen AM, et al. DNA copy number changes in lung adenocarcinoma in younger patients. *Mod Pathol.* 2002;15:372–378.
29. Goeze A, Schluns K, Wolf G, et al. Chromosomal imbalances of primary and metastatic lung adenocarcinomas. *J Pathol.* 2002;196:8–16.
30. Pei J, Balsara BR, Li W, et al. Genomic imbalances in human lung adenocarcinomas and squamous cell carcinomas. *Genes Chromosomes Cancer.* 2001;31:282–287.
31. Mertens F, Johansson B, Hoglund M, et al. Chromosomal imbalance maps of malignant solid tumors: a cytogenetic survey of 3185 neoplasms. *Cancer Res.* 1997;57:2765–2780.
32. Petersen I, Bujard M, Petersen S, et al. Patterns of chromosomal imbalances in adenocarcinoma and squamous cell carcinoma of the lung. *Cancer Res.* 1997;57:2331–2335.
33. Balsara BR, Sonoda G, du Manoir S, et al. Comparative genomic hybridization analysis detects frequent, often high-level, overrepresentation of DNA sequences at 3q, 5p, 7p, and 8q in human non-small cell lung carcinomas. *Cancer Res.* 1997;57:2116–2120.
34. Petersen I, Langreck H, Wolf G, et al. Small-cell lung cancer is characterized by a high incidence of deletions on chromosomes 3p, 4q, 5q, 10q, 13q and 17p. *Br J Cancer.* 1997;75:79–86.
35. Levin NA, Brzoska PM, Warnock ML, et al. Identification of novel regions of altered DNA copy number in small cell lung tumors. *Genes Chromosomes Cancer.* 1995;13:175–185.
36. Levin NA, Brzoska P, Gupta N, et al. Identification of frequent novel genetic alterations in small cell lung carcinoma. *Cancer Res.* 1994;54:5086–5091.
37. Ried T, Petersen I, Holtgreve-Grez H, et al. Mapping of multiple DNA gains and losses in primary small cell lung carcinomas by comparative genomic hybridization. *Cancer Res.* 1994;54: 1801–1806.
38. Schrock E, du Manoir S, Veldman T, et al. Multicolor spectral karyotyping of human chromosomes. *Science.* 1996;273:494–497.
39. Speicher MR, Gwyn Ballard S, Ward DC. Karyotyping human chromosomes by combinatorial multi-fluor FISH. *Nat Genet.* 1996;12:368–375.

13 Proteomics

Larry Fowler and Wieslaw Furmaga

"Omics" is a term designating a complete analysis of biological systems in which entire metabolic pathways are studied. "Omics" changes the methodological approach from small-scale research of one gene, one protein, or one metabolic reaction to the large comprehensive level of synthetic study in which the entire genome, protein composition, or metabolic pathway is studied simultaneously. The main breakthrough to the "omics era" was a confirmation of the complete human genome sequence. Entering this new research territory was possible because of an improvement in scientific instrumentation that has been able to perform high-sensitivity testing at high-throughput mode. The software-directed, "walk-away" DNA and protein microarray or highly sensitive, operator-friendly mass spectrometry have become a driving force for omics research.

These instruments become data factories that produce enormous amounts of information, transcending the capacity of the human brain. The need to store and analyze such a large amount of data has stimulated the advancement of new bioinformatics tools that contribute significantly to the development of "omics" disciplines. Despite its short history, this comprehensive methodology has proved its effectiveness by advancing our understanding of physiological and pathological processes, which brings hope for more accurate diagnosis and treatment of diseases.[1]

Genomics and Transcriptomics

In light of omics as a "method of thinking large," *genomics* can be defined as a comprehensive study of DNA structure and function. Genomics includes intensive efforts to determine the entire DNA sequence of organisms and fine-scale genetic mapping. The field also includes studies of intragenomic phenomena such as heterosis, epistasis, pleiotropy, and other interactions between loci and alleles within the genome.[2]

The analysis of gene expression has evolved into *transcriptomics*, an independent field with specific methodology and instrumentation. *Transcriptomics* examines gene expression on the mRNA level and describes the genes that are being actively expressed at any given time.[3]

The main shortcoming of both genomics and transcriptomics is their inability to predict phenotypic and functional consequences of mutations or single nucleotide polymorphisms. Only analysis of the proteins, on an effector level, is able to fill this gap and bring another "omics" into play – proteomics.

Proteomics

The aim of proteomics is to perform a systematic study of proteins and to present a complete explanation of their structural, functional, and regulatory roles in a particular biological system. Similarly to the other "omics" disciplines, *proteomics* is defined as a comprehensive study of the entire protein composition that is characteristic for a particular physiological or pathological state of the cell or whole organism. Technological and methodological progress has made it possible to go beyond a simple biochemical analysis of a single protein and to progress into an investigation of complex protein mixtures. In conjunction with the constantly growing field of bioinformatics, proteomics is able to explain physiological and pathological processes in both normal and diseased cells, and therefore it has the potential to become an essential component in laboratory diagnostics.

The first tool that assisted the evolution of protein biochemistry into proteomics was two-dimensional gel electrophoresis. Despite its imperfections, two-dimensional gel electrophoresis is still used for protein profiling and/or for quantitation of protein expression. Low sensitivity is a major limitation that makes this method unfeasible for analysis of posttranslational modifications and protein-protein interactions; however, rapid progression in mass spectrometry and protein microarray technology has offered solutions to the two-dimensional electrophoresis limitations. The new generation of user-friendly mass spectrometers is able to analyze a protein mixture with high sensitivity and specificity as well as to perform testing in a high-throughput mode on different

P.T. Cagle and T.C. Allen (eds.), *Basic Concepts of Molecular Pathology*, Molecular Pathology Library 2,
DOI: 10.1007/978-0-387-89626-7_13, © Springer Science+Business Media, LLC 2009

types of specimens (whole blood, serum, body fluids, lavages, and tissue and cellular extracts). The improvements in protein microarray technology and in affinity chromatography permit efficient isolation of protein complexes and study of protein-protein interactions. All these bring proteomics closer to being a useful diagnostic tool for any human disorders, and for hematological disease particularly.

Proteomics: A New Diagnostic Tool for Hematology

Genomics and transcriptomics measure gene expression and transcriptional activities of the cells. It has been widely recognized that the changes in protein expression cannot be predicted at the level of gene amplification or at the level of mRNA production.[4] For this reason, as well as because proteins are the true effectors that change the functional state of cells, it is the proteins which should be examined as the most pertinent markers for cellular function.[5] Performed on cytological or surgical specimens, proteomics shows a complete picture of cellular activity and, along with other omics, provides new information about physiology, pathology, and cellular reaction to treatment. If cytological or histological diagnostic material is not accessible, or if not a cellular activity but discovery of a new marker of disease is of particular interest, protein profiling on biological fluids such as blood, urine, bronchoalveolar lavage (BAL), or serosal transudate or exudate as well as synovial fluid may be performed.[6] The primary objective for protein profiling is the discovery of a protein or its posttranslational modification that is different in the pathological than in the physiological state. This profiling is accomplished by a comparison of an entire collection of proteins present in the specimen derived from healthy as well as diseased organisms, and it is named comparative proteomics.

The Ideal Proteomics

The ideal proteomics can be described as a method, or a series of methods, that can analyze an unmodified specimen in its entirety in both the control and testing groups and is able to detect 100% of the proteins and their modifications present in these sets. The results of proteomics experiments are submitted for an evaluation, which analyzes the entire body of data and recognizes the proteins that are different in testing compared to the control group (proteins of difference). These selected proteins are subsequently used as a characteristic signature for a particular state of a cell or an organism and become a basis for the explanation of biochemical pathways or for diagnosis of diseases.

This simple model of proteomics becomes complicated by what is referred to as *abundance*. There are approximately 35,000 human genes encoding proteins. Most of the proteins undergo posttranslational modifications, so that the actual number of functionally active proteins exceeds 1 million.[7] Therefore, if a particular metabolic process involves 0.1% of the total protein composition (proteome) to fulfill the proteomics requirement that the entire spectrum of proteins must be detected, the protein collection characteristic for this state exceeds 1,000 individual particles.[7]

The problem of abundance affects not only the laboratory (separation and identification) but also the analytical (data analysis) portion of the proteomics process. To efficiently separate and identify proteins, especially those at low concentration, that are present in the unprocessed sample, the highest possible sensitivity is desired which overcomes the "masking effect" created by highly concentrated particles. However, testing with high sensitivity instrumentation detects "artificial peptides" that are not naturally produced. Also, for the data analysis process, this abundance makes it more difficult to obtain sufficient statistical power to prove that the detected differences in protein composition between testing and control groups are systematic, not incidental.[7]

Shotgun and Protein Profiling

For diagnostic purposes, proteomics methods are not restricted to surgically removed bone marrow or lymphatic tissues, but they can also be used to analyze fine-needle aspirates, effusion fluids, urine, and serum. For all these specimens, proteomics provides a unique opportunity to evaluate almost the entire protein content. Because the amount of analyzed proteins is colossal, to overcome "abundance" effects the original specimen is fractionated and divided into portions containing proteins with similar physicochemical properties.

The process of fractionation is performed by gel electrophoresis or by mass spectrometry, and the information received during this process is used in two different ways. The first one is the "shotgun" protocol, in which the protein identification is performed on a chosen fraction that separates the control from the testing group. This fraction of difference is represented by a spot on the electrophoretic gel or by a peak on spectrometric spectra that is unique for the testing or control samples. The proteins present in this fraction and identified by fingerprinting and/or a sequence tag methodology are subsequently evaluated as a potential diagnostic marker.

The second approach is "protein profiling," in which a pattern of all electrophoretic spots or a pattern of all spectrometric peaks obtained from the control and test groups is compared. In this methodology, the pattern that exclusively represents one of these groups (discriminatory pattern) or which is highly specific for a particular cellular state (or state of the organism) becomes a biomarker. Subsequently, this discriminatory pattern is evaluated as a potential marker, such that the identification of its single components is not necessary.

The electrophoretic pattern is defined as the presence or absence of protein spots at particular locations, as well as the intensity of their staining. The mass spectrometry pattern is described as either the presence or absence of spectral peaks and their amplitudes. The discriminatory pattern that separates the control from the testing group is usually related to the small subset of proteins or peptides present in the barely visible electrophoretic spots, or it shows up as small spectrometric peaks hiding in the background of the spectra. To find a spot or a peak of difference and to separate these two groups, a massive amount of data must be analyzed that for most instances exceeds the capacity of the human brain. In these instances, the proteomics practitioner must be assisted by pattern recognition software that helps in the preliminary evaluation of data. Such software is an integral tool of proteomics protocols.

Despite difficulties, the protein profiling approach has noted some successes. The pattern of mass spectrometry peaks has been successfully used as a marker for differentiation between ovarian carcinoma and noncancer cases with sensitivity of 100% [95% confidence interval (CI), 93–100], a specificity of 95% (87–99%), and a positive predictive value of 94% (84–99%).[8,9] The initial success of this novel methodology was strongly criticized. One of the problems was related to the disconnection between the 10 log protein dynamic range of serum and much lower 2 log dynamic range of the Surface-enhanced laser desorption/ionization (SELDI) mass spectrometer that was preferentially used for serum protein profiling.[10] Another critique was based on the observation that biomarkers accumulate in the serum despite the fact that their molecular masses are lower (500–2,500 m/z) than the renal filtration threshold, and so they should be cleared by the kidney.[11] In addition, the validation of cancer markers by liquid chromatography–tandem mass spectrometry (LC-MS/MS) performed on the fraction designated by SELDI has failed.[12] To date, it is widely accepted that proteomics patterns cannot be considered as biomarkers without the identification of its components. The protein profiling as a diagnostic method is accepted only as a part of hybrid strategies that combines the best attributes of pattern- and identity-based methods. This combination provides an optimal approach to the proteomics-based biomarkers discovery.[13]

Protein Identification

There are two general methods of protein identification - database dependent and database independent.[14] In the database-dependent method, a simple protein mixture or a fraction chosen from two-dimensional electrophoresis is enzymatically digested, or chemically cleaved, and submitted for mass spectrometry analysis. As a result, molecular masses of the peptides from the mixture are obtained that are subsequently used for identification of the proteins from which they derive. This identification is made using protein mass fingerprinting[15] and/or peptide sequence tagging[16,17] methodology.

The fingerprinting protein identification approach utilizes the molecular masses of the peptides from the mixture and compares them against a database containing a collection of proteins and peptides derived from them. Based on the similarity of masses, a cluster of protein candidates that may be related to the particular peptide is selected. For each of these protein candidates, a matching score is calculated that reflects the probability that this candidate is a factual protein present in the specimen. The higher the score, the more certain it is that a given protein is truly present in the specimen.

In the peptide sequence tagging identification method, the information is collected from a tandem MS/MS. The first MS measures masses of intact peptides present in the mixture. The role of the second MS is to assess masses of a series of ions generated from a parent peptide in the collision chamber located between mass spectrometers. The derived ions are arranged according to increased molecular masses. The difference of masses between subsequent ions serves as a criterion for establishing which amino acid is added and thus determines the sequence. These sequenced fragments (short sequenced tags) are used for peptide and protein identification, which requires comparison with a sequence database. The protocol for database searching and score calculation varies depending on the goal,[18,19] but the principles of this process are similar for both the fingerprinting and tagging methods. Although each method, fingerprinting and tagging, can be used separately, the combination of the two brings the most efficient results.

A relatively new protein identification method is based on comparing the tandem MS/MS spectra obtained during experimentation with the spectra and corresponding sequences stored in the spectral library. Specially created software performs the search, and as a result a sequence of the potential peptide is attained.[20]

The database-independent method of protein identification is usually performed for those organisms for which little or no sequence information is available. One of them is the de novo sequencing protocol[21] that is used not only in cases of incomplete database information about particular protein but also for validation of the database-dependent identification or for identification of new peptides.

Data Validation

As already mentioned, the amount of data produced by proteomics instrumentation is so sizeable that it requires a special method for its storage and analysis. The need for that method has stimulated a dynamic development of bioinformatics. There are two general levels of proteomics investigation where bioinformatics plays a key role. The first is on the instrument level, where "raw" data are produced. What are referred to as raw data are, in fact, data that are already processed by the instrument, which is referred to as "spectra deconvolution."[14] Spectra deconvolution is the process whereby continuous data obtained by the instrument are transformed into a list of spectral peaks. This list of peaks is subsequently used as input data for the second level of proteomics investigation, which leads to protein identification (database search, score calculation, etc.) and peak validation as a potential marker. The process of "deconvolution" consist of subtraction of background, noise removal, charge state deconvolution, and deisotoping.[22] As a result, the peaks that are extracted from the continuing line of the mass spectrometry recording are obtained and submitted for the next level of elaboration involving peak clustering and identification of the peaks of difference. Peak clustering consists of grouping of peaks that are characteristic for a particular testing and control group (peaks of difference). The peaks of difference are identified by a variety of statistical methods, such as hierarchical clustering and principal component.

All commercial mass spectrometer systems support the first level of data processing, but the process is not standardized, and often the final product differs across different brands of instrumentation. There is thus a real need for standardization in this area, as differences in peak identification affect not only subsequent database identification but also the de novo sequencing process.[22]

The second level of data analysis is also not standardized among different systems. The principles for peak clustering as well as the criteria for identification of the peaks of difference vary significantly from system to system, such that the evaluation and comparison of results from different laboratories is impossible. A new initiative for standardization of these two processes and a requirement for multicenter validation of obtained results slightly improves this situation, but much must yet be done to decrease confusion and inconsistency in this area.

Proteomics Instrumentation

Two-Dimensional Gel Electrophoresis

The first protein separation method that met omics criteria was two-dimensional polyacrylamide gel electrophoresis (2D PAGE). This technique is able to accept minimally prepared specimens and to separate all proteins present in the mixture. Because of its contribution to proteomics and its popularity among researchers, the name *two-dimensional (2D) electrophoresis* has become almost synonymous with the term *proteomics*. Two-dimensional PAGE includes two procedural steps. The first is isoelectric focusing (IEF), which separates proteins based on differences in isoelectric (pI) points. During IEF separation, proteins migrate along a pH gradient until they reach a point where the pH is equal to the pI. At that point, the net charge of the protein becomes zero, and migration stops. At the end of this step, the original mixture is divided into fractions that contain proteins with the same pI. The second step of 2D PAGE separates proteins based on their molecular mass and is performed on a sodium dodecyl sulfate (SDS)-polyacrylamide gel. The final result of these two separation steps is a collection of the fractions that contain proteins with the same pI and the same molecular mass. To visualize separated proteins located in the spots, staining with silver or Coomassie Blue G-250 is performed. In an ideal situation, in which the sensitivity of separation is 100%, each spot would contain one protein. Unfortunately, the sensitivity of 2D PAGE is not perfect, which contributes to several serious methodological problems.

Methodological Obstacles of 2D PAGE

For both shotgun and protein profiling approaches, precise and reproducible placement of spots on the gel is critical. Even minute changes in protocol and/or environment may lead to the misplacement of identical spots, which could be mistakenly evaluated as spots of difference. During IEF, even minimal changes in the pH gradient from one experiment to another deposit the same proteins in different locations on the strip and change their placement on the final gel. This problem was significantly reduced when an immobilized pH gradient was introduced. Despite improvements in the procedure, large hydrophobic proteins failed to be separated by IFE because of their tendency to aggregate and precipitate around their pI. The problem of precipitation also plagues separation of strongly acidic and basic proteins, which occurs on the extreme points of the IEF strip. The second step of the 2D PAGE separation requires a precise placement of IFE strips on the SDS polyacrylamide gel slab because any inconsistency changes the final location of the same protein from run to run. To secure precise and reproducible placement of the IFE strip on the gel slab, new hardware has been introduced that is capable of a consistent transfer of immobilized pH gradient strips onto the SDS-PAGE. If performed on polyacrylamide gel, the second dimension of separation is limited to the proteins with sizes larger than 10 kDa. The smaller proteins, as well as low-abundance proteins with fewer than 1,000 copies per cell, do not separate. Hence, this system is unable to separate membrane proteins because of their large hydrophobic domain, as well as posttranslationally modified proteins because of their low concentration, which is far below the sensitivity of this technique.

Protein Staining

The next step in the 2D PAGE separation is protein visualization. The staining not only reveals individual proteins but also, by its intensity, which is measured by densitometry, provides information about protein concentration. Silver (without glutaraldehyde) and Coomassie Blue G-250 stains are the most commonly used, with a preference for the former because of its significantly higher sensitivity. The drawback of the silver staining is its low sensitivity resulting from the staining of the background. To minimize this effect and to increase the sensitivity for protein detection, glutaraldehyde is used for destaining of the background. There is no end-point for the silver staining reaction, which means that the intensity of the color is directly proportional to time of staining.[23] The silver stain procedure is also not compatible with mass spectrometry because of protein oxidation and covalent modification during contact with silver and formaldehyde, as well as because peptide-peptide binding occurs during background decoloration with glutaraldehyde. The Coomassie Brilliant Blue stains proteins by chemical interaction with arginine residues. The staining process is not time dependent but relates to amino acid composition and stops when all arginine amino acids are saturated. Because the relationship between staining intensity and protein concentration is not linear, and changes from run to run are significant, this stain cannot be used for protein semiquantitation.[24] Compared to silver, the Coomassie Brilliant Blue stains possess lower sensitivity but higher compatibility with mass spectrometry. To obtain high sensitivity and compatibility with MS in a single experiment, both staining techniques should be used in tandem.[25]

The sensitivity of protein visualization has been significantly improved through the use of fluorescent labeling agents such as ruthenium II-Tris, SYPRO dyes, and cyanine dyes, to the extent that posttranslationally modified proteins can be revealed as distinct spots. Another advantage of using fluorescent dyes is related to the intensity of the staining, which is not time dependent and produces minimal background staining. Although special protocols and equipment are required, a fluorescent stain can be used for the detection of low-abundance proteins. This method cannot, however, be used for protein quantitation, because the intensity of staining is proportional to the amino acid composition rather than to the protein concentration. To quantify proteins on a gel, a special staining with stable isotopes has been developed. In this procedure, both light and heavy isotopes of hydrogen, nitrogen, and carbon are used for in vitro or in vivo protein labeling. In vitro labeling is used mostly for relative quantification of the differences in protein concentrations between two states of a cellular culture or an organism.[23] The concentration of proteins that are present at different cellular states can be calculated from the difference of signal intensities detected by mass spectrometry between the heavy and the light forms of proteins. Thus far, in vitro labeling has generally been tested on samples of low complexity, such as commercially available proteins or simple systems such as viruses and prokaryotic cells. Despite its promise, this method requires significant improvement before utilization with more complex systems such as eukaryotic organisms.

Enzymatic and Nonenzymatic Digestion

The spots present on 2D PAGE contain groups of similarly sized and charged proteins. To perform protein identification, the spots are excised and digested using enzymatic or nonenzymatic methods. During this process the proteins are cleaved into smaller peptides suitable for mass spectrometry. The enzyme most commonly used for this process is trypsin. Its popularity is related to its low cost and to generation of peptides of a size that is ideal for mass spectrometry analysis. Other proteases such as Glu-C endoprotease are used sparingly because they produce large amounts of autoproteolytic peptides.[25] Cyanogen bromide is the most commonly used nonenzymatic cleaving agent. It is not very popular because it produces peptides that are too large for mass spectrometry and thus require additional digestion with trypsin.

All the enzymes used in preparatory digestion cleave proteins at well-defined sites. If the sequence of the proteins is known, it is possible to deduce the number, sequence, and molecular weight of all peptides produced by a particular enzyme. These data have been stored in databases; they are readily accessible on the Internet and they are used for protein identification. Similar to any other PAGE-based procedure, enzymatic preparatory digestion carries several methodological problems that need to be resolved. For example, trypsinization of proteins with small lysine and arginine residues produces large fragments not suitable for mass spectrometric analysis. In addition, tryptic fragmentation of proteins is a very imprecise process that not only skips certain cleavage sites but also targets amino acids other than lysine and arginine, which complicates the fingerprinting method of protein identification.[26]

Mass Spectrometry

Mass spectrometry became a potential diagnostic technology at the time when electrospray ionization (ESI) and matrix-assisted laser desorption/ionization (MALDI) instruments were developed. These tools combine both high sensitivity and easy utility, which makes them ideal for commercialization. The new mass spectrometry technology coupled with automated and semiautomated specimen delivery systems provide the ability to perform high-throughput operations, an important aspect of diagnostic tools. Analyzing proteins in their original complex forms and performing this operation in high-throughput mode fulfills the omics criteria and makes modern mass spectrometry useful for proteomics procedures.

Matrix-Assisted Laser Desorption/Ionization

Originally described in 1987,[27] MALDI is a mass spectrometer in which the specimen is embedded into a matrix in a 1:1,000 ratio. The matrix is composed of small molecules that absorb ultraviolet waves, inducing solvent vaporization and matrix crystallization. During crystallization, the specimen is incorporated into matrix crystals and subsequently vaporized. The vaporized clouds of ionized crystals containing the protein particles are directed into a detector, most commonly the time-of-flight (TOF) mass analyzer. The MALDI-TOF is capable of analyzing thousands of different components (not exclusively proteins), but its role in proteomics seems to be the most significant. It is the first instrument that competently detects entire proteins and measures their molecular masses with a precision that allows for fingerprinting identification, even when the proteins are present in the mixture at very low concentrations.[28]

Two modifications of the MALDI mass spectrometer allow for protein identification based on the amino acid sequence: MALDI-Oq-TOF (Oq indicates tandem quadrupole/time-of-flight mass spectrometry) and MALDI-TOF/TOF.[29] These instruments operate with a collision chamber located between the first and second mass analyzers. The first analyzer measures molecular masses of peptides present in the primary mixture, and the second analyzer measures the molecular masses of the fragments derived from the peptide that was chosen for sequencing. This chosen peptide is directed from the first MS to the collision chamber for fragmentation. The molecular masses of these fragments are measured in the second MS, and they are the base for the sequencing process.

Imaging Mass Spectrometry

A new feature of the mass spectrometry development has been attained with an innovative MALDI technology called *imaging mass spectrometry* (IMS). In this technology, the mass spectrometer produces a two-dimensional ionic image of the sample in which the localization of the targeted molecules is tagged.[30] Two techniques are used for obtaining an image. The first one is a blotting technique that requires fresh tissue to be placed on a methanol-wetted organic membrane for protein binding. After the tissue is removed, the blotted area is submitted for MALDI-TOF mass spectrometry analysis. The second technique calls for frozen tissue to be placed on a flat metal MALDI target plate and covered by the matrix. The image is constructed by mapping the m/z value over the target area. A unique feature of IMS technology is that it can reveal a distribution of hundreds of unknown compounds in a single measurement. Conventional methods for visualizing molecules such as chemical staining, immunohistochemistry, and radiolabeling require prior knowledge about targeting molecules and preparation of specific molecular tags. This requirement significantly limits the utility of these methods. MS/MS imaging offers an analysis of the tryptic fragments that are obtained after on-slide enzymatic digestion. The potential benefits promised by this method may be not only an identification of the proteins directly from the slide but also a localization of high molecular weight, membrane-bound or membrane-associated proteins that are difficult to detect with direct MALDI-MS analysis.[31]

In addition to obtaining information about proteins distribution directly from the tissue, IMS allows for measuring their relative concentration and local composition. This technique has been successfully used for protein visualization in glioblastoma of the brain and in non-small cell carcinoma of the lung. In clinical practice, it can be used to identify different populations of cells based on their molecular differentiation, which can be used to clear surgical margins at the molecular level and for discovering of a new marker for ovarian cancer.[30,32]

Although this technology seems extremely attractive, several serious limitations must be overcome before its diagnostic implementation. The size of the analyzing specks is limited by the laser spot, which is 30 μm in diameter. It means that MALDI-IMS produces only about 1,000 discrete images from the cells that contain tens of thousands of proteins,[31] which translates into instrumental sensitivity significantly too low to produce any meaningful results. In addition to low sensitivity, the overshadowing effect limits the usage of this technique in anatomic pathology. In the overshadowing effect, a low signal is suppressed by a higher one, which most often is related to hemoglobin that is present in high concentration on the cut surface. An additional difficulty is formed by an abundance of the produced data. The doubling of the resolution increases the amount of data fourfold with concomitant increases in acquisition time and data file size. To overcome these difficulties, significant improvement in the process of data analysis and storage is required.[31] In addition to the outlined specific limitations, every limitation of the MALDI is also pertinent in this system.[33]

Surface-Enhanced Laser Desorption/Ionization Mass Spectrometry

SELDI mass spectrometry is another spectrometric system that has gained interest for it potential diagnostic usefulness. This system uses microchip technology and is promoted by the Bio-Rad Diagnostic Company. A minimally prepared or unprepared specimen containing a mixture of proteins is applied to an aluminum or steel microchip support platform measuring 1–2 mm in diameter. The surface of the miniplate is coated with chemical matrices ("bites") similar to chromatographic columns and composed of hydrophobic, cationic, or anionic phases. With these types of bites, proteins are separated based on their physicochemical properties such as the isoelectric point (pI). The mixture of proteins and peptides present in the specimen is applied on the microchip, and groups of proteins with similar physicochemical properties bind to the surface. The unbound proteins are removed during the washing process. The coating matrices of the chips can be more specific, such as antibodies, enzymes, ligands, proteins, receptors, or DNA oligonucleotides. These specifically reacting chips are designed to interact with a single target protein and are used to study specific molecular interactions such as protein-protein, protein-antibody, protein-receptor, or protein-DNA.[34] The complexes of "bite" and proteins of interest are subsequently analyzed by MALDI-TOF or MALDI-TOF/TOF mass spectrometry. Microchips are suitable for analyzing all possible specimens, including serum, whole blood, urine, cellular lysate, body fluids, and lavages. To enhance the efficiency of the result analysis, the instrument is equipped with a biomarker discovery software packet. This software packet supports recognition of the spectrometric peaks and combines them into clusters that are characteristic for a particular group. It also extracts the peaks that are characteristic for healthy and diseased cells or for different stages of disease or different states of normal organisms (peaks of difference).

The entire process of the spectrometric spectra evaluation provides a basis for discovering single biomarkers and for a diagnostic strategy called *proteomics pattern diagnostics*.[35] The SELDI mass spectrometer along with provided software was tested during development and validation of the spectral patterns as a diagnostic marker for ovarian,[36] prostate,[37] and breast cancer.[38] As a result, a large amount of very complex data was obtained that could not be evaluated by classical statistics. To validate whether spectrometric peak patterns can be utilized in cancer diagnosis, an artificial neuronal network algorithm was used, integrated with the SELDI software packet. An artificial neural network is a statistical tool that imitates the learning ability of the human brain. PRoPeak software provided by Bio-Rad is based on an artificial neural network and is able to "learn" to recognize universal elements based on repeated analysis of the complex data from at least two experimental groups. Based on the "experience" gained during the training set, the instrument develops criteria and rules that are validated during subsequent test runs. Each run is used to validate and modify the rules, similar to the typical learning process.

The usage of the pattern of mass spectrometric peaks as a marker for disease is a completely new approach in laboratory diagnostics and carries several methodological and technical issues that need to be addressed. The most serious ones are related to the low sensitivity of SELDI instrumentation, its limited reproducibility, and difficulties in protein identification.[39] In addition, several questions have been raised that are not specifically related to the SELDI instrumentation, such as the validation of results among different laboratories, the influence of different sample handling on results, and different experimental designs.[39]

Other Types of Spectrometry

The other type of mass spectrometry used in proteomics is ESI mass spectrometry. This MS operates on soft ionization, which allows the generation of ions from large (>100 kDa) nonvolatile analytes, such as proteins, with no fragmentation. The main difference between MALDI and ESI is that the latter produces multicharged ions and consequently more complex spectra. ESI collects more information; however, its results seem to be more difficult to interpret than the less-complex spectra from MALDI.[40] At the same time, MALDI possesses lower resolution and limited sequence information, which makes it less useful for identification of protein isoforms, posttranslational modifications, and protein complexes. In the recent years, ESI has become a more user-friendly instrument and is more widely used than MALDI in proteomics investigations.

Quadrupole mass spectrometers, which are also named quadrupole mass filters, are another commonly used MS. The wide popularity is accounted for by their flexibility, easy operation, adequate performance for most applications, relatively low cost, and well-developed software. The main functional element consists of four parallel electrically conductive rods arranged into a channel through which the ions from the sample pass. While passing through this channel the mixture of ions is sorted so that only the ion with a selected, very narrow m/z range is able to reach the detector. Because only chosen ions are detected by this system, the quadrupole mass spectrometer is often referred to as the "mass filter." The instrument operates in two basic modes: selective and scanning. The selective mode is used to detect only one ion with precisely defined m/z. In the scanning mode, a wider range of m/z is chosen in which ions across a full range of m/z are detected.[28]

Quadrupole mass spectrometry is predominantly used in target-driven proteomics where a single ion is targeted for identification, as well as for multiple reactions monitoring MS-based biomarker validation.

A "Dream Team" Mass Spectrometry System

The technological challenge for today's proteomics is to develop a system that will allow rapid fractionation of the sample and efficient identification of proteins in the mixture in high-throughput mode. In 2004, Schrader and Klein published a paper describing the ideal proteomics system.[41] This scheme consists of a high performance liquid chromatography-based separation system coupled with Fourier transform ion cyclotron resonance (LC FTICR) mass spectrometry. The Schrader system helps to overcome the limited sensitivity of gel-based separation and limited accuracy of mass spectrometry. High performance liquid chromatography (HPLC) is able to fractionate an entire specimen in a liquid environment into fractions containing proteins with precisely defined m/z. The FTICR detects ions in the mixture by measuring their frequency of resonance. Because frequency of resonance changes with mass, a precise measurement of frequency allows for mass measurement with high resolution and accuracy. It seems to be the solution for problems related to the currently used, less-accurate mass spectrometers. The FTICR is also able to analyze unfractionated samples. It can provide fast fractionation of a complex protein mixture, and it is able to perform mass measurement with precision, allowing for library-based protein identification using very narrow mass tolerances.

At the same time, an efficient fragmentation method for FTICR has been developed: infrared multiphoton dissociation (IRMPD). IRMPD produces peptide fragments similar to those obtained in the collision chamber in a tandem mass spectrometry system. The analysis of fragments is similar to that obtained by a tandem mass spectrometry system, but with significantly higher mass accuracy and efficiency.

The unmatched advantages of this system are high resolution, large dynamic range, and supreme sensitivity, which provide the highest quality data for biological analyses, especially in proteomics. It can be used for both the "bottom-up" method that identifies proteins based on accurate mass measurements of peptides derived from the digested mixture as well as the "top-down" approach which is able to identify intact proteins. Since this technology has been used, especially with the improvement of dissociation techniques, significant progress in providing sequencing and structural information about proteins is noted.[42] Recently, a few papers describing an attempt to couple a desorption electrospray ionization (DESI) technique with a Fourier transform ion cyclotron resonance mass spectrometer have been published.[43] This system combines the unique ability of DESI to ionize a natural and unprepared sample with the high sensitivity of FTICR. In this system, unfractionated samples are directed by DESI to the FTICR for top-down analysis and sequencing of the peptides present in the samples. If successful, the presence of such a system may open new opportunities for sample processing and protein identification, without prepreparation and in high-through output mode.[43]

Proteomics Protocols

Protein Identification

Protein Mass Fingerprinting Identification

The commonly used detecting systems that provide information allowing for protein identification are based on single (MS) or tandem (MS/MS) mass spectrometry. The specimens for these systems consist of prefractionated mixtures of proteins that are enzymatically digested or chemically cleaved into derived peptides. To recognize proteins present in the sample, the molecular masses of the derived peptides are measured with the highest possible precision, and they become a basis for protein identification by the mass fingerprinting procedure.[44]

Protein mass fingerprinting identification is the first and the most often utilized identification method by proteomics investigators. This method is based on the fact that proteolytic enzymes cleave proteins at precisely defined points such that for any known protein the number of derived peptides, their amino acids sequences, and molecular masses can be predicted based on amino acid composition and the characteristics of enzymatic action.

The information about proteins and theoretical sets of peptides derived from them after enzymatic digestion is stored in a database. To identify proteins present in the specimen, molecular masses of experimental peptides are compared against molecular masses present in the database. If they match, the peptide and its "mother" protein are identified. In real life, this perfect picture of protein identification is complicated by several factors. The most important of these are the mass spectrometer's instrumental sensitivity (a degree of precision for molecular mass measurement) and the completeness of the database for these particular proteins. A higher sensitivity and a more detailed database give a better chance for the protein to be identified. To increase the chance for correct protein recognition, it is crucial to measure the mass of the ions with the highest possible accuracy. The number of peptides that differ from each other in the mass range of 0.5 Da is four times higher than the number of different peptides in the mass range higher than 0.5 Da. For the mass spectrometer with a mass resolution higher than 0.5 Da, the probability that dissimilar peptides are differentiated is higher than if the resolution is lower than 0.5 Da. The instrumental sensitivity is even more important for the detection of posttranslational modification or for the uncovering of a single element from a complex eukaryotic proteome.

Another difficulty confront by fingerprinting identification is related to the sizes of the peptide fragments. The general rule for peptide matching is that the shorter fragments are less specific but have a higher chance to be present in the experimental mixture and to be found in the database. On the other hand, the longer peptides are more specific for the "mother" protein, but they are present in the mixture in lower concentrations. To obtain a result of the database search with higher specificity, longer peptides are more desirable. Unfortunately, large peptides not only are present in low concentration, but they also do not ionize as well as small ones do during the spectrometric procedure. Thus, their chances to be detected are diminished. For this reason, the optimal results for protein identification are obtained when a larger number of small peptides containing 8–10 amino acids is detected and submitted for evaluation. The peptides with these characteristics are produced by trypsin, the proteolytic enzyme that is widely used during preparatory digestion. Another problem related to size is that small proteins have a lower chance of being identified by the fingerprinting method than do large ones; this occurs because the number of digested fragments produced from small proteins is low, and consequently the number of peptides available for comparison against the database is small, which decreases the probability that the mother protein will be identified.

Some of the proteins present in the primary mixture may be lost during preparation and/or fractionation. As it was mentioned before, two-dimensional PAGE is still widely used for sample prefractionation. During this process, large proteins with a molecular mass higher than 100 kDa are missed due to precipitation during the first IEF separation, and subsequently they are not identified.

It is important to remember that each mass spectrometer has its own characteristics with variabilities. For that reason, the final product of protein identification can be distorted by the type and internal properties of the particular instrument. A case in point is MALDI-based protein identification. As a general rule for successful protein identification, only the most intense peaks should be analyzed. However, peak intensity is not always related to peptide abundance. An example is MALDI-TOF, which creates peaks whose intensity depends more on arginine content than on peptide concentration. This result indicates that arginine-rich proteins have a greater chance of being identified by MALDI-TOF spectrometry than proteins with low arginine content.

A highly advanced tool for fingerprint identification is "iterative analysis," which analyzes in depth the 2D PAGE protein distribution and protein composition within a single two-dimensional spot. The first step for interactive analysis is the removal of masses of contaminants such as keratins, products of trypsin autolysis, and dyes. The second step of this analysis consists of removing the peaks caused by neighboring spots. The final results of protein identification are reevaluated for the major components and contaminants.[45]

Sequence-Based Identification

It is possible that, in a digested protein mixture, especially one derived from eukaryotic organisms, there are a number of peptides with the same molecular mass but with different amino acid compositions. It is also possible that the same mixture contains peptides with slightly different masses that are not discriminated by the mass spectrometer. For these cases, molecular mass alone cannot be the basis for protein identification, and to increase scoring and certainty for protein recognition, amino acid sequencing is added to the protocol. Peptide sequencing is performed via tandem mass spectrometry, in which the mass spectrometers are separated by a collision chamber. When the mixture of ions reaches the first mass spectrometer, the molecular masses are measured and one of them is chosen for future sequencing. This selected ion is subsequently directed to the collision chamber, where it is fragmented into a unique spectrum of derived ions whose masses are measured in the second mass spectrometer. As a result, a series of increasing m/z is obtained, progressing from a single C amino acid to the entire peptide, where each subsequent ion differs from the previous one by the mass of one amino acid.[46] The key elements for this sequencing process are precise measurements of the molecular masses produced in the collision chamber, and their arrangement from lowest to highest one allows for construction of amino acid sequencing. For protein recognition, the spectra and sequenced fragments are compared with a comprehensive protein sequence database by using one of the identification algorithms.

The simplest algorithm starts with a match of the experimental ion mass with the masses of peptides present in the database. Usually, a wide window of mass tolerance is chosen, and a broad range of peptides is selected. In the next step, a theoretical sequence for each peptide candidate is created, which is then compared with the sequence of the experimental one. The identification process by collision-induced spectra is very efficient, because sequence analysis brings supportive information to the identification process performed by the fingerprinting method.

Database Searching

Several available databases allow for comprehensive searches for molecular masses and for sequence matching. No matter which database is used, the peaks produced by internal standards and the peaks of autolytic peptides must be excluded from the search. The search process and data interpretation require precise information about search parameters, such as specific instrumentation and fragmentation or cleavage methods. There are a few search engines that support multiple fragmentations. For comprehensive fingerprinting and sequence tag protein identification, the RefSeq, NCBI Entrez Protein, UniProt, TrEMBL, and Swiss-Prot databases are most widely used. The Swiss-Prot is not so complete as the others, but it contains several tools for sequence analysis.

The NCBI is an example of a sequence repository database that contains sequence data translated from the nucleotide sequences of the DDBJ/EMBL/GenBank. The NCBI is constantly updated by sequence information from other protein database such as Swiss-Prot, Protein Information Resource (PIR), and Protein Data Bank (PDB).[47] To be efficient, the search process requires some specific information to be entered, such as keratin contamination, the proteolytic enzyme used, the number of allowed missed cleavages, the usage of iodoacetamide for cysteine modification, and the fixative modification in cysteine residues. This information narrows the spectrum of the search and reduces searching time.[48]

The RefSeq collection's main feature includes explicitly linked nucleotide and protein sequences, which are automatically updated to reflect current knowledge of sequence data, data validation, and review status in each record.[47]

The UniProt database was developed with the aim to be the most comprehensive protein database. It merges other collections such as Swiss-Prot, TrEMBL, and PIR-PSD. All suitable PIR-PSD sequences that are missing from Swiss-Prot or TrEMBL were incorporated into UniProt. To allow fast tracking of any PIR-PSD entries, cross references between Swiss-Prot, TrEMBL and PIRPSD have been created. This comprehensive UniProt database is an initiative that provides a foundation for uniform information flow among researchers.[47]

For database searching, there are several commercial and publicly available engines, but not all provide the same quality of search, and only a few cover the fragmentation method that is used by a particular instrument. Commercial search engines such as MASCOT and Phenyx support most instrumentation with its fragmentation methods.

MASCOT provides a complex search report expressed by numerical scores. A score of 100 identifies a protein, but any score between 70 and 100 indicates that a chosen protein is a good candidate for identification. MASCOT also provides a list of unassigned masses. The most common reasons for unmatched peaks are tryptophan oxidation, methylation of aspartic and glutamic acids, and (not as often, but most excitingly) a new protein.

Publicly available search engines such as VEMS, OMSSA, and S!Tandem return results of a similar quality as those attained by MASCOT or Phenyx.[47]

There are several reasons for obtaining unassigned peaks. If investigation proves that the presence of unassigned peaks is the result of the presence of an unknown or posttranslationally modified protein, it must be sequenced by tandem mass spectrometry or retested with one of the modification targeted methods such as multiple reaction monitoring (MRM).

Validation of Protein Match

To validate matching accuracy for each candidate protein, various scoring methods are available. In the simplest one, the peak intensities of the experimental peptides are compared with those of the theoretical group. The most serious bias for this method is that the intensity of peaks depends on the collision energy. Any inconsistency in collision energy significantly changes the matching results. The most popular scoring systems using signal intensity produced by tandem mass spectrometry is MASCOT. More advanced scoring methods calculate the matching score based on intensity of the collision energy and internal properties of the peptides.[49] The popularity of these analytical methods is very low, and the simpler MASCOT and SEQUEST systems still dominate the market.[50]

The sequence-based identification process may be obscured if the reference database for a particular organism is not available; if the identified protein underwent posttranslational modification not listed in the database; or if spectrometric peak intensity is doubtful because of a low signal-to-noise ratio. These situations can be overcome by using a de novo sequencing method. There are three basic approaches to de novo sequencing method: the exhaustive listing method, the subsequencing method, and the "graph theoretical" approach. Detailed descriptions of these methods are available elsewhere.[51] A simpler version of de novo sequencing identification is the partial sequence method. It is based on the mass sequence-mass protocol, which matches a molecular mass and establishes a sequence for an experimental peptide with a theoretical partial sequence and the masses of peptide candidates.[50] The other group of methods is based on the calculating of theoretical mass from the peptide sequences in the database and comparing the results with observed peaks.

The method that has recently gained popularity is a calculation of the false discovery rate (FDR) for reverse sequences obtained from a real database. The popularity of the FDR scheme is related to its simplicity and to the fact that it can be used with any search engine and scoring system. It is a method that offers a practical way to establish cutoff values for searching scores, and is especially valuable for validation of assignments for complex posttranslational modifications.[52]

Ascore is another computational scoring system that works in most instances except in cases of ambiguous phosphorylation sites, when the potential score error is too high, so that manual validation is still preferred by most researchers.[53]

In the instances when phosphorylated peptides are validated, the VEMS program may be used. It offers the option to generate all possible sites of phosphorylation for a particular peptide and rank them using the VEMS score.

Although a large number of methods is used for validation of sequence-based matching, it is difficult to assess which one is the best because there is no consistent evaluation protocol that can be used to compare these methods.[51] As of today, there is no method that would satisfy all possible situations, and it seems obvious that the comprehensive identification and validation approach is the best choice. It would include fingerprinting and sequencing identification with computing validation followed by manual verification. This comprehensive tactic increases the chance for correct identification not only of common cases but also of proteins with extremely low and extremely high molecular masses, posttranslationally modified as well as unknown ones.

Protein Quantification

For higher organisms, the old concept of "one gene, one protein" no longer holds for the obvious reason that a single gene is the source of a large number of protein isoforms that subsequently undergo extensive posttranslational modifications. The different isoforms of the same protein play important regulatory roles and may have the opposite effect on cellular function. The idea that one protein participates in cellular metabolism as a single factor has been revised and changed to the theory that a multiple protein network governs cellular activities. The function of these complexes is regulated not only by the actions of different isoforms but also by different concentrations of the participating proteins. Although the importance of different protein concentrations for cellular metabolism is well known, to this day there is no good method that measures absolute concentrations of proteins or their isoforms. The only well-established method uses a stable isotope and measures a relative concentration of a protein in prokaryotic cells. This strategy is based on the ability of mass spectrometry to separate chemically identical proteins that differ by a stable isotope. In recent years, some hope has been brought to this field by the development of a new technology that is based on the multiphoton detection methodology. The sensitivity of this method allows not only for detection of as few as 600 molecules per liter but also for separation of different protein isoforms.[54]

The general methods for protein quantitation can be classified into densitometry that is used for 2D electrophoresis, relative quantitation by chemical and metabolic isotopic labeling, and spectral intensity profiling.

On 2D PAGE the protein concentration on stained spots can be quantified by measuring the absorbance of each gel fraction by a densitometer. The currently available densitometers are equipped with a microprocessor that calculates the area under each peak of absorbance, which represents a protein spot, and reports the peak as a percent of the total or as an absolute concentration. Reliable densitometric quantitation requires an appropriate light wavelength, a linear relation between absorbance and concentration, and a transparent background of the protein-free gel.

Quantitation techniques by stable isotope labeling are based on comparing the measured intensity of the mass of the unlabeled "light peptide" and the intensity of the mass of the labeled "heavy peptides." A more advanced form of the same method is called "multiple encoding," and it uses multiple stable isotopes that produce multiple "heavy peptides." This method allows for the comparison of more than two samples in the same MS experiment. Although different quantitative methods are based on the same principles, the calculating algorithm and interpretation of the results are different and specialized. ASAPratio, RelEx, MSquant, and VEMS are examples of the tools used for these calculations.[14]

In the spectral intensity profiling method, the intensities of spectral peaks received from different LC-MS are compared. Because of the low reproducibility of LC, the bias must be corrected by duplicate measurement and calculation of the experimental variances[55] or by adding an internal standard to the mixture.

There is no single method that can be used for simultaneous protein identification and quantification. The most widely used tactic is to identify and quantify individual components separately and subsequently integrate these results using analytical tools. The 2D PAGE is an example for this approach, where concentration of proteins is assessed by measuring stain intensity of the separated spots. For protein identification, the selected spot is excised, digested, and submitted for mass spectrometry testing. The gel-based approach carries several limitations described earlier, of which the most important are inability to detect low-abundance proteins, to recognize posttranslational modification, and to separate protein complexes.[51]

The non-gel approach for protein identification and quantification is a mass spectrometry-based protocol in which an entire protein mixture is submitted for analysis. Although more effective than the gel-based approach, mass spectrometry protocols are still imperfect. For example, the analytical sensitivity of single-dimension mass spectrometry is too low to detect low-abundance proteins. In addition, the relationship between peak intensity and the amount of the tested substances in the mixture is too poor to be used for assessing their concentration. To overcome these problems, an internal standard must be used for a more precise assessment of the protein concentrations, and a "dream team" mass spectrometry system ought to be employed, such as HPLC coupled with LC-FTICR for surmounting low instrumental sensitivity.

Detection of Posttranslational Modification

Posttranslational modification is a reversible event serving as a regulatory process that modifies protein activity. During this process, proteins become mature and ready for their physiological function. It is possible that the only cellular reaction to diseases or environmental modification is an alteration in the concentration of modified proteins without simultaneous changes in protein expression.

There is no single and efficient method to determine posttranslational modification of proteins in a biological sample or in any protein mixture. The most popular method of posttranslational modification detection is the use of different enzymes interacting with posttranslationally modified proteins/peptides and subsequent analysis of the differences in molecular masses. Proteins undergo more than 300 different posttranslational modifications, of which the most commonly observed are phosphorylation, splicing, and disulfide bridges.

Phosphorylation is an important regulatory mechanism and occurs at the serine, threonine, and tyrosine residues. During this modification, 80 Da is added to the molecular mass of the peptide, which can be detected by comparative mass measurement before and after treatment with alkaline phosphatase. However, this method is complicated by a low consistency in peptide ionization and by unpredictable protein fragmentation.[56] The nonenzymatic approach for detection of phosphorylation and other posttranslational modifications is based on a specific database search aimed at the evaluation of modified and unmodified peptides.[57] The main complication of this analysis is technical difficulty of detecting phosphoproteins by mass spectrometry in the presence of nonphosphorylated species.

Splicing is a regulatory mechanism for gene expression present in higher organisms that leads to the production of multiple proteins from a single transcript. Splicing can be detected by the amino acid sequencing as a complementary identification method to the fingerprinting protocol. Disulfide bridging is formed between cysteine residues and can be produced in vivo or during sample preparation. It can occur within the same protein, producing a change in the tertiary structure, or between molecules with the formation of multimers. This modification changes the molecular mass and subsequently the behavior during electrophoretic separation. Formation of multimers is suspected if the same amino acid sequence is detected in several bands or spots on the electrophoretic gel and can be confirmed by the presence of a molecular mass two or more times higher than that of a basic peptide.

Several methods are alternatively used for the detection of posttranslational modification. The immunoprecipitation method is most commonly used to detect phosphoproteins, and it uses antibodies against phosphorylated amino acids. The precipitated proteins are subsequently separated via gel electrophoresis and analyzed by MS. Another method is chemical derivatization with alkali solution, which eliminates *p*-serine and *p*-threonine, and subsequent tagging with biotin. The phosphoproteins after being tagged with biotin are separated and identified by affinity chromatography.[58,59] Other methods are immobilized metal affinity chromatography and affinity purification, both of which use the high affinity of phosphoproteins to positively charged metal ions such as iron, aluminum, or zirconium. Because of this high affinity, the method is biased toward multiply phosphorylated proteins, and also it does not give quantitative information.[58]

Modification of proteins not only occurs in vivo as a regulatory process of protein activity but can also be artificially produced during laboratory procedures. For example, during electrophoresis, proteins react with nonpolymerized acrylamide monomers and produce unusually modified amino acids such as alkalysated cysteine or propionamide. Similarly, during the staining procedure in the presence of acetic acid and methanol, asparic and glutamic acids are methylated.[60] For practical reasons it is important to be aware of these artificial modifications and not misinterpret them as naturally occurring.

The methodology recently implemented into proteomics, multiple reaction monitoring (MRM), gives a significant boost for mass spectrometry in its role to detect and monitor the posttranslational status of proteins. The MRM experiment requires mass spectrometry that runs at high sensitivity to be able to detect targeted components. This method has been developed and used by the pharmaceutical industry to detect and quantify drugs and their metabolites. The potential value for the proteomics investigation was shown by Cox et al,[61] who identified a low level of phosphorylated MEF2A by using a highly sensitive MRM experiment and a hybrid quadrupole linear ion-trap instrument for identification of targeted ions. The phosphorylated targets were deduced based on specific information for MEF2A, which included the primary sequence, the type of phosphorylation (serine/threonine vs. tyrosine), and the acceptor sites for phosphorus (consensus peptide that is targeted by a kinase). The principle of MRM can be applied to studying other types of posttranslational modifications that occur on proteins with known molecular masses and deduced derived peptides. This information about the protein molecular masses and derived peptides can be used to design an MRM experiment specifically for the identification of potential sites of posttranslational modifications that have become a target for mass spectrometry.

Protein Complexes and Protein Interactions

In a physiological environment, proteins do not exist as single entities but rather as a complex system responsible for cellular function and regulation. The methods of studying protein complexes must be able to evaluate not one but multiple elements, as one-on-one interaction between proteins is more an exception than a rule. Proteins interact not only with other proteins but also with nucleic acids, small molecules, and other components, such as drugs. The classic low-throughput system for studying interactions between proteins or between a protein and other molecules is a design in which one component is used as bait for the isolation of its binding partner. The most serious obstacle to this system is its low reproducibility, even if the same bait is used for protein binding.[62]

The protein–protein interaction depends on variations in protein affinity, which is regulated by changes in the cytoplasmic environment. Because the environment of mass spectrometry is completely different from that in the cytoplasm, mass spectrometry is not a good method for studying protein-protein interactions. In recent years, new and more advanced methods for protein complex study have been developed. One is the "complex walking" strategy, in which the complex is studied sequentially. The other method uses stable isotopes for labeling the different components of the protein complexes. This method allows for the identification of dynamic changes in protein composition relative to different cellular states.

There are also several MS-based methods that use modified tagged proteins as a bait, forming physiological complexes with other proteins. These complexes are created in vivo and are subsequently pulled out. Extracted proteins with the tagged bait are identified using the MALDI MS. This method has been used on a large scale for the characterization of protein complexes in yeasts and has been named TAP (tandem affinity purification). The high-throughput modification of this method is called the high-throughput mass spectrometric protein complex identification (HMS-PCI).[63]

Another method of mapping protein-protein interaction is with microchip technology. This method uses a large number of purified proteins that are attached to a microchip surface at known locations and form baits for protein complexes. These bait proteins are preserved in their folded conformation so that they can interact with other proteins. The specimen that contains the mixture of proteins to be tested is incubated with a chip. If specific interaction between bait proteins and protein(s) from the solution occurs, it is detected and characterized by the position on the chip. A substantial advantage of microchip technology over the conventional methods is its high sensitivity and small amount of sample needed for a run. In addition, short turnaround time for a single sample and a large amount of samples that can be tested simultaneously, making this technology attractive, especially in the perspective of diagnostic usage.[64]

All these high-throughput protein-protein interaction experiments produce large amounts of data, which become the foundation for new biological discoveries. However, simultaneously a great challenge to bioinformatics is created to manage, analyze, and model these data.

References

1. OmicsWorld. http://www.omicsworld.com/?gclid = CJ3e6_yU05 ICFR5uIgodNgI0BA.
2. Wikipedia. http://en.wikipedia.org/wiki/Genomics.
3. Wikipedia. http://en.wikipedia.org/wiki/Transcriptomics.
4. Gygi SP, Rochon Y, Franza BR, Aebersold R. Correlation between protein and mRNA abundance in yeast. *Mol Cell Biol.* 1999;19: 1720–1730.
5. Karin DR. Proteomics and cancer diagnosis: the potential of mass spectrometry. *Clin Biochem.* 2004;37:579–583.
6. Patterson DS. Proteomics: beginning to realize its promise? *Arthritis Rheum.* 2004;12:3741–3744.
7. Check E. Proteomics and cancer: running before we can walk? *Nature*. 2004;429:496–497.
8. Petricoin FE III, Ardekani MA, Hitt AB, et al. Use of proteomics patterns in serum to identify ovarian cancer. *Lancet.* 2002;359:572–577.
9. Conrads TP, Fusaro VA, Ross S, et al. High resolution serum proteomic features for ovarian cancer detection. *Endocr Relat Cancer*. 2004;11:163–178
10. Anderson NL, Anderson NG. The human plasma proteome: history, character and diagnostic prospects. *Mol Cell Proteomics*. 2002;1:845–867.
11. Diamandis E. Analysis of serum proteomics pattern for early cancer diagnosis: drawing attention to potential problems. *J Natl Cancer Inst.* 2004;96:353–356.
12. Lowenthal MS, Mehta AI, Frogale K, et al. Analysis of albumin-associated peptides and proteins from ovarian cancer patients. *Clin Chem.* 2005;51:1933–1945.
13. Gillette MA, Mani DR, Carr SA. Place of pattern in proteomics biomarker discovery. *J Proteome Res.* 2005;4:1143–1154.
14. Matthiesen R. Methods, algorithms and tools in computational proteomics: a practical point of view. *Proteomics*. 2007;7: 2815–2832.
15. Henzel WJ, Stults JT, Wong SC, et al. Identifying proteins from two-dimensional gels by molecular mass searching of peptide fragments in protein sequence databases. *Proc Natl Acad Sci U S A*. 1993;90:5011–5015.
16. Mann M, Wilm M. Error-tolerant identification of peptides in sequence databases by peptide sequence tags. *Anal Chem.* 1994; 66:4390–4399.
17. Eng JK, McCormack AL, Yates JR. An approach to correlate tandem mass spectral data of peptides with amino acid sequences in a protein database. *J Am Soc Mass Spectrom.* 1994;5:976–989.
18. Hjero K. Protein identification by PMF. *Methods Mol Biol.* 2006;367:61–76.
19. Hjero K, Hojrup P. Calibration of matrix-assisted laser desorption/ionization time-of-flight peptide mass fingerprinting spectra. *Methods Mol Biol.* 2006;367:49–60.
20. Matthiesen R, Bunkenborg J, Stensballe A, Jensen ON. Database independent, database-dependent and extended interpretation of peptide mass spectra in VEMS V2.0. *Proteomics*. 2004;4: 2583–2593.
21. Craig R, Cortens JC, Fenyo D, Beavis RC. Using annotated peptide mass spectrum libraries for protein identification. *J Proteome Res.* 2006;5:1843–1849.
22. Matthiesen R. Extracting monoisotopic single-charge peaks from liquid chromatography-electrospray ionization-mass spectrometry. *Methods Mol Biol.* 2006;367:37–48.
23. Moritz B, Meyer HE. Approaches for the quantify cation of protein concentration ratios. *Proteomics*. 2003;3:2208–2220.
24. Consoli L, Damerval C. Quantification of individual zein isoforms resolved by two-dimensional electrophoresis: genetic variability in 45 maize inbred lines. *Electrophoresis.* 2001;22:2983–2989.
25. Thiedea B, Höhenwarterb W, Kraha A, et al. Peptide mass fingerprinting. *Methods*. 2005;35:237–247.
26. Godovac-Zimmermann J, Kleiner O, Brown RL, Drukier KA. Perspectives in spicing up proteomics with splicing. *Proteomics*. 2005;5:699–709.
27. Karas M, Bachnann D, Bahr U, Hillenkamp F. Matrix-assisted ultraviolet laser desorption of non-volatile compounds. *Int J Mass Spectrom Ion Process*. 1987;78:53–68.
28. Burtis AC, Ashwood RE, Burns D. *Tietz Textbook of Clinical Chemistry and Molecular Diagnosis*, 4th edn. Philadelphia: Elsevier; 2006:170.
29. Loboda AV, Krutchinsky AN, Bromirski M, et al. A tandem quadrupole/time-of-flight mass spectrometer with a matrix-assisted laser desorption/ionization source: design and performance. *Rapid Commun Mass Spectrom* 2000;14:1047–1057.
30. Chaurand P, Caprioli MR. Direct profiling and imaging of peptides and proteins from mammalian cells and tissue sections by mass spectrometry. *Electrophoresis*. 2002;23:3125–3135.
31. Groseclose MR, Andersson M, Hardesty WM, Caprioli RM. Identification of proteins directly from tissue: in situ tryptic digestions coupled with imaging mass spectrometry. *J Mass Spectrom*. 2007;42:254–262.
32. Lemaire R, Menguellet S, Stauber J, et al. Specific MALDI imaging and profiling for biomarker hunting and validation: fragment of the 11S proteasome activator complex, Reg alpha fragment, is a new potential ovary cancer biomarker. *J Proteome Res.* 2007;6: 4127–4134.
33. Stoeckli M, Farmer TB, Caprioli RM. Automated mass spectrometry imaging with a matrix-assisted laser desorption ionization time-of-flight instrument. *J Am Soc Mass Spectrom.* 1999;10:67–71.
34. Poetz O, Schwenk MJ, Kramer S, et al. Protein microarrays: catching the proteome. *Mech Ageing Dev.* 2005;126:161–170.
35. Petricoin E, Paweletz C, Liotta L. Clinical applications of proteomics: proteomic pattern diagnostics. *J Mammary Gland Biol Neoplasia*. 2002;7:433–440.
36. Petricoin EF, Ardekani AM, Hitt BA, et al. Use of proteomic patterns in serum to identify ovarian cancer. *Lancet.* 2002;359: 572–577.
37. Petricoin EI, Ornstein D, Paweletz C, et al. Serum proteomic patterns for detection of prostate cancer. *J Natl Cancer Inst.* 2002;94:1576–1578.
38. Zhang L, Rosenzweig J, Wang Y, Chan D. Proteomics and bioinformatics approaches for identification of serum biomarkers to detect breast cancer. *Clin Chem.* 2002;48:1296–1304.
39. Rodland DK. Proteomics and cancer diagnosis: the potential of mass spectrometry. *Clin Biochem.* 2004;37:579–583.
40. Reinders J, Lewandrowski U, Moebius J, et al. Challenges in mass spectrometry-based proteomics. *Proteomics.* 2004;4:3686–3703.
41. Schrader W, Klein WH. Liquid chromatography/Fourier transform ion cyclotron resonance mass spectrometry (LC-FTICR MS): an early overview. *Anal Bioanal Chem.* 2004;379: 1013–1024.

42. Page JS, Masselon CD. Smith RD. FTICR mass spectrometry for qualitative and quantitative bioanalysis. *Curr Opin Biotechnol.* 2004;15:3–11.
43. Takats Z, Kobliha V, Sevcik K, et al. Characterization of DESI-FTICR mass spectrometry: from ECD to accurate mass tissue analysis. *J Mass Spectrom.* 2008;43:196–203.
44. Johnson RS, Davis MT, Taylor JA, Patterson SD. Informatics for protein identification by mass spectrometry. *Methods* 2005;35: 223–236.
45. Schmidt F, Schmid M, Jungblut PR, et al. Iterative data analysis is the key for exhaustive analysis of peptide mass fingerprints from proteins separated by two-dimensional electrophoresis. *J Am Soc Mass Spectrom.* 2004;14:943–956.
46. Papayannopoulos IA. The analysis of progressive spectra yields information about the amino acid sequence for the chosen peptide. *Mass Spectrom Rev.* 1995;14:49–73.
47. Apweiler R, Bairoch A, Wu CH. Protein sequence databases. *Curr Opin Chem Biol.* 2004;8:76–80.
48. Thiedea B, Höhenwarterb W, Kraha A, et al. Peptide mass fingerprinting. *Methods.* 2005;35:237–247.
49. Tabb LD, Smith LL, Breci AL, et al. Statistical characterization of ion trap tandem mass spectra from doubly charged tryptic peptides. *Anal Chem.* 2003;75:1155–1163.
50. Shadforth I, Crowther D, Bessant C. Protein and peptide identification algorithms using MS for use in high-throughput, automated pipelines. *Proteomics.* 2005;5:4082–4095.
51. Gygi SP, Corthals GL, Zhang Y, et al. Evaluation of two dimensional gel electrophoresis-based proteome analysis technology. *Proc Natl Acad Sci U S A.* 2000;97:9390–9395.
52. Lopez-Ferrer D, Martinez-Bartolome S, Villar M, Campillos M. Statistical model for large scale peptide identification in databases from tandem mass spectra using SEQUEST. *Anal Chem.* 2004;76:6853–6860.
53. Beausoleil SA, Villen J, Gerber SA, Ruch J, Gygi SP. A probability-base approach for high throughput protein phosphorylation analysis and site localization. *Nat Biotechnol.* 2006;24:1285–1292.
54. Kleiner O, Price DA, Ossetrova N, et al. Ultra-high sensitivity multi-photon detection imaging in proteomics analyses. *Proteomics.* 2005;5(9):2322–2330.
55. Listgarten J, Emili A. Statistical and computational methods for comparative proteomic profiling using liquid chromatography-tandem mass spectrometry. *Mol Cell Proteomics.* 2005;4:419–434.
56. Mann M. Analysis of protein phosphorylation using mass spectrometry: deciphering the phosphoproteome. *Trends Biotechnol.* 2002;20:261–268.
57. MacCoss MJ. Shotgun identification of protein modifications from protein complexes and lens tissue. *Proc Natl Acad Sci U S A.* 2002;99:7900–7905.
58. Mirza SP, Olivier M. Methods and approaches for the comprehensive characterization and quantification of cellular proteomes using mass spectrometry. *Physiol Genomics.* 2008;33:3–11.
59. Goshe MB, Conrads TP, Panisko EA, Angell NH, Veenstra TD, Smith RD. Phosphoprotein isotope-coded affinity tag approach for isolating and quantitating phosphopeptides in proteome-wide analyses. *Anal Chem.* 2001;73:2578–2586.
60. Galvani M, Hamdan M, Herbert B, Righetti GP. Alkylation kinetics of proteins in preparation for two-dimensional maps: a matrix assisted laser desorption/ionization - mass spectrometry investigation. *Electrophoresis.* 2001;22:2058–2065.
61. Cox DM, Zhong F, Du M, Duchoslav E, Sakuma T, McDermott JC. Multiple reaction monitoring as a method for identifying protein posttranslational modifications. *J Biomol Tech.* 2005;16(2): 83–90.
62. Rotheneder H, Geymayer S, Haidweger E. Transcription factors of the Sp1 family: interaction with E2F and regulation of the murine thymidine kinase promoter. *J Mol Biol.* 1999;293:1005–1015.
63. Yu C, Dong X. Computational analyses of high-throughput protein-protein interaction data. *Curr Protein Peptide Sci.* 2003;4:159–181.
64. Zhu H, Klemic JF, Chang S, Bertone P, Casamayor A, Klemic K.G, Smith D, Gerstein M, Reed MA, Snyder M. Analysis of yeast protein kinases using protein chips. *Nat Gene.* 2000;26:283–289.

14
Immunohistochemistry of Biomarkers

Patrick L. Fitzgibbons and Kumarasen Cooper

Introduction

Immunohistochemistry (IHC) has an essential role in the diagnostic evaluation of cancers, most commonly to help identify and subclassify tumors, but its utility in assessing biomarkers that are predictive of benefit or lack of benefit from specific chemotherapies is becoming increasingly important. When IHC is used for predictive marker testing, the assessment of the extent of protein expression may be even more important than the presence or absence of expression, but many variables affect the quantitative measurement of protein expression in routinely processed tissues, and these can have profound influences on the test results.

Immunohistochemistry (IHC) has been a critical tool for pathologists for more than 20 years and is still used most frequently to determine cell lineage and tumor type. Because of its far greater specificity, IHC has essentially replaced most traditional histochemical stains once used for this purpose. IHC is also used to identify specific cellular constituents (e.g., basal and myoepithelial cells), which may help to determine the presence of malignancy and/or invasion and to diagnose infectious diseases, but its role in predictive marker testing to select or exclude patients for specific therapies is growing rapidly in frequency and importance.

Until recently, only a few predictive markers were routinely assessed in all patients, but an increasing number are becoming a regular part of clinical management. Such tissue-based biomarkers, many assessed solely by IHC, may predict responsiveness (or lack of response) to specific drugs or to entire classes of chemotherapeutic agents; however, along with the discovery of new biomarkers and their clinical significance, we are increasingly recognizing problems related to variability in laboratory assessment.

Role of Immunohistochemistry in Biomarker Testing

For some markers, such as steroid hormone receptors in breast cancer, IHC is the standard method of identifying patients likely to respond to specific drugs. Patients with tumors that are estrogen receptor positive [ER(+)] (about 70% of breast cancers) are more likely to respond to selective estrogen receptor modulators (e.g., tamoxifen) or aromatase inhibitors than those with ER(−) tumors. Indeed, ER expression as assessed by IHC has been shown to be the single most important biomarker in breast cancer: tamoxifen treatment of ER(+) patients reduces annual breast cancer death rates by nearly one-third.[1] Similarly, responsiveness to the drug trastuzumab is much more likely in patients with tumors that overexpress HER2 protein (or have amplification of the HER2 gene). Other examples of IHC-based identification of potential therapeutic responsiveness include CD20 expression in B-cell non-Hodgkin lymphoma (rituximab), CD117 staining of gastrointestinal stromal tumors (imatinib mesylate), and topoisomerase II alpha in breast cancer (anthracyclines).

More recently, detection of the excision repair cross-complementation group 1 (ERCC1) protein by IHC has been shown to be useful as both a prognostic indicator and a predictor of response to chemotherapy in patients with non-small cell lung cancer. Patients with ERCC1(+) tumors survive longer than those whose tumors do not express ERCC1. More importantly, patients with ERCC1(−) tumors may benefit from adjuvant cisplatin whereas those with ERCC1(+) tumors appear resistant to platinum-based chemotherapy.[2]

For some tumors, there is little correlation between overexpression of a specific protein and responsiveness to agents

P.T. Cagle and T.C. Allen (eds.), *Basic Concepts of Molecular Pathology*, Molecular Pathology Library 2,
DOI: 10.1007/978-0-387-89626-7_14, © Springer Science+Business Media, LLC 2009

targeting that same molecule. Cetuximab is a monoclonal antibody that has a high affinity for the epidermal growth factor receptor (EGFR). The drug was initially shown to be effective in some patients with EGFR(+) colorectal carcinoma, but similar response rates were subsequently described in patients whose tumors were negative for EGFR by IHC.[3] Similarly, response to the EGFR tyrosine kinase inhibitors gefitinib and erlotinib in non-small cell lung cancer is related to the presence of activating mutations in the epidermal growth factor receptor, but not to the presence of EGFR positivity by IHC.[4] In such cases, assessment of the EGFR status of the tumor by IHC seems to have little or no value in terms of predicting response to therapy.

In addition to its utility in identifying potential therapeutic response, IHC can sometimes be used as a surrogate test to identify underlying molecular abnormalities. As previously noted, IHC can reveal overexpression of cell-surface or cytoplasmic proteins because of gene amplification, for instance, HER2 and TOP2A in breast cancer, but can also provide evidence of specific translocations and mutations, which may be of great value for diagnosis and patient management. For example, follicular lymphoma is classically associated with a t(14,18)(q32;q21) translocation that brings the BCL2 gene adjacent to the immunoglobulin heavy chain gene. This change results in aberrant overexpression of functionally active bcl-2 protein, which can be detected by IHC and used diagnostically.

Soft tissue tumors are also commonly associated with specific translocations that result in proteins detectable by IHC. Desmoplastic small round cell tumor, for example, has a specific chromosomal translocation, t(11;22)(p13;q12), that fuses EWS on chromosome 22 to WT1 on chromosome 11; this results in a EWS/WT1 fusion protein that can be detected by immunostaining for WT-1 (usually with antibodies to the C-terminal of WT-1). Similarly, about 90% of primitive neuroectodermal tumor/Ewing's sarcomas have a t(11;22)(q24;q12) translocation that results in a EWS/FLI-1 fusion gene and overexpression of the FLI-1 protein.

Immunohistochemistry can also be used to identify abnormal protein expression resulting from gene mutations. Mutations in the tumor suppressor gene p53 are one of the most common genetic changes found in human cancers. Although the protein produced by wild-type p53 has a very short half-life and is undetectable by IHC, p53 gene mutation results in an inactive but stable protein that is easily detected by IHC stains. Similarly, mutations in the adenomatous polyposis coli (APC) gene result in abnormal intranuclear accumulation of β-catenin protein, and IHC stains for β-catenin are very useful in diagnosing tumors such as mesenteric fibromatosis.

The opposite staining pattern is seen with other markers. Hereditary nonpolyposis colon carcinoma (HNPCC) most often arises because of a germline mutation in one or more genes for DNA mismatch repair, principally MLH1 and MSH2. Although the proteins produced by these genes are detectable by IHC in normal colonic epithelial cells, mutation in tumor cells results in loss of functionality of the corresponding proteins and, in most cases, absence of staining by IHC (a mutant protein is sometimes produced that is detected by IHC). As such, IHC can be used as a screening tool to identify patients likely to have mutations and for whom genetic sequencing may be useful.[5,6]

Identification of abnormal mismatch repair status in colorectal carcinoma provides useful prognostic information[7] and may be of predictive value in identifying patients who are resistant to 5-fluorouracil-based chemotherapy.[8] Although IHC is a sensitive method for identifying defects in mismatch repair proteins, the results are not specific for HNPCC. About 15% of sporadic colorectal cancers also have microsatellite instability, usually because of hypermethylation of the MLH1 promoter and silencing of the gene, and such tumors exhibit loss of expression of the corresponding protein similar to that seen in HNPCC.[9]

Importance of Method Standardization in Biomarker Testing

IHC stains that are used to determine cell lineage are almost always done as part of an overall diagnostic evaluation and interpretation, and the data obtained are generally meaningful only in conjunction with the clinical and histological features. The significance of any one lineage stain depends on the context in which it is assessed, and this must be determined by the pathologist evaluating the entire case. Because such immunostains are usually done in panels, minor staining differences caused by variations in antigenicity, fixation, or laboratory methods may have little consequence. In this setting, the presence or absence of staining is usually much more important than the extent or intensity of staining.

Predictive biomarker testing, by contrast, has several important differences. Although the significance of a lineage stain cannot be separated from the overall diagnostic evaluation, a predictive marker stain can be considered a discrete laboratory test that provides independent information leading directly to a treatment decision. In this setting, the extent or intensity of staining may be more important than its presence or absence and can be of paramount importance in determining whether a patient is eligible for drug therapy. Thus, minor differences in staining may have a major impact on patients.

Few data were available on the overall accuracy of IHC testing for various biomarkers until recently, but it was generally assumed that experienced laboratories using modern equipment and methods were providing reliable results. This perception changed with the publication of studies showing significant discordances in the results of HER2 testing among different laboratories. In these studies, central testing failed to confirm the results of testing done in local hospital laboratories in up to a quarter of cases.[10,11] This finding was alarming because of the impact on patients with inaccurate test results. A false-negative HER2 test would result in denial of a potentially life-saving therapy, whereas a false-positive test might lead to very expensive but unnecessary treatment with a drug that has significant cardiotoxicity in some patients.

These papers drew attention to the underappreciated reality that relatively common variations in preanalytical and analytical phases of routine testing can lead to a significant number of incorrect test results. Not surprisingly, this situation was deemed unacceptable and led directly to the creation of new HER2 testing guidelines by the College of American Pathologists and the American Society of Clinical Oncology.[12] Lessons learned from HER2 have had an impact on predictive marker testing in general and underscore the need for greater stringency when testing for biomarkers. Some of the most important issues that must be addressed include fixation, assay validation, adherence to standardized methods, and ongoing laboratory performance assessment.

Preanalytical Factors

The single most important preanalytical factor influencing consistency of biomarker testing is variability in fixation (time of fixation and fixative type). Incomplete fixation produces heterogeneous staining across tissue sections whereas prolonged fixation can result in partial or complete loss of immunoreactivity. Goldstein et al, in a study of the effect of time of formalin fixation on the results of estrogen receptor staining, showed that tissues must be fixed at least 6–8 h to achieve reliable results.[13] This 6-h minimum is irrespective of specimen size; it is not the rate of tissue penetration by formalin that is important, but rather the time needed to complete the chemical reactions that comprise formalin "fixation." Although penetration of aqueous formalin into tissues is relatively fast (about 1 mm per hour), the biochemical cross-linking needed for complete fixation is much slower (up to 24 h).[14,15,16] Thus, all specimens, including needle biopsies, need to be fixed for at least 6 h before processing. Placing specimens in the tissue processor before they have been completely formalin fixed leads to completion of fixation in alcohol (a coagulation fixative). This event is an added source of variability that can result in false-positive and false-negative IHC test results.[16]

The adverse effects of underfixation on IHC have been shown to be greater than those of overfixation.[13,17,18] While problems with overfixation can be overcome to a significant extent by heat-induced antigen retrieval,[19,20] there are no such remedies for underfixed tissues.

When using an in vitro diagnostic assay that specifies specimen handling conditions, one must follow the manufacturer's instructions on fixation and processing. Most IHC assays have been validated on tissues fixed in 10% neutral buffered formalin, and use of an alternative fixative may invalidate the predictive aspects of the assay. Laboratories that choose to use another fixative must validate that fixative's performance by testing the same tissues that have also been fixed in buffered formalin and showing equivalent results.[12]

Analytical Factors

Formalin fixation alters the three-dimensional configuration of protein as a result of protein–protein cross-links, thereby potentially altering antibody-antigen interaction. The introduction of antigen retrieval techniques in the early 1990s revolutionized IHC by drawing attention to methodologies to unmask these protein cross-links. Microwave pretreatment of sections immersed in a buffer enhanced immunostaining reactions several fold. Subsequently, other techniques employing heat (e.g., pressure cooking, steam heating, and high-temperature water bath) and proteolytic enzyme digestion (e.g., proteinase K, trypsin, pepsin, etc.) and combinations thereof have been employed to great success[21]; however, caution with respect to overdigestion must be exercised, because excessive tissue digestion may result in destruction of antigenicity with subsequent false-negative immunoreactions. Conversely, extreme tissue digestion may also unmask "unwanted" antigens, resulting in false positivity. Both pitfalls can be avoided with cautious optimization and monitoring of antigen retrieval techniques plus use of appropriate positive and negative controls.

Antibodies used to assess biomarkers can vary significantly in sensitivity and specificity. Until recently, most antibodies used in clinical practice were rabbit polyclonal or mouse monoclonal reagents. Newer rabbit monoclonal antibodies generally have greater affinity and specificity than mouse monoclonal antibodies. As a result, these antibodies are often quickly adopted for routine practice[22]; however, because the intensity or extent of staining is of such critical importance in predictive biomarker testing, conclusive evidence of superiority over existing reagents in predicting outcome should be required before changing methods.[23]

The choice of an immunodetection system is of critical importance in accurate and reproducible biomarker testing. The most widely employed detection system is the avidin–biotin IHC technique, but this is being gradually replaced by more sensitive polymer-based detection systems. The latter involve a two-step procedure, comprising a dextran polymer "backbone" to which are bound peroxidase molecules and secondary antibodies (to bind to the unlabeled primary antibody).[24] This highly recommended system enjoys the advantage of being avidin-biotin free, thus avoiding potential nonspecific binding with endogenous biotin in some tissues.[25] Newer techniques employing tyramide-based amplification technology (or catalyzed reporter deposition) provide even more sensitive methods of antigen detection. Although these latter techniques are useful for detecting low levels of antigenicity, they do carry the potential problem of spurious false-positive immunoreactions, and, hence, cautious interpretation with appropriate positive and negative controls is always mandatory.

When assessing biomarkers by IHC, one must always consider the possibility of intratumoral heterogeneity. A significant portion of cases with equivocal immunohistochemical staining for HER2 also show heterogeneity of gene amplification. Lewis et al,[26] in a study of cases that were equivocal (Score 2) by IHC, found that when staining multiple additional blocks from the same case, more than half of cases had a score that was different from the initial score. This study confirmed that equivocal staining is often the result of intratumoral genetic heterogeneity and/or low-level gene amplification.

All laboratory assays must be validated before results can be reported on patient specimens, but validation is particularly important for biomarker assays. For initial IHC test validation, labs should show that their assay agrees with a reference method in a high percentage of cases. The ASCO/CAP HER2 Guidelines require at least 95% concordance between IHC and fluorescence in situ hybridization (FISH) (or with another lab's validated IHC assay) when testing the same blocks. The validation study must be done on at least 25–100 cases, which include negative and positive samples. Assays must be revalidated whenever there is a significant change to test methods (new antibody type, new detection system, new automated platform, etc.).[27]

Postanalytical Factors

Besides strict adherence to specimen handling guidelines and assay performance specifications, ongoing assessment of laboratory performance is essential. This step includes training and regular competency assessment of all laboratory personnel, including the pathologists who interpret the stain slides.

Importantly, some form of continuing quality assessment is needed for each biomarker assay offered by the laboratory. This assessment is usually done by participating in an external proficiency testing program,[28] but if such a program is not available, for example, for new or rare analytes, an alternative form of proficiency testing may be done. This effort may take the form of split sample analysis with other labs or comparing the results of staining with an alternative test method (e.g., FISH). Such steps must be done on a regular, ongoing basis (e.g., several challenges two to four times per year).

Conclusions

Molecular tests that directly analyze nuclei acids can provide highly specific diagnostic and predictive information, but IHC continues to play an essential role in biomarker testing and clinical decision making. Because the quantitative aspect of predictive marker testing by IHC is subject to significant analytical variability, steps must be taken to ensure that standardized methods are used throughout the testing process.

References

1. Early Breast Cancer Trialists' Collaborative Group (EBCTCG). Effects of chemotherapy and hormonal therapy for early breast cancer on recurrence and 15-year survival: An overview of the randomised trials. *Lancet.* 2005;365(9472):1687–1717.
2. Olaussen KA, Dunant A, Fouret P, et al. DNA repair by ERCC1 in non-small-cell lung cancer and cisplatin-based adjuvant chemotherapy. *N Engl J Med.* 2006;355:983–991.
3. Chung KY, Shia J, Kemeny NE, et al. Cetuximab shows activity in colorectal cancer patients with tumors that do not express the epidermal growth factor receptor by immunohistochemistry. *J Clin Oncol.* 2005;23:1803–1810.
4. Pinter F, Papay J, Almasi A, et al. Epidermal growth factor receptor (EGFR) high gene copy number and activating mutations in lung adenocarcinomas are not consistently accompanied by positivity for EGFR protein by standard immunohistochemistry. *J Mol Diagn.* 2008;10:160–168.
5. Lynch HT, Boland CR, Rodriguez-Bigas MA, et al. Who should be sent for genetic testing in hereditary colorectal cancer syndromes? *J Clin Oncol.* 2007;25:3534–3542.
6. Shia J. Immunohistochemistry versus microsatellite instability testing for screening colorectal cancer patients at risk for hereditary nonpolyposis colorectal cancer syndrome: part I. The utility of immunohistochemistry. *J Mol Diagn.* 2008;10:293–300.
7. Lanza G, Gafa R, Santini A, et al. Immunohistochemical test for MLH1 and MSH2 expression predicts clinical outcome in stage II and III colorectal cancer patients. *J Clin Oncol.* 2006;24: 2359–2367.
8. Jover R, Zapater P, Castells A, et al. Mismatch repair status in the prediction of benefit from adjuvant fluorouracil chemotherapy in colorectal cancer. *Gut.* 2006;55:848–855.
9. Kane MF, Loda M, Gaida GM, et al. Methylation of the hMLH1 promoter correlates with lack of expression of hMLH1 in sporadic colon tumors and mismatch repair-defective human tumor cell lines. *Cancer Res.* 1997;57:808–811.
10. Paik S, Bryant J, Tan-Chiu E, et al. Real-world performance of HER2 testing: National Surgical Adjuvant Breast and Bowel Project experience. *J Natl Cancer Inst.* 2002;94:852–854.
11. Roche PC, Suman VJ, Jenkins RB, et al. Concordance between local and central laboratory HER2 testing in the breast intergroup trial N9831. *J Natl Cancer Inst.* 2002;94:855–857.
12. Wolff AC, Hammond ME, Schwartz JN, et al. American Society of Clinical Oncology/College of American Pathologists guideline recommendations for human epidermal growth factor receptor 2 testing in breast cancer. *Arch Pathol Lab Med.* 2007;131:18–43.
13. Goldstein NS, Ferkowicz M, Odish E, et al. Minimum formalin fixation time for consistent estrogen receptor immunohistochemical staining of invasive breast carcinoma. *Am J Clin Pathol.* 2003;120:86–92.
14. Fox CH, Johnson FB, Whiting J, Roller PP. Formaldehyde fixation. *J Histochem Cytochem.* 1985;33:845–853.
15. Helander KG. Kinetic studies of formaldehyde binding in tissue. *Biotech Histochem.* 1994;69:177–179.

16. Jacobs TW, Gown AM, Yaziji H, et al. Specificity of HercepTest in determining HER-2/neu status of breast cancers using the United States Food and Drug Administration-approved scoring system. *J Clin Oncol.* 1999;17:1983–1987.
17. De Marzo AM, Fedor HH, Gage WR, Rubin MA. Inadequate formalin fixation decreases reliability of p27 immunohistochemical staining: probing optimal fixation time using high-density tissue microarrays. *Hum Pathol.* 2002;33:756–760.
18. Oyama T, Ishikawa Y, Hayashi M, et al. The effects of fixation, processing and evaluation criteria on immunohistochemical detection of hormone receptors in breast cancer. *Breast Cancer.* 2007;14:182–188.
19. Boenisch T. Effect of heat-induced antigen retrieval following inconsistent formalin fixation. *Appl Immunohistochem Mol Morphol.* 2005;13:283–286.
20. Shi SR, Liu C, Taylor CR. Standardization of immunohistochemistry for formalin-fixed, paraffin-embedded tissue sections based on the antigen-retrieval technique: from experiments to hypothesis. *J Histochem Cytochem.* 2007;55:105–109.
21. Leong TY-M, Leong AS-Y. How does antigen retrieval work? *Adv Anat Pathol.* 2007;14:129–131.
22. Cheang MC, Treaba DO, Speers CH, et al. Immunohistochemical detection using the new rabbit monoclonal antibody SP1 of estrogen receptor in breast cancer is superior to mouse monoclonal antibody 1D5 in predicting survival. *J Clin Oncol.* 2006;24:5637–5644.
23. Dowsett M. Estrogen receptor: methodology matters. *J Clin Oncol.* 2006;24:5626–5628.
24. Sabbattini E, Bisgaard K, Ascani S, et al. The EnVision++ system: a new immunohistochemical method for diagnostics and research. Critical comparison with the APAAP, ChemMate, CSA, LABC and SABC techniques. *J Clin Pathol.* 1998;51:506–511.
25. Cooper K, Haffajee Z, Taylor L. Comparative analysis of biotin intranuclear inclusions of gestational endometrium using the APAAP, ABC and the PAP immunodetection systems. *J Clin Pathol.* 1997;50:153–156.
26. Lewis JT, Ketterling RP, Halling KC, et al. Analysis of intratumoral heterogeneity and amplification status in breast carcinomas with equivocal (2+) HER-2 immunostaining. *Am J Clin Pathol.* 2005;124:273–281.
27. Goldstein NS, Hewitt SM, Taylor CR, et al. Recommendations for improved standardization of immunohistochemistry. *Appl Immunohistochem Mol Morphol.* 2007;15:124–133.
28. Fitzgibbons PL, Murphy DA, Dorfman DM, et al. Interlaboratory comparison of immunohistochemical testing for HER2: results of the College of American Pathologists HER2 Immunohistochemistry Tissue Microarray Program. *Arch Pathol Lab Med.* 2006;130:1440–1445.

15

Basic Principles of Flow Cytometry

Youli Zu, Munir Shahjahan, and Chung-Che Chang

Introduction

The concept of flow cytometry has been in existence for more than five decades. With the emergence of antibodies and fluorochrome chemistry and the advances in laser and computer technology, flow cytometry has become a powerful and indispensable tool for the diagnosis of hematopoietic and lymphoid disorders. Although limited, it can be also used for determining cell types of solid tissue neoplasms. Current flow cytometers have the capability of simultaneously measuring multiple parameters of individual cells in a cell suspension. Thus, a large number of cell specimens can be processed with a quick turnaround time. In addition, flow cytometry is also highly sensitive and can detect immunophenotype of cells in a specimen with thousands of cells. The parameters analyzed by flow cytometry include physical properties of cells; the size, cytoplasmic granularity, and amount of DNA contents; and cell antigens/markers (surface, cytoplasmic, and nuclear) that can be recognized by specific antibodies. By using appropriate antibody panels, flow cytometry can reveal the cell type (hematopoietic, lymphoid, or nonhematopoietic), cell lineage (B- and T cells, natural killer cells, myeloid/monocytic cells, neuro/neuroendocrine cells, and epithelial cells), cell maturation stage (precursors vs. matured cells), and B-cell clonality of expressing immunoglobulin light chains. Moreover, flow cytometry allows us to sort cells based on their distinct parameters for clinical and research purposes. In this chapter, we outline the basic principles of flow cytometry for understanding when flow cytometry will be helpful and how to interpret flow cytometric data for disease diagnosis. Importantly, flow cytometry analysis results must be correlated with clinical presentations and pathological findings for accurate diagnosis.

Basic Principles of Flow Cytometry Analysis

Flow cytometry has a variety of clinical applications include phenotyping analysis (diagnosis) of leukemia and lymphoma, following-up of residual disease, assessment of cell size and membrane integrity (viability), detection of antigens used as therapeutic targets, monitoring CD4/CD8 T-cell ratio in human immunodeficiency virus (HIV)-infected individuals, detection of DNA/RNA content (cell-cycle analysis), assessment of reticulocyte count, detection of platelet antibody, diagnosis of paroxysmal nocturnal hemoglobinuria, detection of nonhematopoietic neoplastic cells, transplant applications (HLA typing), and estrogen receptor analysis.[1–14] Flow cytometry analysis mainly relies on identification of cell antigens/markers by using specific antibodies (immunophenotyping study). However, the quantity and quality of the diagnostic markers may differ in different cells. Thus, the diagnostic value of flow cyto-metry varies among diseases. Generally speaking, flow cytometry is

- Most powerful for diagnosis and classification of B-cell lymphoma/leukemia
- Very powerful for diagnosis of plasma cell neoplasm
- Very useful for diagnosis of precursor B- and T-cell neoplasms
- Very useful for diagnosis and classification of acute myeloid leukemia
- Useful for diagnosis of T-cell lymphoma/leukemia
- Helpful in diagnosis of myelodysplastic syndromes and chronic myeloproliferative disorders
- Not useful for diagnosis of nonhematopoietic/lymphoid disorders

P.T. Cagle and T.C. Allen (eds.), *Basic Concepts of Molecular Pathology*, Molecular Pathology Library 2,
DOI: 10.1007/978-0-387-89626-7_15, © Springer Science+Business Media, LLC 2009

In comparison to immunohistochemical techniques, flow cytometry analysis has its incomparable advantages including simultaneous multiparametric analysis of individual cells, shorter turnaround time (minutes to hours), and the need for many fewer specimens (thousands of cells) for diagnosis (Table 15.1). Flow cytometry also provides quantitative results with less subjective interpretation. However, sampling bias without morphological correlation may diminish its accuracy. All these factors need to be taken into consideration when ordering a flow cytometry analysis.

Technical Aspects

Antibodies

One of the most significant advancements in the immunophenotyping analysis of cell-surface antigens/markers is the development of highly specific monoclonal antibodies. Individual monoclonal antibodies are produced by cloned antibody-secreting cells and specifically recognize the antigen that was used in immunolization for the antibody production. The cluster of differentiation (abbreviated as CD) was initially a protocol used for the identification and distinction of cell-surface molecules (antigens) present on leukocytes. In 1982, the CD nomenclature was formally established in the first International Workshop and Confe-rence on Human Leukocyte Differentiation Antigens (HLDA). This system was used for the classification of the monoclonal antibodies based on their recognized CD molecules. Each surface molecule that is recognized by antibodies is assigned a CD number individually. Using the CD system, we can identify cells by the presence or absence of particular surface molecules. Cells are usually defined using a "+" or a "–"' symbol to indicate whether they express or lack the CD molecules. For example, a "CD3+ and CD20–" cell is one that expresses CD3 but not CD20. Currently, the CD system has been expanded and is widely used in all types of cells.[15–17]

Fluorochromes and Fluorescent Dye

The characterization of cells by flow cytometry is mainly based on the presence of distinct cell antigens/markers and cell contents, which can be detected through the fluorescent signals directly or indirectly. Fluorescent dyes can directly bind to certain cellular content, such as DNA and RNA, and allow us to perform quantitative analysis on individual cells. However, in most cases fluorochromes are conjugated with monoclonal antibodies, which specifically target cellular antigens/markers. Fluorochromes are substances that can be excited by certain light source (such as laser) and emit a fluorescent signal at a single wavelength.[18] Table 15.2 provides a list of commonly used fluorochromes and dyes. Interestingly, although some of them can be excited by the same light source, the different fluorochromes may emit fluorescent signals with different wavelengths/colors. Thus, multiple fluorochromes can be simultaneously excited by a light source and detected by their emission fluorescent signals with different wavelengths, respectively. Under the current consensus guidelines, a minimum of three fluorochromes should be used in a multicolor flow cytometry for the analysis of leukemia and lymphoma.[13] As illustrated in Fig. 15.1, by utili-zing different fluorochrome-conjugated antibodies, multiple antigens/markers in the same cells can be simultaneously detected.

Flow Cytometers

For cell marker detection and analysis, the basic components of a flow cytometer include (1) the flow system or fluidics, (2) an optical system to stimulate and detect fluorescent signals, (3) an electronic system, and (4) a computer system (Fig. 15.2). The *flow system* has a coaxial streaming nature that transports and presents single cells in suspension to the laser beam(s) and the optical path for the detectors through hydrodynamic focusing. The coaxial stream consists of the inner sample fluid surrounded by an outer sheath of isotonic fluid that creates a laminar flow, allowing the cells in the

TABLE 15.1. Comparison of immunophenotypic techniques.

Flow cytometry	Immunohistochemistry
Shorter turnaround time (minutes to hours)	Longer turnaround time (hours to days)
Less subjective result interpretation	Subjective result interpretation
Quantitative results	Semiquantitative results
Good to detect immunoglobulin light chain restriction	High background staining for immunoglobulin light chains
Multiple antibodies/fluorochromes per test	Usually limited to a single antibody per slide
Greater antibody selection	Fewer antibodies available
Data/results can be electronically transferred	Slides can be shipped by mail or courier service
Need fresh cells or tissue	Can use fixed/archived tissue
Limited morphologic correlation	Architectural and cytologic correlation
Cannot assess nonviable cells	Can assess nonviable "ghost" cells

TABLE 15.2. Characteristics of fluorochromes commonly used in flow cytometry.

Fluorochromes conjugated to antibodies	Excitation wavelength (nm)	Emission wavelength (nm)
Fluorescein isothiocyanate (FITC)	488[a]	530
Phycoerythrin (PE)	488	580
PE-Texas Red	488	615
PE-Cy5	488	670
Peridinin chlorophyl protein (PerCP)	488	670
Allophycocyanin (APC)	633[a]	670
APC-Cy7	633	767
Propidium iodide (dye to stain DNA)	488	620

[a]Argon laser is commonly used for 488 nm and HeNe laser for 633 nm excitation wavelength

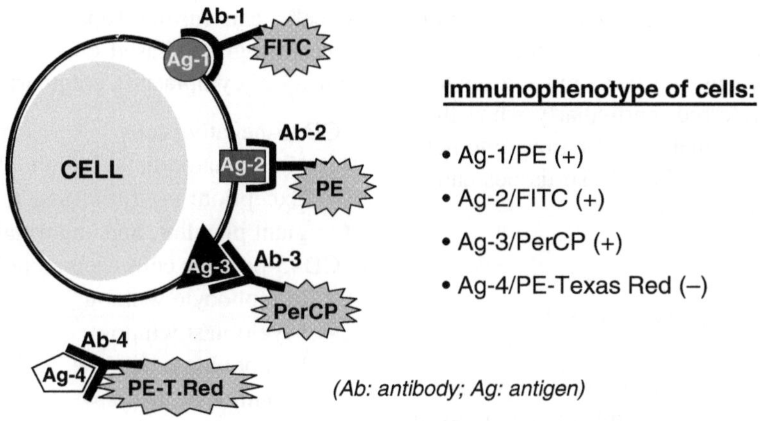

FIG. 15.1. Simultaneous detection of multiple cell antigens/markers. Multiple cell antigens (*Ag*) are recognized by fluorochrome-conjugated specific antibodies (*Ab*). Because different fluorochromes have different emission wavelengths/colors, they can be simultaneously detected by a flow cytometer. *FITC* fluorescein isothiocyanate; *PE* phycoerythrin; *PerCP* peridinin chlorophyll protein; *PE-T Red* PE-Texas Red.

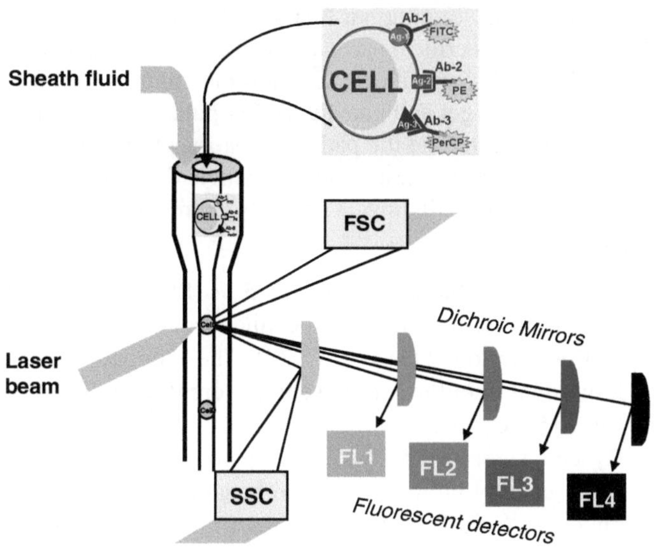

FIG. 15.2. Schematic diagram of the basic components of a flow cytometer. For immunophenotyping, analysis cells are incubated with fluorochrome-conjugated antibodies, which can specifically bind to different cell antigens. The flow system transports and presents single cells to the laser beam(s). Resultant forward light scatter (*FSC*) and side light scatter (*SSC*) from cells and emitted fluorescent signals from individual fluorochromes with different wavelengths/colors are simultaneously detected by fluorescent detectors (*FL*).

center to pass individually through the interrogation point of the laser beams. The *optical system* carries laser beam(s) to excite the cells and fluorochromes and uses a series of optical devices including filters, mirrors, and detectors to collect the incident light, reflected or scattered light, and emitted fluorescent signals (see Fig. 15.2). The most commonly used lasers, argon ion lasers, produce a laser beam of 488-nm wavelength that is capable of exciting several fluorochromes and dyes with different excitation wavelengths/colors, as listed in Table 15.2. The multicolored fluorescent signals with different wavelengths are then separated by the combined usage of filters, lenses, and dichroic mirrors. In the *electronic system*, after traveling through the optical system, the light and fluorescent signals are collected by the corresponding detectors and then converted into electronic impulses proportional to the number of photons. Although these impulses/signals are analog, they can be further amplified and digitized for computer analysis. An external *computer system* can be directly connected to the flow cytometer and interfaced with it in two ways. As a navigator, it controls many operating functions of the flow cytometer such as establishing the instrument setting, setting the compensation, monitoring operation conditions, gating cell populations, and directing the cell-sorting controls. The computer system also plays an important role in acquisition, storage, anal-ysis, and presentation of cytometric data. Most often data are graphically displayed as dot plots or two-parameter histograms, although a variety of other data presentation formats, such as contour plot and isometric plot, are available. Contemporary cytometers are generally capable of analyzing between 5 and 13 parameters (forward scatter, side scatter, and as many as 11 different fluorescent wavelengths/colors).[19]

Specimens Suitable for Flow Cytometry

Theoretically, any specimens from which a single cell suspension can be generated are suitable for flow cytometry analysis. However, a lack of distinct antigens or markers in the cells of interest or tissues limits the diagnostic value of flow cytometry. Common specimens suitable for flow cytometry analysis include peripheral blood, bone marrow, body fluids, cerebrospinal fluid, urine, lymph node (cells or fresh tissues), any fine-needle aspirates, and fresh tissues suspicious for hematopoietic and lymphoid disorders. For blood and bone marrow specimens, anticoagulants such as ethylenediaminetetraacetic acid (EDTA), heparin, or acid citrate dextrose are needed. Fresh tissue specimens are best transported and stored in sterile tissue culture medium such as RPMI1640. Although specimens may be stored at room temperature, refrigeration is preferred, particularly when there is a delay for flow cytometric analysis. For flow cytometry analysis, single-cell suspensions of the fresh tissues can be achieved by mechanical dissociation.

Data Analysis and Interpretation

At the time of flow cytometry analysis, it is important to collect data on sufficient events (cell numbers) and all available parameters to ensure detection of the cells of interest. The immunophenotyping characterization of hematopoietic and lymphoid neoplasms relies solidly on the principle that neoplastic cells frequently exhibit abnormal or aberrant antigenic expression that differs from normal cells.[20–22] Abnormal/aberrant antigenic expression can be grouped into four basic categories:

- Abnormally increased or decreased levels of antigenic expression (aberrant expression)
- Gain of antigens not normally expressed in the cell type or lineage
- Expression of antigens not synchronized with normal development and maturation stage of the cell type or lineage
- Homogeneous expression of antigen(s) by a cell population that normally show more heterogeneous expression

Basic Parameters and Windows of Cell Populations

Forward light scatter (FSC) and side light scatter (SSC). FSC collects light at 180° from the point at which the laser beam intersects the cells, usually on a linear scale. It is correlated with cell size, and thus can distinguish normal lymphocytes (small), monocytes (intermediate), and neoplastic cells (generally they are large in size). SSC collects right-angle light at 90° and is correlated with cytoplasmic granularity and nuclear configuration. The combination of both FSC and SSC can distinguish normal lymphocytes, granulocytes, and monocytes (Fig. 15.3a-c). The detection of lymphocytes and monocytes provides a reliable internal control to evaluate the size of the cells of interest.

CD45 and SSC. As the first step, it is most important to determine whether the cells of interest are hematopoietic/lymphoid. Generally speaking, all hematopoietic/lymphoid cells express CD45 antigens (CD45+). Thus, a histogram of CD45 on a logarithmic scale vs. SSC on a linear scale is indispensable as a starting point of flow cytometry analysis (Fig. 15.3d-f). Based on antigen expression, cells are divided into CD45+ and CD45– groups. Among the CD45+ group, the cells can further separated into subgroups (windows in the histogram) based on expression levels of CD45 and intensity of cytoplasmic granularity.[23–25]

- CD45-negative cells
  - Nonhematopoietic/lymphoid cells
  - Exception: erythroid and megakaryocytic precursors, giant platelets, and abnormal plasma cells
- CD45-positive cells (hematopoietic/lymphoid)
  - Lymphocyte window
  - Myeloblast window
  - Lymphoblast window
  - Granulocyte window
  - Monocyte window

Gating Cell Populations and Immunophenotyping Analysis

Gating is the most important first step in immunophenotyping analysis. It is critical particularly in a specimen that contains mixed cell populations, such as bone marrow aspirate. To accurately target and gate the cell population(s) of interest, one must correlate the patient's history, clinical presentation, and laboratory tests, and the clinician's concerns and possible pathology findings. Generally, there are three ways to gate cells:

1. By ce\ll distribution in the CD45 vs. SSC. histogram: This is most useful in a specimen containing mixed cell populations (see Fig. 15.3d). The grouped cells in individual windows represent different cell lineages (Fig. 15.4).
2. By cell size: In FSC vs. SSC histograms, neoplastic cells (usually large in size) can be gated by using lymphocytes (small) and monocytes (intermediate) as an internal size control (see Fig. 15.3a). Once the cells of interest are gated, further analysis of cell lineage can be performed.
3. By cell lineage-specific antigens (immunophenotype): If cells are CD45+ but do not fit into particular windows in the CD45 vs. SSC histogram (see Fig. 15.3e), identification of lineage-specific antigen expression is needed (Table 15.3).

Immunophenotyping Analysis and Interpretation

When the cell populations of interest have been accurately gated, detailed immunophenotyping analysis can be carried out by the presence and absence of specific cell antigens/markers. Figures 15.4–15.8 outline the basic principles of data interpretation for individual cell windows/cell lineage step by step. Generally, analysis can start with the CD45 vs. SSC histogram, and the cells of interest are gated for further characterization based on clinical indication. In specimens

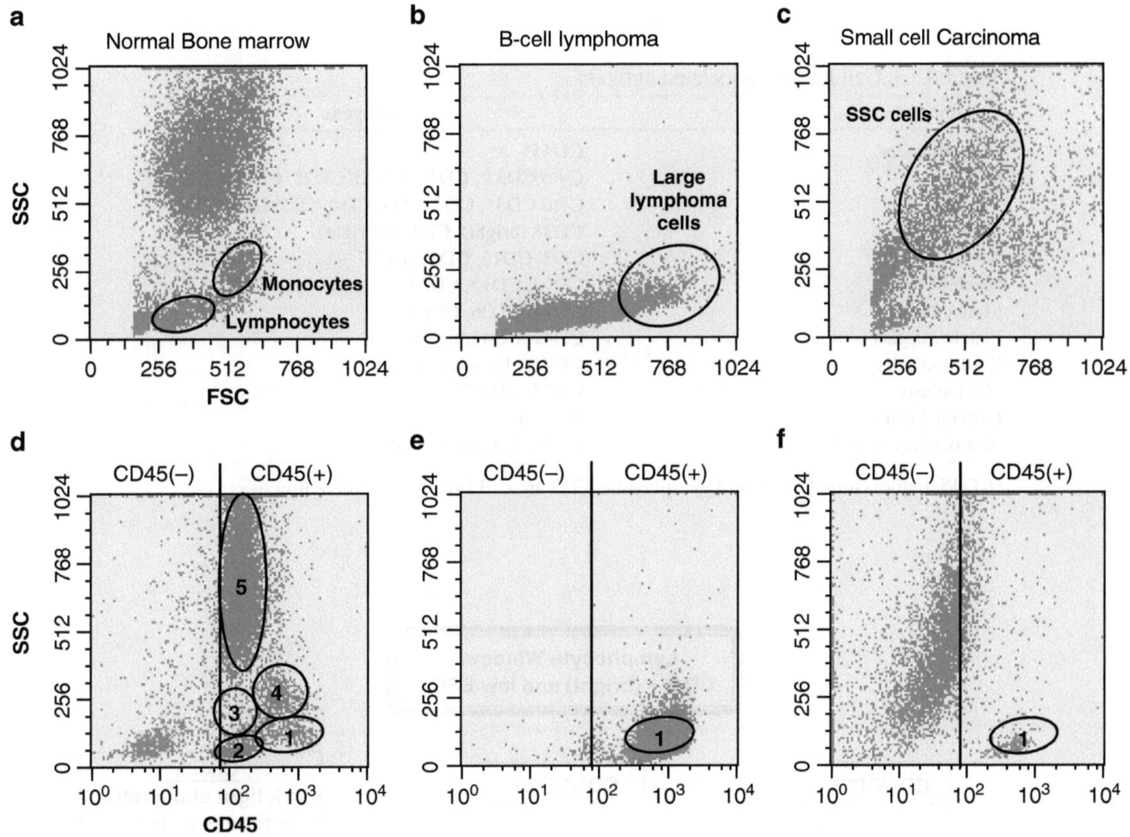

FIG. 15.3. FSC vs. SSC and CD45 vs. SSC histograms. (**a, d**) Normal bone marrow aspirate. In the FSC vs. SSC histogram, identification of monocytes (*intermediate size*) and lymphocytes (*small size*) provides a useful internal size control. In the CD45 vs. SSC histogram, cells are initially divided into CD45(−) nonhematopoietic/lymphoid cells and CD45(+) hematopoietic/lymphoid cells. The CD45(+) cells can be further classified by their distribution in individual windows: (*1*) lymphocyte window; (*2*) lymphoblast/hematogone window; (*3*) myeloblast window; (*4*) monocyte window; and (*5*) granulocyte window (see Fig. 15.4 for details). (**b, e**) B-cell lymphoma biopsy. In the FSC vs. SSC histogram, cell size can be determined by evaluating cell FSC as indicated. In addition, the CD45 vs. SSC histogram reveals that cells are CD45(+) hematopoietic/lymphoid cells and present within the lymphocyte window. (**c, f**) Small cell carcinoma (SCC) tissue biopsy. The FSC vs. SSC histogram demonstrates that SCC cells are intermediate in size (intermediate FSC) and have a complicated nuclear configuration (intermediate to high SSC). In the CD45 vs. SSC histogram, SSC cells are CD45(−) and a small group of background lymphocytes are CD45(+).

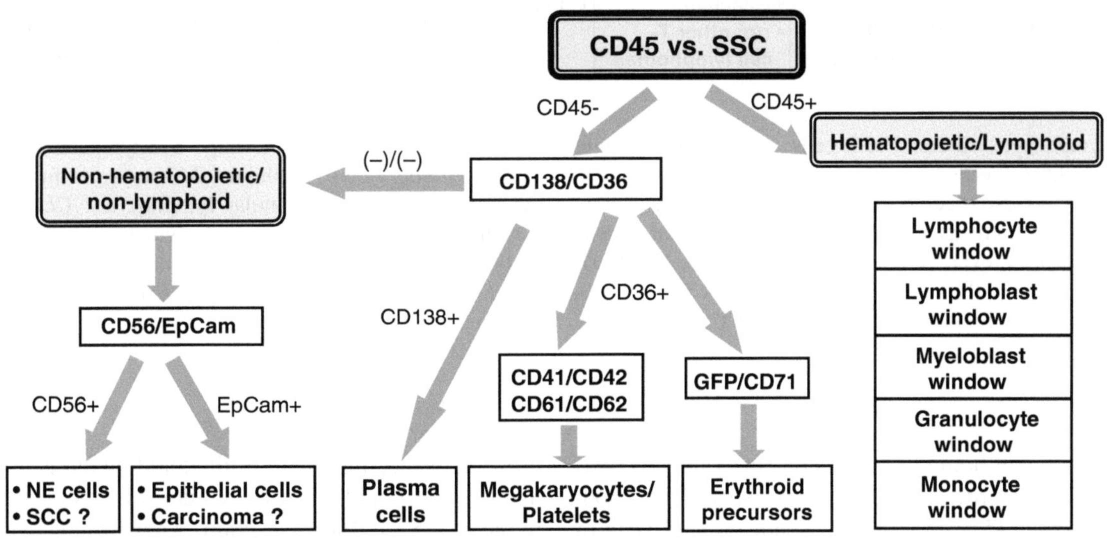

FIG. 15.4. Identification of cell windows/lineages in CD45 vs. SSC histogram. As the first step, cells are divided into CD45(+) and CD45(−) groups. The majority of CD45(−) cells are nonhematopoietic/lymphoid, with rare exceptions. CD45(+) hematopoietic/lymphoid cells can be further divided into different windows/lineages as illustrated in Fig. 15.3. Detailed immunophenotyping analysis of the cells in each window is shown in Figs. 15.5–15.8.

TABLE 15.3. Cell lineage associated antigens.

Cell lineaget	Antigens
All leukocytes	CD45[a]
B cells	Cyto CD22[a], CD19, CD20, CD22, CD79a, CD79b
T cells	Cyto CD3*, CD2, CD3, CD4, CD5, CD7, CD8
Plasma cells	CD38 (bright), CD138 (bright)
NK cells	CD2, CD16, CD56, CD57
Myeloid cells	CD13, CD15, CD33
Monocytes	CD14, CD36, CD68
Platelet/megakaryocyte	CD41, CD42, CD61
Erythroid cells	CD71, glycophorin A
Myeloblasts	CD34, CD117
Epithelial cells	EpCam
Neuroendocrine cells	CD56, EpCam, CD138

[a] CD45, cytoplasmic CD22, and cytoplasmic CD3 are cell lineage specific antigens

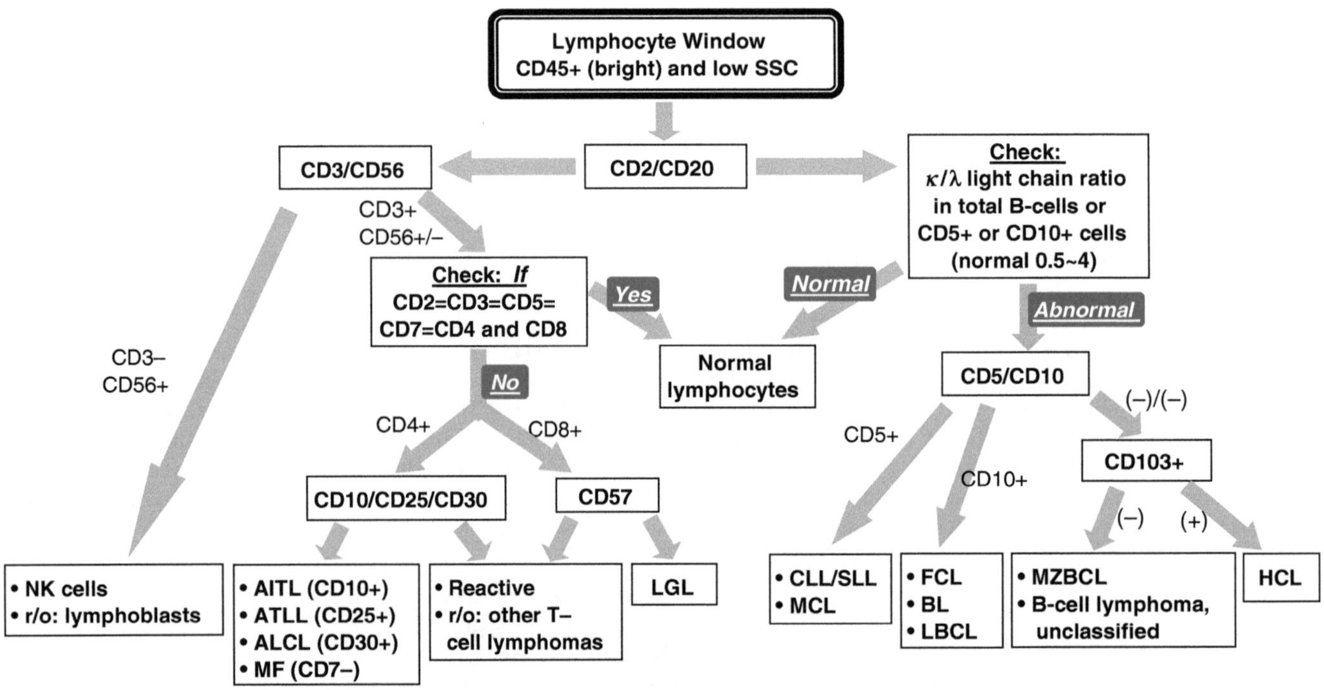

FIG. 15.5. Lymphocyte window. For B cells, initially the κ/λ light chain ratio should be checked in total cells (normal range, 0.5–4), as well as in CD5+ or CD10+ cell populations. For T cells, the expression level of T-cell antigens should be examined (normal T cells show same or similar expression as CD2, CD3, CD5, and CD7, which are also equal to the sum of CD4+ and CD8+ T cells). *r/o*, rule out; *NK cells*, natural killer cells; *ATLL*, adult T-cell leukemia/lymphoma; *ALCL*, anaplastic large cell lymphoma; *AITL*, angioimmunoblastic T-cell lymphoma; *MF*, mycosis fungoides/Sézary syndrome; *LGL*, large granular lymphocytes; *CLL/SLL*, chronic lymphocytic leukemia/small lymphocytic lymphoma; *MCL*, mantle cell lymphoma; *FCL*, follicular lymphoma; *BL*, Burkitt lymphoma; *LBCL*, large B-cell lymphoma; *MZBCL*, marginal zone B-cell lymphomas; *HL*, hairy cell leukemia.

containing relatively homogeneous cells, analysis may start with cell lineage identification by using the specific antibodies listed in Table 15.3. In cases in which the cell type of interest is unknown, gating large cells in the FSC vs. SSC histogram might be helpful for cell lineage determination (neoplastic cells are usually large). This step is particularly useful in a specimen with limited and mixed cells, by which distinct cell populations or windows cannot be established in the CD45 vs. SSC histogram. Again, to make an accurate interpretation of flow cytometry data, clinical correlation and discussion with clinicians are indispensable. Consultation with experts is also valuable.

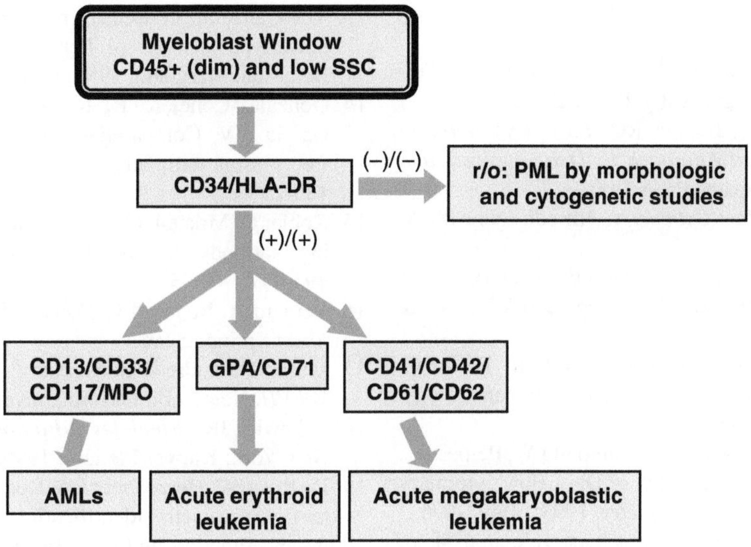

FIG. 15.6. Myeloblast window. *r/o*, rule out; *PML*, acute promyelocytic leukemia; *AMLs*, acute myeloid leukemias excluding acute erythroid and megakeryoblastic leukemias.

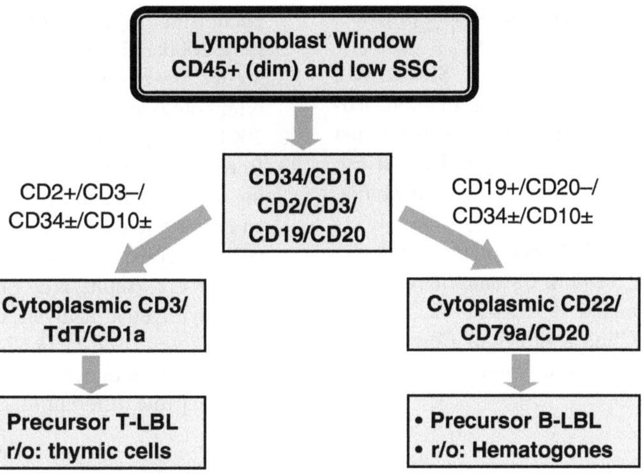

FIG. 15.7. Lymphoblast window. *r/o*, rule out; *LBL*, lymphoblastic leukemia/lymphoma.

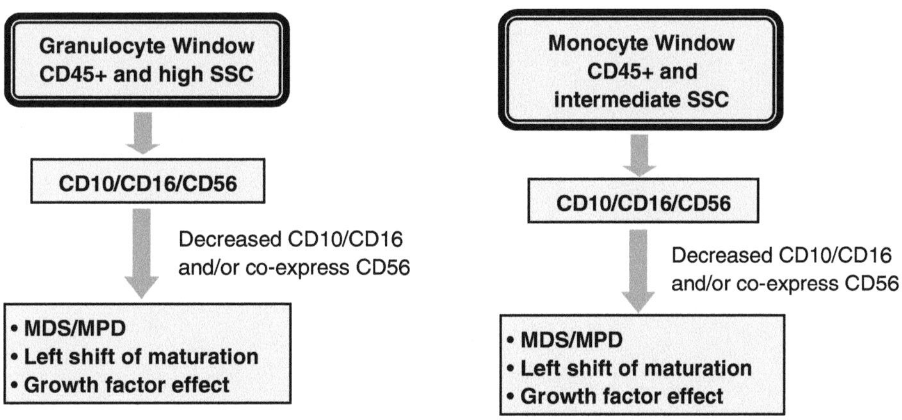

FIG. 15.8. Granulocyte and monocyte windows. *MDS*, myelodysplastic syndromes; *MPD*, myeloproliferative diseases.

References

1. Carey JL III, McCoy JP Jr, Keren DF. *Flow Cytometry in Clinical Diagnosis*. 4th ed. Chicago: ASCP Press; 2007.
2. Nguyen D, Diamond LW, Braylan RC. *Flow Cytometry in Hematopathology: A Visual Approach to Data Analysis and Interpretation*. Totowa: Humana Press; 2003.
3. Shapiro HM. *Practical Flow Cytometry*. 4th edn. New York: Wiley-Liss; 2003.
4. Sun T. *Flow Cytometry and Immunohistochemistry for Hematologic Neoplasms*. Philadelphia: Lippincott Williams & Wilkins; 2008
5. Orfao A, Ruiz-Arguelles A, Lacombe F, Ault K, Basso G, Danova M. Flow cytometry: its applications in hematology. *Haematologica*. 1995;80:69-81.
6. Duque RE, Andreeff M, Braylan RC, Diamond LW, Peiper SC. Consensus review of the clinical utility of DNA flow cytometry in neoplastic hematopathology. *Cytometry*. 1993;14:492-496.
7. Ware RE, Rosse WF, Hall SE. Immunophenotypic analysis of reticulocytes in paroxysmal nocturnal hemoglobinuria. *Blood*. 1995, 86:1586-1589.
8. Garratty G, Arndt PA. Applications of flow cytofluorometry to red blood cell immunology. *Cytometry*. 1999;38:259-267.
9. Saag MS, Holodniy M, Kuritzkes DR, O'Brien WA, Coombs R, Poscher ME, et al. HIV viral load markers in clinical practice. *Nat Med*. 1996;2:625-629.
10. Khalidi HS, Medeiros LJ, Chang KL, Brynes RK, Slovak ML, Arber DA. The immunophenotype of adult acute myeloid leukemia: high frequency of lymphoid antigen expression and comparison of immunophenotype, French-American-British classification, and karyotypic abnormalities. *Am J Clin Pathol*. 1998;109:211-220.
11. Ruiz-Arguelles A, Duque RE, Orfao A. Report on the First Latin American Consensus Conference for Flow Cytometric Immunophenotyping of Leukemia. *Cytometry*. 1998;34:39-42.
12. DiGiuseppe JA, Borowitz MJ. Clinical utility of flow cytometry in the chronic lymphoid leukemias. *Semin Oncol*. 1998; 25:6-10.
13. US–Canadian consensus recommendations on the immunophenotype analysis of hematologic neoplasia by cytometry. Bethesda, Maryland, 1995. *Cytometry*. 1997;30:213-274.
14. Gonzalez-Conejero R, Rivera J, Rosillo MC, Lozano ML, Garcia VV. Comparative study of three methods to detect free plasma antiplatelet antibodies. *Acta Haematol*. 1996;96: 135-139.
15. Kohler G, Milstein C. Continuous cultures of fused cells secreting antibody of predefined specificity. *Nature*. 1975;256: 495-497.
16. Bernard A, Bounsell L, Dausset J, eds. *Leukocyte Typing*. New York: Springer; 1984.
17. Mason D, Andre P, Bensussan A, et-al, eds. *Leukocyte Typing. Vol VII*. New York: Oxford University Press; 2002.
18. Lakowicz JR. *Principles of Fluorescence Spectroscopy*. 2nd ed. New York: Kluwer/Plenum; 1999.
19. De Rosa SC, Herzenberg LA, Roederer M. 11-color, 13-parameter flow cytometry: identification of human naïve T-cells by phenotype, function, and T-cell receptor diversity. *Nat Med*. 2001;7: 245-248.
20. Wood BL. Multicolor immunophenotyping: human immune system hematopoiesis. *Methods Cell Biol*. 2004;75:559-576.
21. Loken MR, Wells DA. Normal antigen expression in hematopoiesis: basis for interpreting leukemia phenotypes. In: Stewart C, Nicholson JKA, eds. *Immunophenotyping*. New York: Wiley-Liss; 2000:133-160.
22. Stetler-Stevenson M, Braylan RC. Flow cytometric analysis of lymphomas and lymphoproliferative disorders. *Semin Hematol*. 2001;38:111-123.
23. Borowitz MJ, Guenther KI, Shults KE, et al. Immunophenotyping of acute leukemia by flow cytometric analysis. Use of CD45 and right-angle light scatter to gate on leukemic blasts in three-color analysis. *Am J Clin Pathol*. 1993;100(5):534-540.
24. Stelzer GT, Shults KE, Loken MR. CD45 gating for routine flow cytometric analysis of human bone marrow specimens. *Ann NY Acad Sci*. 1993;677:265-280.
25. Rainer RO, Hodges L, Stelzer GR. CD45 gating correlates with bone marrow differential. *Cytometry*. 1995;22:139-145.

16
Stem Cells

Louise Hecker, Victor J. Thannickal, and Sem H. Phan

Introduction

Stem cells are defined by their potential to self-renew and differentiate into more specialized cell types within a given tissue. Stem cells can be found in the embryo and in adult tissue and are termed embryonic stem cells and adult stem cells, respectively. Embryonic stem cells are responsible for the development of an entire organism, whereas adult stem cells provide a mechanism for maintenance of tissue homeostasis by replacing damaged cells throughout the life of the organism. Stem cell biology has provided a platform for addressing important scientific questions in the areas of development, repair/regeneration, and disease. Here, we discuss some basic concepts regarding stem cells, including definitions of commonly used terminology and the current status of the stem cell field. A glossary of terms defined in this chapter is given in Table 16.1 (Table 16.1).

Embryonic Stem Cells

Embryonic stem cells are derived from the inner cell mass of a 4- to 5-day-old embryo (blastocyst), before implantation of the embryo in the uterus (Fig. 16.1). These embryonic stem cells are pluripotent, defined as having the potential to differentiate into any of the tissues derived from the three major germ layers—endoderm, ectoderm, and mesoderm—and ultimately into all the cells that make up the fully developed organism. They express telomerase to maintain telomere length, which would account for their extended capacity to replicate and self-renew. In addition to certain cell-surface markers, embryonic stem cells characteristically express Oct-4 and Nanog, as part of a group of transcription factors that are considered to be important in suppressing gene expression and maintenance of their pluripotent and undifferentiated state. Another distinctive characteristic is the absence of, or highly abbreviated, G_1 checkpoint in the cell cycle.

Adult Stem Cells

Adult stem cells are undifferentiated cells that can be found throughout the body of a mature organism. Adult stem cells can self-renew as do embryonic stem cells; however, they are multipotent rather than pluripotent. Adult stem cells have been identified in tissues that have developed from all three germ layers, including the brain, bone marrow, lungs, peripheral blood, blood vessels, skeletal muscle, skin, digestive tract, cornea, retina, dental pulp, liver, and pancreas. Therefore, adult stem cells have been identified in tissues that develop from all three germ layers (see Fig. 16.1). However, none of these adult stem cell populations is capable of differentiating into cells from all three germ layers. Instead, the major function of adult stem cells is to serve as a source of cells to regenerate the tissues in which they reside, both for homeostasis and for reparative responses to injury.

Although the origin of embryonic stem cells is well defined, much less is known with respect to adult stem cells, partly because they are present in very small numbers within defined tissue niches and are difficult to identify because of the absence of specific markers. This lack of distinct cell-surface markers also contributes to the technical challenges for their isolation and purification. A combination of markers, both negative and positive, is often required to define a particular stem cell population for purification; in the absence of specific stem cell markers, negative selection based on the absence of markers that define more differentiated cells is necessary. Hematopoietic stem cells (HSCs) represent an example of adult stem cells that have been well characterized, as they were the first to be discovered and isolated. By using fluorescence-activated cell sorting (FACS) and a panel of monoclonal antibodies that recognize cell-surface antigens specific to HSCs, a nearly homogeneous population of HSCs has been isolated in mice[1] and in humans.[2] Further research aimed to better characterize cell-surface markers and/or a gene expression profile that identify specific adult stem cell populations will allow for improved isolation techniques and better characterization of tissue-resident adult stem cells and their origins.

P.T. Cagle and T.C. Allen (eds.), *Basic Concepts of Molecular Pathology*, Molecular Pathology Library 2,
DOI: 10.1007/978-0-387-89626-7_16, © Springer Science+Business Media, LLC 2009

TABLE 16.1. Glossary of terms.

Determination: The irreversible commitment of a cell to follow a specific developmental pathway.

Differentiation: The overall process by which stem/progenitor cells are activated to become more specialized cell types.

Fate: Term that describes what cells will become as a result of differentiation.

Multipotent stem cells: Cells that self-renew and differentiate into several different specialized cell types, often within a tissue (e.g., hematopoietic stem cells).

Plasticity: The ability of an adult stem cell from one tissue to develop into differentiated cell types from a different tissue; this remains largely unproven as an in vivo physiological process.

Potential: Term that describes what cells are able to become.

Progenitor cell: An undifferentiated precursor cell with the capacity to undergo differentiation into specialized cell types; unlike putative stem cells, they do not retain the capacity for self-renewal.

Pluripotent stem cells: Cells capable of self-renewing and differentiating into any of the three germ layers (endoderm, ectoderm, and mesoderm); unlike totipotent cells, they do not give rise to embryonic components of the trophoblast and placenta.

Totipotent stem cells: Cells derived from the first few divisions of the fertilized egg; these cells have the potential to give rise to all the differentiated cells of the fully developed organism.

Self-renewal: A defining property of stem cells that allows them to undergo repeated mitotic cell divisions to create at least one daughter cell equivalent to the mother cell that retains latent capacity for differentiation.

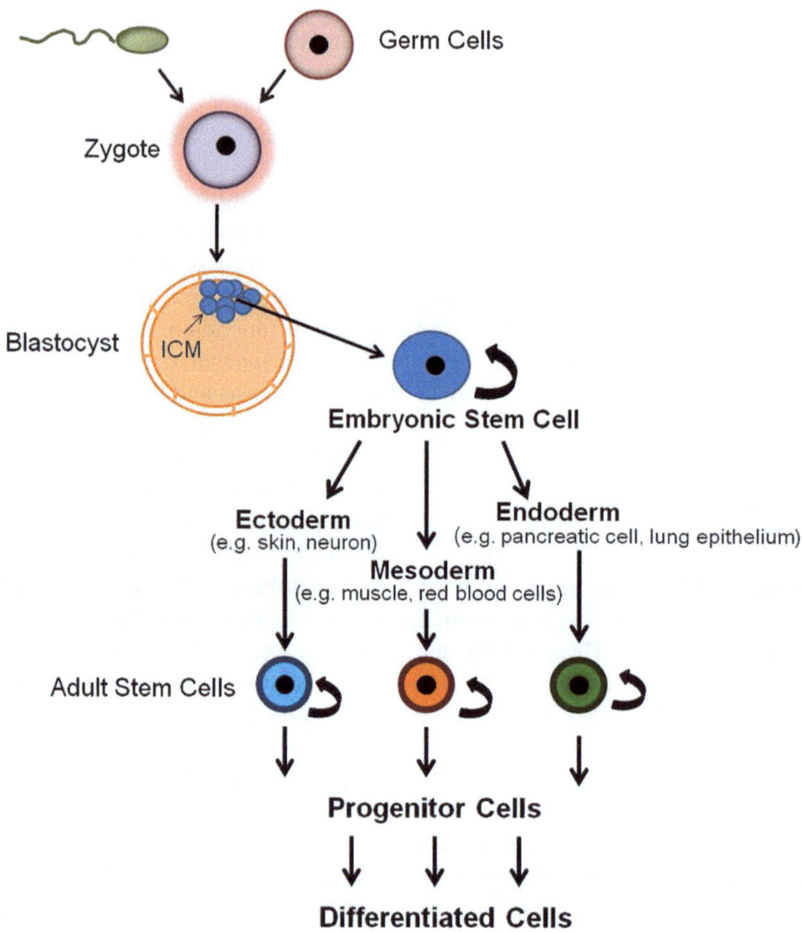

FIG. 16.1. Schematic depicting the hierarchy and differentiation potential of embryonic and adult stem cells. Pluripotent embryonic stem cells are derived from the inner cell mass (ICM) of a 4- to 5-day-old embryo (blastocyst) and can differentiate into the tissues that comprise the endoderm, ectoderm, and mesoderm. Multipotent stem cells that may also derive from all three germ layers have been identified in adult tissues; these adult stem cells self-renew within tissue niches, maintain tissue homeostasis, and participate in repair/regeneration following injury. Both embryonic and adult stem cells undergo asymmetrical mitotic cell division to form progenitor cells, which eventually differentiate into specialized cell types.

Hierarchy of Stem Cell Differentiation

Both embryonic and adult stem cells can undergo mitotic cell division to form progenitor cells (sometimes referred to as a transit-amplifying population) with the loss of some degree of multipotency before reaching their specialized fate. The convergence of intrinsic cellular signals and extrinsic microenvironmental cues from the surrounding niche regulates differentiation, but the specific signals involved are poorly understood. For example, the mechanisms by which the daughter cells of a stem cell either remain a stem cell or become a progenitor cell are not entirely clear, although it can involve the inheritance of cell fate determinants from the mother cell and/or niche factors. In contrast to stem cells, progenitor cells have not been shown to possess unlimited self-renewal capacity. There are, however, some poorly defined elements with respect to this and other aspects (e.g., degree of plasticity) of the definition of a progenitor cell versus a stem cell that have been used interchangeably to reflect this uncertainty. Progenitor cells will eventually differentiate into mature specialized cell types. During differentiation, activation of a program of gene expression orchestrated by a set of transcription factors and epigenetic changes results in the phenotypic maturation of a cell with specialized functions. For example, bone marrow stromal cells or mesenchymal stem cells (MSCs) have the capacity to differentiate into adipocytes, bone, and cartilage in culture and upon transplantation into donors.[3–5] In general, there exists a hierarchy of stem cell differentiation, where stem cell division yields progenitor cells that become progressively more restricted in their developmental potential, assuming the phenotype and function of specialized cells (see Fig. 16.1).

Stem Cell Plasticity

Stem cells remain quiescent within their niche, but possess extraordinary proliferative capacity compared to other cell types. For example, a single hematopoietic stem cell (HSC) can supply all the blood cells throughout the life of a mouse via self-renewal and multilineage differentiation capabilities.[6] Some adult stem cells appear to exhibit extended plasticity, in that they can differentiate into tissues other than the tissue from which they originated.[7,8] However, there is also controversy with respect to the degree of plasticity of many of these adult stem cells, including HSCs, which have been claimed to give rise to parenchymal cells of many organs, including the liver, lung, and heart.[8–10] An added complication is that fusion of these stem cells with the target differentiated cell at a distal site may give the appearance of transdifferentiation, a phenomenon that is known to occur.[9,10] Further research is necessary to conclusively establish the extent of plasticity of these adult stem cells in vivo. A noteworthy circulating cell of limited plasticity has been described with the capacity to differentiate to fibroblasts, consequently referred to as fibrocytes.[11,12] Presumably derived from HSCs, these cells are thought to contribute to the fibroblast populations at sites of injury and repair. With a more restricted plasticity are the resident stem cells found in many tissues that may give rise to the differentiated parenchymal elements within the tissue or organ itself.[13] Further studies are required to determine the origin(s), niche, cell-specific markers, and plasticity of these tissue-resident stem/progenitor cells.

Mechanisms of Homeostasis and Response to Injury

A primary function of adult tissue-resident stem/progenitor cells is to maintain cellular homeostasis within the tissue/organ and replace damaged cells following injurious insults. The mechanisms for repair/regeneration in humans are organ specific, with distinct resident stem cell populations giving rise to the differentiated elements of the tissue of origin. This mechanism may involve the potential for differentiated cells to act as precursors to other differentiated cell types in response to microenvironmental cues, particularly in organs that turn over slowly, such as the liver, lung, and pancreas. In skeletal muscle tissues, a population of stem cells that reside between the plasma membrane of multinucleated skeletal muscle cells and the basal lamina that surrounds each muscle cell are called "satellite stem cells" and function as classic adult tissue-resident stem cells.[14,15] Satellite cells are normally nondividing, quiescent cells unless they are triggered to proliferate as a result of injury. Transit-amplifying satellite cells can produce myoblasts that fuse with preexisting multinucleated skeletal muscle help repair damaged tissue. Similarly, progenitors of lung and intestinal epithelium have been described that reside within or subjacent to the epithelium itself and serve as the precursors for the differentiated epithelium to replace cells lost by normal sloughing or toxic injury.

Stem Cells and Disease

Recent studies have begun to implicate stem cells in diverse human diseases. In carcinogenesis, for example, work by several groups has led to the notion that cancer stem cells may be responsible for the growth and maintenance of tumors. The cancer stem cell hypothesis suggests that malignant tumors are initiated and maintained by a population of cancer stem cells that share biological properties similar to those of normal adult stem cells.[16] A cancer stem cell has been defined as a self-renewing cell within a tumor that has the capacity to regenerate the phenotypic diversity of the original tumor.[17] The first evidence for this concept was provided by the finding that a subpopulation of $CD34^+/CD38^-$ cells from acute myeloid leukemia patients initiated leukemia when transplanted into NOD/SCID mice.[18] Cancer stem cells have been implicated in a growing list of cancers including those in the breast, brain, pancreas, colon, and ovary. Current efforts are focused on dissecting the molecular pathways by which cancer stem cells are initiated and propagated. Specific genes and pathways, such as Shh and Bmi-1, are just beginning to be identified in these processes. The ability to selectively target cancer stem cells in human patients would revolutionize the field of cancer treatment.

The role of adult stem cells in other diseases is also receiving increasing interest. For example, discovery of the fibrocytes suggests that bone marrow-derived cells may contribute to the pathogenesis of organ fibrosis[11,12,19]; this is in addition to the well-known role of inflammation in disease and the HSC origin of the inflammatory cell participants. In addition to their role as the source of cells that directly participate in disease, deficiencies in stem cell function or numbers may play a role in pathogenesis. Moreover, a number of age-related degenerative disorders, and even aging itself, may be related to a deficiency in stem cell responses and tissue-renewal capacity. Deficient stem cell function or replicative capacity has significant consequences on the ability to produce essential cells for normal function of the affected organ. For example, dyskeratosis congenita is characterized by bone marrow stem cell failure caused by impaired telomerase function and consequent associated aplastic anemia.[20]

Significance and Future Research

Embryonic stem cell research will likely uncover fundamental events during development and lead to a better understanding of the causes of birth defects and novel strategies to allow us to prevent or correct them. Adult stem cells have recently received a great deal of research interest because of their potential for future clinical applications. Two obvious advantages for clinical applications include these: (a) adult stem cells may be isolated from the patient's own tissues and thus provide an autologous cell source that obviates issues of immune incompatibility; and (b) ethical concerns surrounding the use of embryonic stem cells can be avoided. More recently, another alternative is provided by the discovery that fibroblasts can be reprogrammed or induced to revert to a phenotype comparable in many respects to that of a pluripotent stem cell.[21,22] These so-called induced pluripotent stem cells (iPSCs) can be artificially derived from a nonpluripotent somatic cell by inducing expression of three or four specific genes that are normally expressed in pluripotent stem cells. Thus, human iPSCs are obtained by inducing expression of Oct3/4, Sox2, Klf4, and c-Myc with a retroviral system in fibroblasts,[22] while another study achieved a comparable outcome by inducing expression of Oct4, Sox2, and Nanog using a lentiviral system.[21] The hope is that these iPSCs and other stem cells will provide novel cell-based therapies that facilitate or directly contribute to the regeneration of damaged or lost tissues following injury or as a result of aging. In this regard, embryonic and adult stem cells, and potentially iPS cells, have been shown to contribute to the repair or replacement of damaged tissues in animal models with varying degrees of success.[23,24] Recent observations of the protective effects of MSCs in acute tissue injury and inflammation represent another example of the potential utility of stem cells in therapy.[25] To realize the full potential of cell-based therapies for regenerative medicine, we must take into account alterations to tissue niches in disease states and in aging that may influence the capacity for successful differentiation of transplanted cells, much of which remains to be discovered and elucidated.

References

1. Spangrude GJ, Heimfeld S, Weissman IL. Purification and characterization of mouse hematopoietic stem cells. *Science.* 1988;241(4861):58–62.
2. Baum CM, Weissman IL, Tsukamoto AS, Buckle AM, Peault B. Isolation of a candidate human hematopoietic stem-cell population. *Proc Natl Acad Sci USA.* 1992;89(7):2804–2808.
3. Caplan AI, Bruder SP. Mesenchymal stem cells: building blocks for molecular medicine in the 21st century. *Trends Mol Med.* 2001;7(6):259–264.
4. da Silva ML, Chagastelles PC, Nardi NB. Mesenchymal stem cells reside in virtually all post-natal organs and tissues. *J Cell Sci.* 2006;119(pt 11):2204–2213.
5. Bruder SP, Jaiswal N, Ricalton NS, Mosca JD, Kraus KH, Kadiyala S. Mesenchymal stem cells in osteobiology and applied bone regeneration. *Clin Orthop Relat Res.* 1998;355(suppl): S247–S256.
6. Osawa M, Hanada K, Hamada H, Nakauchi H. Long-term lymphohematopoietic reconstitution by a single CD34-low/negative hematopoietic stem cell. *Science.* 1996;273(5272):242–245.
7. Ratajczak MZ, Kucia M, Reca R, Majka M, Janowska-Wieczorek A, Ratajczak J. Stem cell plasticity revisited: CXCR4-positive cells expressing mRNA for early muscle, liver and neural cells "hide out" in the bone marrow. *Leukemia.* 2004;18(1):29–40.
8. Wagers AJ, Weissman IL. Plasticity of adult stem cells. *Cell.* 2004;116(5):639–648.
9. varez-Dolado M, Pardal R, Garcia-Verdugo JM, et al. Fusion of bone-marrow-derived cells with Purkinje neurons, cardiomyocytes and hepatocytes. *Nature.* 2003;425(6961):968–973.
10. Wagers AJ, Sherwood RI, Christensen JL, Weissman IL. Little evidence for developmental plasticity of adult hematopoietic stem cells. *Science.* 2002;297(5590):2256–2259.
11. Hashimoto N, Jin H, Liu T, Chensue SW, Phan SH. Bone marrow-derived progenitor cells in pulmonary fibrosis. *J Clin Invest.* 2004;113(2):243–252.
12. Phillips RJ, Burdick MD, Hong K, et al. Circulating fibrocytes traffic to the lungs in response to CXCL12 and mediate fibrosis. *J Clin Invest.* 2004;114(3):438–446.

13. Lama VN, Smith L, Badri L, et al. Evidence for tissue-resident mesenchymal stem cells in human adult lung from studies of transplanted allografts. *J Clin Invest.* 2007; 117(4):989–996.
14. Bailey P, Holowacz T, Lassar AB. The origin of skeletal muscle stem cells in the embryo and the adult. *Curr Opin Cell Biol.* 2001;13(6):679–689.
15. Raff M. Adult stem cell plasticity: fact or artifact? *Annu Rev Cell Dev Biol.* 2003;19:1–22.
16. Tan BT, Park CY, Ailles LE, Weissman IL. The cancer stem cell hypothesis: a work in progress. *Lab Invest.* 2006;86(12): 1203–1207.
17. Lobo NA, Shimono Y, Qian D, Clarke MF. The biology of cancer stem cells. *Annu Rev Cell Dev Biol.* 2007;23:675–699.
18. Bonnet D, Dick JE. Human acute myeloid leukemia is organized as a hierarchy that originates from a primitive hematopoietic cell. *Nat Med.* 1997;3(7):730–737.
19. Moore BB, Kolodsick JE, Thannickal VJ, et al. CCR2-mediated recruitment of fibrocytes to the alveolar space after fibrotic injury. *Am J Pathol.* 2005;166(3):675–684.
20. Calado RT, Young NS. Telomere maintenance and human bone marrow failure. *Blood.* 2008;111(9):4446–4455.
21. Yamanaka S. Induction of pluripotent stem cells from mouse fibroblasts by four transcription factors. *Cell Prolif.* 2008; 41(suppl 1):51–56.
22. Yu J, Vodyanik MA, Smuga-Otto K, et al. Induced pluripotent stem cell lines derived from human somatic cells. *Science.* 2007;318(5858):1917–1920.
23. Hanna J, Wernig M, Markoulaki S, et al. Treatment of sickle cell anemia mouse model with iPS cells generated from autologous skin. *Science.* 2007; 318(5858):1920–1923.
24. Wernig M, Zhao JP, Pruszak J, et al. Neurons derived from reprogrammed fibroblasts functionally integrate into the fetal brain and improve symptoms of rats with Parkinson's disease. *Proc Natl Acad Sci USA.* 2008;105(15):5856–5861.
25. Morigi M, Introna M, Imberti B, et al. Human bone marrow mesenchymal stem cells accelerate recovery of acute renal injury and prolong survival in mice. *Stem Cells.* 2008;26(8): 2075–2082.

17
Animal Models of Lung Disease

Roberto Barrios

Introduction

Animal models afford the opportunity for investigators to study the study the mechanisms of human disorders by experimentally manipulating a number of controlled variables. Experimental models can be arbitrarily divided into (a) spontaneous disease models, which are mutant animals that carry a disease similar to a human condition; (b) models obtained through genetic manipulation, in other words, gene-modified models including: transgenic and knock out animals and (c) chemically or physically induced changes. Researchers have to chose from a number of possibilities such as species and strain of animal, environment, and the genome to investigate the molecular interactions involved in the pathogenesis of many diseases.[1] Experimental models also provide a unique opportunity to test potential therapeutic intervention. Well-developed animal models should share features with specific human disorders. Although the basic molecular biology terminology is covered in other chapters of this book, it is convenient to briefly review the basic technology for the development of transgenic mice.

Transgenic Mouse Models

Transgenic mice technology has provided a powerful tool for both neoplastic and non-neoplastic disease investigation. Injection of exogenous DNA into fertilized eggs (zygotes) is known as trangenesis and is the most common method to modify the germline of mice. The offspring of transgenic founders are usually examined for a specific phenotype and are used to characterize genes and gene regulatory elements. The procedure involves a construct made of a segment of DNA that contains the sequence of interest and that has been cut with restriction enzymes. This segment may be ligated to a promoter that will induce expression of the transgene. Injection of the construct into mouse embryos results in a random integration and variable copy numbers into the mouse genome. The result will be expression of the transgene contained in the construct. Random integration and the copy number of the transgene are factors that cannot be controlled in the classical transgenic mouse technology. The procedure necessitates the generation and characterization of several founders to ensure that the phenotype is dependent on the transgene and not a result of alterations in the function of an endogenous gene. The standard protocol involves injection of the DNA construct into the pronucleus of fertilized eggs (0.5-day-postcoital embryos). These injected eggs are then implanted into the oviduct of a pseudopregnant foster mother. The injected DNA integrates randomly into the genome in a fraction of the implanted embryos. Founders are mated to demonstrate germline transmission and to establish a transgenic line.[3]

We describe here some examples of experimental models to illustrate the powerful tool that they represent.

A good example of the role of lymphocytes in host defense against tumor development was provided in the paper by McCabe et al.[2] They observed an interesting role for B lymphocytes in the surveillance of two experimental tumors in the rat. Following intravenous (i.v.) injection of MADB106 tumor cells into syngeneic WAG rats, most animals develop lung tumors. Previously, these tumors had been shown to be under control by natural killer (NK) cells.[3] It was noted that after i.v. injection of these tumors, a rapid influx of B cells into the lung occurred. The study demonstrated that homing of B lymphocytes to the potential site of tumor development plays an important role in defense against these tumors. Another area of a common human disease that gained knowledge from animal models is chronic obstructive pulmonary disease (COPD). COPD is one of the most common pulmonary diseases seen in clinical practice. Smoking is by far the most important risk factor for COPD. Because COPD is produced by a combination of environmental and genetic factors, animal models are ideal to study the pathogenesis of this condition. A number of species to study COPD have been proposed in the literature.[4] In recent years, mice have been shown to be excellent for these studies because mouse and human genomes are very similar,[5] and because

P.T. Cagle and T.C. Allen (eds.), *Basic Concepts of Molecular Pathology*, Molecular Pathology Library 2,
DOI: 10.1007/978-0-387-89626-7_17, © Springer Science+Business Media, LLC 2009

of the ability to modify the genetic constitution of the mouse by either inducing a protein, knocking out genes, or creating new mutations. In addition, the physiology and general aspects of mouse biology are well known and a large number of inbred strains are commercially available.[6,7] There are several experimental models of COPD and emphysema.[8–10] Some depend on endobronchial perfusion of proteolytic enzymes in guinea pigs, hamsters, and mice.[11] Others depend in tobacco smoke.[12,13] There are also mouse strains that develop emphysema spontaneously.[9]

An imbalance between collagen synthesis and degradation and between proteases and their inhibitors has been proposed for many years as a possible pathogenetic mechanism for the development of emphysema. It is not surprising that this hypothesis has been studied in experimental models. In theory, this imbalance may be caused by an excessive production of proteases by inflammatory cells or by reduced synthesis or increased breakdown of antiproteases. This hypothesis generated a number of experimental models that included intratracheal instillation of papain, trypsin, porcine elastase, and human neutrophil elastase as well as a large number of other proteases.[14,15]

More recently, transgenic mice have allowed expression or overexpression of proteases, such as human interstitial collagenase (MMP-1).[16,17] These mice develop emphysema that is very similar to that seen in human alpha-1 antitrypsin deficiency. The emphysematous changes are induced by degradation of type III collagen. Overexpression of interleukin 13 (IL-13) or interferon-gamma increases the expression of matrix metalloproteinases (MMPs) that are most likely responsible for the emphysematous changes. Knockout of TIMP-3 and, in other models, knockout of surfactant protein D, induces "spontaneous," age-related emphysema.[18,19]

A number of the previously mentioned models provided support to the hypothesis of imbalance between protein formation and degradation, a hypothesis that has also been proposed for the development of interstitial fibrosis.

Asthma is another common disease that has been studied through a large number of animal models.[20–23] These models have aided in the study of mechanisms and pathways leading to the development of airway hyperreactivity and inflammation. The majority of described models depend on previous sensitization and challenge with the offending antigen. Most of these models are reminiscent of the changes seen in individuals with airway hyperreactivity; however, the morphological changes seen in these animals do not resemble the changes that the pathologist sees in patients with asthma. Sensitization and challenge with interleukins such as IL-13 have provided new knowledge related to the role of leukotrienes and the glutathione pathway in this disease.[24]

With the discovery of the cystic fibrosis (CF) gene in 1989,[25–28] a number of models have been attempted. Three years later, the first CF mouse model was described.[29] Approximately 11 CF mouse models have been described.[30,31] CF, the most common genetic disease among Caucasians, is caused by mutations in the gene encoding CFTR (cystic fibrosis transmembrane conductance regulator). However, CFTR null mice did not develop pathology resembling human CF lesions.[32] A more promising model was that of Mall et al,[33] in which they enhanced sodium absorption in mouse airways by manipulating the epithelial sodium channel (ENaC). This model has provided new insight in the pathogenesis of CF and the study of ion transport defects.

An important clinical problem is the development of fibrosis of unknown causes in several organs, such as pulmonary interstitial fibrosis, some examples of "cryptogenic" cirrhosis, and even focal segmental glomerulosclerosis in the kidney. All these are common diseases that are associated with high morbidity and mortality, and their pathogenesis is not clear. To understand the pathogenesis and possible therapeutic interventions in interstitial diseases that progress to fibrosis, animal models provide valuable help. Pathologists have known for decades that the pathogenesis of fibrosis was associated with a number of conditions, such as infections, collagen vascular disease, allergic alveolitis, and trauma. However, the origin and mechanisms of development of idiopathic interstitial pulmonary fibrosis remain uncertain, probably in part from a lack of a good experimental model analogous to human interstitial fibrosis. Despite the number of experimental models, we still lack a model that recreates all the characteristics of human disease. Similar to the approach for emphysema, researchers have attempted to instill fibrogenic agents intratracheally in various experimental models. While these approaches attempt to elucidate some mechanisms of fibrogenesis, it is uncertain whether they really can be extrapolated to human disease. Nevertheless, our understanding of the basic pathology of the development of fibrosis has increased dramatically thanks to many of these models. A number of pioneer studies demonstrated that mice and other species were susceptible to the development of fibrosis after bleomycin (a profibrotic drug) instillation.[34–37] These models have been well studied; however, they do not resemble human idiopathic pulmonary fibrosis.[38] Paraquat-induced fibrosis has also been well described in experimental models,[39] but the lesions tend to be heterogeneous. With the development of transgenic technology, induced abnormal expression of extracellular matrix proteins, cytokines, and proteases has provided additional information.[40] Although some of these models mimic pulmonary fibrosis, interpretation of the results requires a critical analysis. At the present time, it is impossible to reproduce the natural history of human pulmonary fibrosis in an experimental model. Recent advances such as microarray analysis, however, are providing better scrutiny of the initial mechanisms of disease. For example, it has been shown that bleomycin alters the gene transcription pattern in the mouse lung by increasing genes associated with inflammation, which reached maximum levels at 5 days after bleomycin administration, whereas genes involved in the development of fibrosis increased gradually up to 14 days after bleomycin treatment. These changes in gene expression signature were well correlated with observed histopathological changes.[41] This approach will provide a sensitive method to assess gene expression and may help identify genes involved in clinical pulmonary fibrosis.

An interesting example of transgenic mice that help elucidate mechanisms of development of glomerular lesions was published by Brantley et al.[42] They found that increased expression of Cux-1, a murine homeobox gene, results in cell proliferation and mesangial expansion. In addition, these changes are potentially related to disruption of podocyte architecture leading to loss of filtration. These results suggest that expression of Cux-1 is sufficient to induce the early events of mesangioproliferative glomerulonephritis.

A large number of animal models of neoplastic diseases have been developed. Urethane-induced lung tumors in some strains of mice is a well-known model that has been studied for many years,[43–48] and the mechanisms for the increased susceptibility of some mouse strains have been studied[49–51] with and without transgenic manipulation.[52] There are, however, many complications in establishing animal models of lung cancer, among them a good correlation between histological patterns of human and animal malignancies, natural strain susceptibility, and time frames.

The first oncogene targeted specifically to the lung was the Simian virus large T antigen (Tag).[53] A model of pulmonary adenocarcinomas has been produced in transgenic mice harboring a chimeric gene comprising the SV40 large T antigen under the control of a transcriptional region derived from the human surfactant protein C (SP-C) gene.[54] In these studies, transgenic mice succumbed with pulmonary tumors within 4–5 months of age. By histology, the tumors were adenocarcinomas with lepidic, papillary, and solid growth patterns that were indistinguishable from adenocarcinomas occurring in humans. This model has been useful in our understanding of regulatory pathways disrupted during tumor progression.[55]

SP-C promoter has also been used to express c-*myc*, epidermal growth factor, the recepteur d'origine nantais (RON), receptor tyrosine kinase (a member of the MET proto-oncogene family), and Raf-1. These models have been very useful for specifically targeting certain cell types such that other lung cell types are not directly affected.[56,57] Animal models for squamous cell carcinomas and small cell carcinomas are needed. Failure to develop specific tumor types is probably the result of variability of the transgene expression early in lung development. Another limiting factor in the study of these models is that once transcription of the transgene is initiated, it is irreversible. Apparently, conditional transgenic models have overcome these limitations. The ligand-inducible binary transgenic systems provide effective regulatory models; they consist of at least two transgene constructions, a regulatory transgene and a target transgene to provide regulated expression of a specific gene. The regulator transgene encodes a transcription factor whose activity is determined by the administration of an exogenous compound. The regulator is placed under the control of a tissue-specific promoter to express the transcription factor in the tissue of interest. This regulator does not activate transcription of the target transgene until the animal receives an exogenous compound. When this compound is administered, the regulator activates only the target transgene. The other construct that contains the target gene contains the sequence of a protein of interest under the transcriptional control of *cis*-acting DNA elements that are responsive to the DNA-binding domain of the regulator transgene. The currently used ligand-inducible binary transgenic systems are the tetracycline transactivator inducible system, the mifepristone gene switch, and the acdysone regulatory system.[58]

Knockout Mouse Models

Knockout mouse production implies mutation or ablation of an endogenous gene by homologous recombination in embryonic stem (ES) cells. Basically, the ES cells with the appropriate target are injected into the blastocyst of a mouse embryo; mice born contain cells from both the host embryo and the targeted ES cells. If these ES cells incorporate themselves into the germline, the mutation can be transmitted to future generations.[59] There are limitations for the use of these models in lung cancer because the mutated or knockout gene either provides a silent phenotype or will not allow the study of new neoplastic transformation in the adult mouse.

Ligand-dependent Cre recombinase provides a clever mechanism to the production of controlled lesions, which is possible by using Cre reombinases. The technique depends on the Cre protein, which is encoded by the coliphage P1. It is a 38-kDa protein that efficiently promotes both intra- and intermolecular synapsis and recombination of DNA both in *Escherichia coli* and in vitro. Recombination occurs at a specific site, called lox, and does not require any other protein factors. The Cre protein causes synapsis of DNA and site-specific recombination in a mammalian cell line. Cre protein activity is directly regulated by a ligand, which binds to the Cre recombinase and causes changes in conformation that allow the recombinase to edit floxed genes. Cre, the product of the *Cre* (cyclization recombination) gene of bacteriophage P1, catalyzes the reciprocal recombination of genomic segments that are flanked by loxP sequences. The recombinase Cre acts on the DNA site *loxP*. If there are two *loxP* sites in the same orientation near each other, Cre can act to loop out the sequence between the two sites, leaving a single *loxP* site in the original DNA and a second *loxP* in a circular piece of DNA containing the intervening sequence. Therefore, a properly designed targeting construct containing *loxP* sites can be used for introducing subtle mutations or for a temporally or spatially controlled knockout[59] (see also Chap. 1 in this volume). Although this technology appears promising, it has not been applied successfully to the study of lung malignancies.[60,61]

Malignant mesothelioma has been linked to asbestos exposure and generally has a poor prognosis because it is often diagnosed in advanced stages and is refractory to conventional therapy. Mouse models of pleural and peritoneal mesotheliomas have been produced by exposure to asbestos fibers, radionuclides, particulate nickel compounds, and chemicals such as 3-methylcholanthrene. The role of SV40 virus as a cofactor with asbestos fibers in the development of diffuse malignant mesotheliomas in humans has also been explored in animal models. Some models have shown that SV40 virus alone induces mesotheliomas in hamsters. Because human malignant mesotheliomas frequently show hypermethylation or deletions at the Cdkn2a/Arf and Cdkn2b gene loci and deletions or mutations at the NF2 gene locus, heterozygous Nf2 (+/–) experiments with mice exposed to crocidolite asbestos fibers have been published. These mice exhibited accelerated development of malignant mesotheliomas compared to wild-type littermates.

An interesting study by Altomare et al in a mouse model of mesothelioma demonstrated that mesotheliomas from asbestos-treated *Nf2* (+/–) mice show somatic genetic changes, including homozygous deletion of the tumor suppressor genes *p16(INK4A)*, *p14(ARF)/p19(Arf)*, and/or *p15(INK4B)* that are very similar to events found in human malignant mesotheliomas. Moreover, they show that a similar reciprocal pattern of ARF loss versus p53 alteration is present in both mouse and human malignant mesotheliomas. Taken together, these data implicate a common set of cellular perturbations in both human and mouse malignant mesotheliomas. Thus, from these studies it seems that alterations of the p53/ARF and p16(INK4A) cell-cycle regulatory pathways and the AKT and p21-activated kinase-merlin signal transduction pathways are critical events that cooperate to drive malignant mesothelioma tumorigenesis in both human and murine malignant mesotheliomas. These findings are consistent with the view that cancer is a multistep process involving the accumulation of somatic genetic changes that enable tumor cells to override fail-safe mechanisms regulating normal cell proliferation.[62,63]

Conclusions

Animal models of lung disease have provided extraordinary information to help us understand human disease. They are powerful tools that enable the study of the mechanisms and natural history of human diseases. Several species have provided good models for certain diseases; however, murine models have several intrinsic advantages compared to other animal models including lower cost, easier maintenance, and rapid reproduction rate. Transgenic or knockout mice can be generated in the laboratory in a relative short time compared to other species. Nevertheless, anatomic and immunological differences between mice and humans mean that murine models have limitations that must be considered when interpreting the results obtained from experimental models and applying these to the pathogenesis of human diseases. The methodology is limited by a number of factors including species differences, lack of models that truly resemble human disease, and strain variations. Although transgenic and knockout mice have been used in research for many years, the sequencing of the mouse and human genomes and high-density DNA expression analysis has recently added powerful tools to researchers using animal models. Furthermore, new knowledge of certain pathways altered in some diseases, obtained from animal models, may provide an opportunity for pharmacological intervention.

References

1. Kumar RK. Experimental models in pulmonary pathology. *Pathology*. 1995;27(2):130–132.
2. Barlozzari T, et al. Direct evidence for the role of LGL in the inhibition of experimental tumor metastases. *J Immunol.* 1985;134(4): 2783–2789.
3. McCabe PM, et al. Animal models of disease. *Physiol Behav.* 2000;68(4):501–507.
4. Brusselle GG, et al. Murine models of COPD. *Pulm Pharmacol Ther.* 2006;19:155–165.
5. Paigen K. A miracle enough: the power of mice. *Nat Med.* 1995;1(3):215–220.
6. Glasser SW, et al. Transgenic models for study of pulmonary development and disease. *Am J Physiol.* 1994;267(5 pt 1):L489–L497.
7. Ho YS. Transgenic models for the study of lung biology and disease. *Am J Physiol.* 1994;266(4 pt 1):L319–L353.
8. Shapiro SD. Animal models for COPD. *Chest.* 2000;117(5 suppl 1): 223S–227S.
9. Shapiro SD. Animal models for chronic obstructive pulmonary disease: age of klotho and marlboro mice. *Am J Respir Cell Mol Biol.* 2000;22(1):4–7.
10. Lucey EC. Experimental emphysema. *Clin Chest Med.* 1983;4(3):389–403.
11. Snider GL, Lucey EC, Stone PJ. Animal models of emphysema. *Am Rev Respir Dis.* 1986;133(1):149–169.
12. Guerassimov A, et al. The development of emphysema in cigarette smoke-exposed mice is strain dependent. *Am J Respir Crit Care Med.* 2004;170(9):974–980.
13. Bartalesi B, et al. Different lung responses to cigarette smoke in two strains of mice sensitive to oxidants. *Eur Respir J.* 2005;25(1):15–22.
14. Valentine R, et al. Morphological and biochemical features of elastase-induced emphysema in strain A/J mice. *Toxicol Appl Pharmacol.* 1983;68(3):451–461.
15. Gross P, et al. Experimental emphysema: its production with papain in normal and silicotic rats. *Arch Environ Health.* 1965;11:50–58.
16. Shiomi T, et al. Emphysematous changes are caused by degradation of type III collagen in transgenic mice expressing MMP-1. *Exp Lung Res.* 2003;29(1):1–15.
17. Foronjy RF, et al. Progressive adult-onset emphysema in transgenic mice expressing human MMP-1 in the lung. *Am J Physiol Lung Cell Mol Physiol.* 2003;284(5):L727–L737.
18. Zheng T, et al. Inducible targeting of IL-13 to the adult lung causes matrix metalloproteinase- and cathepsin-dependent emphysema. *J Clin Invest.* 2000;106(9):1081–1093.
19. Wang Z, et al. Interferon gamma induction of pulmonary emphysema in the adult murine lung. *J Exp Med.* 2000;192(11): 1587–1600.

20. Elwood W, et al. Characterization of allergen-induced bronchial hyperresponsiveness and airway inflammation in actively sensitized brown-Norway rats. *J Allergy Clin Immunol.* 1991;88(6): 951–960.
21. Nagai H, et al. Effect of anti-IL-5 monoclonal antibody on allergic bronchial eosinophilia and airway hyperresponsiveness in mice. *Life Sci.* 1993;53(15):PL243–PL247.
22. Renz H, et al. Specific V beta T cell subsets mediate the immediate hypersensitivity response to ragweed allergen. *J Immunol.* 1993;151(4):1907–1917.
23. Kung TT, et al. Characterization of a murine model of allergic pulmonary inflammation. *Int Arch Allergy Immunol.* 1994;105(1): 83–90.
24. Chavez J, et al. Interactions between leukotriene C4 and interleukin 13 signaling pathways in a mouse model of airway disease. *Arch Pathol Lab Med.* 2006;130(4):440–446.
25. Kerem B, et al. Identification of the cystic fibrosis gene: genetic analysis. *Science.* 1989;245(4922):1073–1080.
26. Buchwald M, Tsui LC, Riordan JR. The search for the cystic fibrosis gene. *Am J Physiol.* 1989;257(2 pt 1):L47–L52.
27. Riordan JR, et al. Identification of the cystic fibrosis gene: cloning and characterization of complementary DNA. *Science.* 1989;245(4922):1066–1073.
28. Rommens JM, et al. Identification of the cystic fibrosis gene: chromosome walking and jumping. *Science.* 1989;245(4922): 1059–1065.
29. Snouwaert JN, et al. An animal model for cystic fibrosis made by gene targeting. *Science.* 1992;257(5073):1083–1088.
30. Davidson DJ, Rolfe M. Mouse models of cystic fibrosis. *Trends Genet.* 2001;17(10):S29–S37.
31. Guilbault C, et al. Cystic fibrosis mouse models. *Am J Respir Cell Mol Biol.* 2007;36(1):1–7.
32. Frizzell RA, Pilewski JM. Finally, mice with CF lung disease. *Nat Med.* 2004;10(5):452–454.
33. Mall M, et al. Increased airway epithelial Na+ absorption produces cystic fibrosis-like lung disease in mice. *Nat Med.* 2004;10(5): 487–493.
34. Fleischman RW, et al. Bleomycin-induced interstitial pneumonia in dogs. *Thorax.* 1971;26(6):675–682.
35. Adamson IY, Bowden DH. The pathogenesis of bloemycin-induced pulmonary fibrosis in mice. *Am J Pathol.* 1974;77(2):185–197.
36. Adamson IY, Bowden DH. Bleomycin-induced injury and metaplasia of alveolar type 2 cells. Relationship of cellular responses to drug presence in the lung. *Am J Pathol.* 1979;96(2):531–544.
37. Snider GL, Hayes JA, Korthy AL. Chronic interstitial pulmonary fibrosis produced in hamsters by endotracheal bleomycin: pathology and stereology. *Am Rev Respir Dis.* 1978;117(6):1099–1108.
38. Borzone G, et al. Bleomycin-induced chronic lung damage does not resemble human idiopathic pulmonary fibrosis. *Am J Respir Crit Care Med.* 2001;163(7):1648–1653.
39. Popenoe D. Effects of paraquat aerosol on mouse lung. *Arch Pathol Lab Med.* 1979;103(7):331–334.
40. Yoshida M, et al. A histologically distinctive interstitial pneumonia induced by overexpression of the interleukin 6, transforming growth factor beta 1, or platelet-derived growth factor B gene. *Proc Natl Acad Sci U S A* 1995;92(21):9570–9574.
41. Katsuma S, et al. Molecular monitoring of bleomycin-induced pulmonary fibrosis by cDNA microarray-based gene expression profiling. *Biochem Biophys Res Commun.* 2001;288(4): 747–751.
42. Brantley JG, et al. Cux-1 transgenic mice develop glomerulosclerosis and interstitial fibrosis. *Kidney Int.* 2003;63(4):1240–1248.
43. Driessens J, et al. The urethane-induced experimental pulmonary adenoma in the mouse. I. Histological study. *C R Seances Soc Biol Fil.* 1962;156:655– 657.
44. Mirvish SS. The carcinogenic action and metabolism of urethan and N-hydroxyurethan. *Adv Cancer Res.* 1968;11:1–42.
45. Brooks RE. Pulmonary adenoma of strain A mice: an electron microscopic study. *J Natl Cancer Inst.* 1968;41(3):719–742.
46. Gargus JL, Paynter OE, Reese WH Jr. Utilization of newborn mice in the bioassay of chemical carcinogens. *Toxicol Appl Pharmacol.* 1969;15(3):552–559.
47. Shabad LM. Dose-response studies in experimentally induced lung tumours. *Environ Res.* 1971;4(4):305–315.
48. Snyder C, et al. Urethan-induced pulmonary adenoma as a tool for the study of surfactant biosynthesis. *Cancer Res.* 1973;33(10): 2437–2443.
49. Cazorla M, et al. Ki-ras gene mutations and absence of p53 gene mutations in spontaneous and urethane-induced early lung lesions in CBA/J mice. *Mol Carcinog.* 1998;21(4):251–260.
50. Lin L, et al. Additional evidence that the K-ras protooncogene is a candidate for the major mouse pulmonary adenoma susceptibility (Pas-1) gene. *Exp Lung Res.* 1998;24(4):481–497.
51. Avanzo JL, Mesnil M, Hernandez-Blazquez FJ, et al. Altered expression of connexins in urethane-induced mouse lung adenomas. *Life Sci.* 2006;79:2202–2208.
52. Umemura T, et al. Susceptibility to urethane carcinogenesis of transgenic mice carrying a human prototype c-Ha-ras gene (rasH2 mice) and its modification by butylhydroxytoluene. *Cancer Lett.* 1999;145(1-2):101–106.
53. Furth PA. SV40 rodent tumour models as paradigms of human disease: transgenic mouse models. *Dev Biol Stand.* 1998;94: 281–287.
54. Wikenheiser KA, et al. Simian virus 40 large T antigen directed by transcriptional elements of the human surfactant protein C gene produces pulmonary adenocarcinomas in transgenic mice. *Cancer Res.* 1992;52(19):5342–5352.
55. Wikenheiser KA, Whitsett JA. Tumor progression and cellular differentiation of pulmonary adenocarcinomas in SV40 large T antigen transgenic mice. *Am J Respir Cell Mol Biol.* 1997;16(6): 713–723.
56. Kwak I, Tsai SY, DeMayo FJ. Genetically engineered mouse models for lung cancer. *Annu Rev Physiol.* 2004;66:647–663.
57. Zhao B, et al. Transgenic mouse models for lung cancer. *Exp Lung Res.* 2000;26(8):567–579.
58. Albanese C, et al. Recent advances in inducible expression in transgenic mice. *Semin Cell Dev Biol.* 2002;13(2):129–141.
59. Argmann C, Dierich A, Auwrex J. *Current Protocols in Molecular Biology*. New York: Wiley; 2006 [unit 29.A.1].
60. Matsuda I, Aiba A. Receptor knock-out and knock-in strategies. *Methods Mol Biol.* 2004;259:379–390.
61. Sato Y, et al. Establishment of Cre/LoxP recombination system in transgenic rats. *Biochem Biophys Res Commun.* 2004;319(4): 1197–1202.
62. Altomare DA, et al. A mouse model recapitulating molecular features of human mesothelioma. *Cancer Res.* 2005;65(18): 8090–8095.
63. Altomare DA, et al. Human and mouse mesotheliomas exhibit elevated AKT/PKB activity, which can be targeted pharmacologically to inhibit tumor cell growth. *Oncogene.* 2005;24(40): 6080–6089.

18

Tissue Culture Models

Roger A. Vertrees, Jeffrey M. Jordan, Travis Solley, and Thomas J. Goodwin

Introduction

The use of tissue cultures as a research tool to investigate the pathophysiological bases of diseases has become essential in the current ag of molecular biomedical research. Although it will always be necessary to translate and validate the observations seen in vitro to the patient or animal, the ability to investigate the role(s) of individual variables free from confounders is paramount toward increasing our understanding of the physiology and their role in disease. Additionally, it is not feasible to conduct certain research in humans because of ethical constraints, yet investigators may still be interested in the physiological response in human tissues; in vitro characterization of human tissue is an acceptable choice.

Tissue culture techniques have been utilized extensively to investigate questions pertaining to physiology and disease. The isolation and propagation of human epithelial cells has allowed investigators to begin to characterize the interactions and reactions that occur in response to various stimuli. Moreover, the culture of other human tissue has allowed researchers to investigate pathological cascades as well as other physiological responses. Combining cultured endothelial cells and leukocytes together in vitro under laminar flow conditions has helped elucidate the critical interactions that occur in rolling and emigration of leukocytes during the inflammatory response. Cultured embryonic stem cells that had been subjected to various growth conditions have advanced our understanding of cellular differentiation and growth.

Numerous cancer cell lines have been established to investigate their responses to chemotherapy and determine their biological properties. Overall, the use of cultured human tissue has provided a windfall of information on the pathogenesis of diseases that affect the body. In light of this, we briefly discuss the history and principles behind the utilization of tissue culture. We then discuss the current use of tissue culture to examine many of the unanswered questions involved in physiology and pathology.

History of Tissue Culture

The technique of tissue culture is generally accepted to have arisen following the experiment of Ross Harrison, around the turn of the twentieth century. In 1907, Harrison began by adapting a previously established bacteriology technique, the "hanging-drop" method, to culture a frog neuron.[1] In 1912, Alexis Carrel[2] built upon this work by successfully culturing small tissue samples from an 18-day-old chick embryo heart, thereby becoming the first scientist to propagate mammalian cells in vitro.[3] Carrel's demonstration that cells could be passaged 18 times, remain viable over 3 months, and continue to maintain cardiac rhythm was the first to show that cardiac tissues in vitro could retain normal characteristics for a prolonged period of time. These elegant studies, conceived by Carrel, initiated the modern day art of *histoculture* as it is now known.[4,5]

Although only a small "sect" of researchers embraced early tissue culture as a methodology to investigate the pathogenesis of disease, it is appropriate to describe Carrel as the father of mammalian tissue culture. In fact, Sven Gard, in his presentation speech for the Nobel Prize in Physiology or Medicine in 1954, referred to tissue culture as a "tissue cult. . . with Carrel as their high priest." In that year, Drs. Weller, Enders, and Robbins shared the prize for their work in propagating poliovirus in tissue culture. This work was the first Nobel Prize awarded in medicine and physiology for work accomplished primarily utilizing tissue culture as a methodology.

An additional important milestone in the use of tissue culture in biomedical research was the establishment of the first human cell line. In 1951, cervical cancer cells from Henrietta Lacks were cultivated into the first immortal cell line – "HeLa."[6] HeLa cells are still one of the most widely used cell lines today. Since the 1950s, tissue culture has become firmly established as a mechanism to answer many questions in biomedical research. Today, tissue culture is widely used to investigate diseases that affect the body, and through this work we have been able to increase our understanding of the pathological cascades that occur in diseases, as well as normal physiologies.

P.T. Cagle and T.C. Allen (eds.), *Basic Concepts of Molecular Pathology*, Molecular Pathology Library 2,
DOI: 10.1007/978-0-387-89626-7_18, © Springer Science+Business Media, LLC 2009

Types of Tissue Culture

Tissue culture is a commonly used generic term for the in vitro cultivation of cells, attributed to the early cultures that generally consisted of heterogeneous cultures of crudely disaggregated tissues. Currently, many terms are used that can be encompassed by the term: organ culture, cell culture, primary explants, and ex vivo propagation all concern the in vitro cultivation of cells or tissues. *Cell culture* in general can be applied either to primary cells (e.g., those with a finite lifespan) or to cell lines (e.g., HeLa cells). Additionally, these cultures can be either a homogeneous or a heterogeneous group of cells.

Primary cell culture involves the isolation of cells from a tissue by disaggregation. Single-cell suspensions from tissues can be completed through either enzymatic digestion of extracellular matrix surrounding the cells – such as with ethylenediaminetetraacetic acid, trypsin, or collagenase – or mechanical disaggregation. These disaggregation procedures have the disadvantage of possibly injuring cells. If the cells of interest are adherent viable cells, they will be separated from nonviable cells when the medium is changed. Alternatively, viable cells can be separated from nonviable cells before culture by subjecting the single-cell suspension to density-gradient centrifugation (e.g., Hypaque). Primary cells have an advantage of possessing many of the biological properties that they possessed in vivo because they are not transformed. Primary cells, unlike cell lines, are not immortal and have only a finite survival time in culture before becoming senescent. Variant cells, however, as well as those obtained from neoplastic tissue, may proliferate infinitely, thus becoming immortal in vitro. This development will eventually allow the immortal cell to take over the culture and can be thought of as a cell line. In general, primary human cultures will survive for 30–80 passages in vitro, although this number is dependent on cell type, conditions, and possibly other unknown factors. Primary cells are widely used to examine the effects of toxins, infectious agents, or other cellular interactions that would not be feasible in vivo. Primary cells have a disadvantage of being a heterogeneous mixture of cells upon primary isolation, with the type of cell obtained generally a component of the disaggregation method used. The most common contaminant seen following isolation of primary cells are cells of mesenchymal origin (e.g., fibroblasts). However, advances have been made that allow the culture of homogeneous populations of cells. For instance, cell-surface molecules specific for the cells of interest may be tagged with monoclonal antibodies. Techniques such as fluorescence-activated cell sorting or the use of magnetic beads can be utilized to enrich the single-cell suspension for the cell type of interest. Additionally, some investigators have recently exploited unique characteristics of certain cells, such as the presence of P-glycoprotein or multidrug resistance-associated proteins expressed on endothelial cells, to poison other contaminating cells in culture.[7]

Another type of primary cell culture are "primary explants." This type of culture is not subjected to a disaggregation procedure as is the primary cell technique described earlier. Therefore, single-cell suspensions do not occur. Briefly, tissue samples are dissected and finely minced. These tissue pieces are then placed onto the surface of a tissue culture plate. Following plating of tissue pieces, cells have been shown to migrate out of the tissue and onto the tissue culture surface.[8] This technique is useful when cells of interest may become damaged or lost in the disaggregation technique described earlier and is often used to culture human bronchial epithelial cells.[9]

Cell lines are another useful source of cells to investigate questions in biomedical research. These cells have the advantage of being immortal as opposed to the finite lifespans that primary cells possess. Additionally, they are generally well studied and characterized, leaving few experimental variables to worry about. These cells however, are prone to dedifferentiation, a process by which they lose the phenotypic characteristics of the cell from which they began. Many of the early cell lines were established from tumor tissue and as such possess abnormal growth characteristics. Newer cell lines have been established by molecular techniques such as inserting a telomerase gene into a cell to allow it to replicate infinitely.[10] Because of the phenotypic changes that allow cell lines to replicate infinitely in culture, they are often a first choice for experiments; however, they are also highly criticized in light of their nonnatural phenotype.

Organ culture, as the name implies, involves ex vivo culture of the whole or significant portion of the organ. The main advantage to this type of culture is the retention and preservation of the original cell-cell interaction and extracellular architecture. This type of culture may be particularly important when experimental design necessitates the use of an ex vivo system, but researchers still need to retain the original organ architecture to answer questions posed. These types of cultures do not grow rapidly, however, and are therefore not suitable for experiments needing large numbers of a particular cell type.[11]

Advantages and Limitations of Tissue Culture

Tissue culture has become the penultimate tool of the reductionist biologist. The utilization of tissue culture as a research methodology has allowed investigators to study isolated interactions in its near-normal environment. These experiments by their very nature introduce artifacts; however, they do minimize the number of confounding variables that may affect a particular experiment. For instance, tissue culture allows investigators to determine the effects of one particular treatment on a particular cell type, which would not be feasible in vivo. Additionally, tissue culture models of disease allow investigators to obtain samples and make observations more readily than those done in vivo. However, it is the relative simplicity of experiments done in vitro that allows models of disease or physiology to come under frequent and warranted criticism. These models do not take into consideration the complexity of biological systems. Diminishing possible confounding variables by culturing cells in vitro brings up the constant criticism of how applicable results are because of alterations of the normal cellular environment in vivo. For example, cell–cell interactions in vitro are reduced and unnatural. Moreover, the culture does not contain the normal heterogeneity and three-dimensional architecture that are seen in vivo. This said, however, tissue culture biology has proved to be successful in many ways.

Cell Culture and the Study of Disease Processes

We have discussed the advantages that experimental systems using tissue culture afford researchers studying physiology and pathogenesis. Because of its ability to isolate individual variables and determine their role(s) in physiology, cell culture has become an integral tool in deciphering the pathological cascades that occur in human disease.

Many diseases that affect humans are multifactorial. This begs the question how can cell culture, because of its reductionist nature only dealing with a minimal number of variables, help to solve the unknown questions and decipher the components involved in disease? Often, clinical observations, and the questions arising therein, have been the launching pad for investigation.

For instance, observations of massive inflammation in the bronchoalveolar lavage samples of patients with acute respiratory disease syndrome (ARDS), consistent with damage seen in histological samples, prompted investigators to determine the role(s) of inflammation in the etiology of ARDS. Through the use of cell culture, investigators were able to determine individual interactions that occurred in the disease process. Investigators have utilized culture models employing microcapillary endothelial cells under flow conditions to understand the role of proinflammatory cytokines in the cytokinesis and emigration of neutrophils in disease. Using a model of pulmonary endothelium under flow conditions allowed investigators to demonstrate the importance of certain proinflammatory cytokines in ARDS.[12]

The role of inhaled toxicants in lung injury, and the mechanism(s) by which they cause disease, is another area of investigation that has utilized cell culture. Scientists have developed diverse and unique tissue culture systems that contain air–liquid barriers of lung epithelium and subjected these cells to various gaseous toxicants to determine what occurs following inhalation of various chemicals. Utilizing these types of systems, investigators are able to control the exposure time and other variables that may be difficult when determining inhaled toxicant effects in vivo. Moreover, the use of tissue culture, as opposed to an animal model, allows investigators to observe effects kinetically, without undue changes (e.g., sacrifice) and expense in the experimental model.[12]

A tissue culture model also permits an investigator to observe multiple changes in real time, such as cellular integrity, cell signaling and intracellular trafficking, protein expression changes, and oxidant-induced cellular damage. Deciphering each of these changes in an animal model would be extremely difficult; by employing a tissue culture model, researchers are able to tightly control the experimental system while isolating the events of interest. Further examples of how tissue culture models are currently being used to elucidate questions in physiology and disease are discussed later.

Biology of the Cultured Cell

Culture Environment

Maintaining cells in vitro was initially a very difficult task. Many characteristics need to be fulfilled before a successful cell culture occurs. Some of these characteristics are dependent on the type of tissue being studied; others may depend on specific requirements of the individual cells. Various chemically defined media are now available commercially to support the growth and differentiation of numerous cell types. The creation of defined media has allowed investigators to culture a multitude of cell types while controlling the local environment to answer pertinent questions. For example, glucose can be removed from a culture medium to study its effects on cellular metabolism, relative position in the cell cycle, and many other effects. Each chemical component is known in these media. Additionally, investigators can add growth factors to nourish their cell cultures.

The medium chosen when culturing cells in tissue culture must fit two main requirements: (1) it must allow cells to continue to proliferate in vitro, and (2) it must allow the preservation of the certain specialized functions of interest.[8] The most common medium formulations used currently are Dulbecco's modified Eagle's medium, minimum essential medium, RPMI 1640, and Ham's F-12. Occasionally, investigators develop new medium types to attain a formulation that optimizes their own experimental conditions. Fetal bovine serum is a common additive to most tissue culture media, although some investigators choose to forgo this additive for more defined supplementation. Additionally, others may choose sera from other sources such as human serum when culturing cells of human origin. Inactivation of complement by heat treating serum for 1 h at 56°C was initially very popular in tissue culture. However, it has become clear that this treatment may in fact damage some of the proteinaceous growth factors present in the medium, rendering it less effective. Currently, many experts recommend heat inactivation only if the cell type of interest is particularly sensitive to complement.[13]

When deciphering if the current culture conditions are sufficient for the experimental design, the investigator must determine which cellular characteristics are important. Not only are the general characteristics, such as adhesion, multiplication, and immortalization of cell types important, but so are tissue-specific characteristics.

Cell Adhesion

Nearly all normal or neoplastic human epithelial cells will attach with relative ease to tissue culture surfaces. However, for culture cells that may loosely adhere, or may not adhere at all, scientists coat tissue culture surfaces with extracellular matrix proteins. Incubating tissue culture surfaces with serum, as well as laminin, fibronectin, or collagen, before culture has been shown to improve attachment of finicky cells.[9] These treatments also help in replicating the normal attachment of cells to extracellular matrix proteins in vivo.

Development of Continuous Cell Lines

The development of continuous cell lines may be serendipitous, as was the development of early cell lines. In brief, many investigators would continue splitting primary cell cultures until one or more cell clones became immortal. Unfortunately, the changes that generally occurred in culture led to cells with abnormal phenotypes that had undergone dedifferentiation. Today, many investigators choose to use molecular biology techniques, exploiting our current knowledge of oncogenic viruses and enzymatic processes of cellular aging to transform primary cells in vitro to an immortal phenotype. It is known that the large T antigen present in the SV (Simian virus) 40 virus is capable of transforming cells to an abnormal phenotype.[12,14,15] Moreover, transfection of primary cells with a transposase enzyme has also been shown to induce an immortal phenotypic change while preserving most normal cellular functions and phenotypes.[12]

Dedifferentiation

A commonly encountered problem in tissue culture is dedifferentiation. This loss of phenotype may be insignificant to the research at hand or it may be critical, and it must be dealt with on a case-by-case basis. When a cell culture undergoes dedifferentiation, it is often unclear whether undifferentiated cells took over the culture of terminally differentiated cells or whether a primary cell of interest became immortal under the culture conditions.

Functional Environment

The functional environment in which cells are cultured is critical when correlating experimental results to those seen in vivo. We previously alluded to the importance of the environment in which cells are cultured when discussing the advantages and limitations of tissue culture. Investigators have frequently striven to replicate integral in vivo environments in vitro to increase the significance of their experimental results.

The development of cell culture insert wells (e.g., Transwells, Corning) has allowed investigators to culture bronchial or alveolar epithelial cells at an air-liquid interface. This ability allows investigators to begin to replicate a significant aspect of these cells' functional environment in vitro, thereby increasing their understanding of the effects of gaseous particles on pulmonary epithelial cells. Alternatively, scientists have also cultured epithelial cells on a roller bottle apparatus. This method allows investigators to determine the amount of time the apical epithelial cell surface is in contact with the air.

Capillary cell cultures have also come under frequent criticism when cultured in a monolayer in a tissue culture plate. Investigators have been able to utilize gel matrices in which capillary cells form tubule-like structures, more closely replicating the architecture these cells maintain in vivo. Additionally, endothelial cells are constantly under flow conditions in vivo. Addressing this condition in vitro has allowed investigators to look at the role of endothelial cells during inflammation, helping to increase the understanding of the role endothelium plays in various diseases.

At times, researchers may also choose to determine the effects of soluble factors (e.g., cytokines, hormones, neurotransmitters) from acute patients or animal models in a cell culture model. The milieu of soluble factors present in the serum that may play a role in a disease state is considerable. Moreover, these factors may have actions alone that are different when combined with other soluble factors. Reconstituting every factor presents a difficulty in vitro and leaves the possibility that an unknown factor may be missing. To address this, investigators have harvested sera from patients or animal models and used these samples as additives in their media formulations. For instance, through the use of serum samples from an animal model of smoke- or burn injury-induced acute lung injury, investigators have demonstrated that use of arteriovenous CO_2 removal in acute lung injury significantly reduces apoptotic cell death in epithelial cells.[16]

Epithelial Cell Culture Models

Epithelial cells provide many important and equally diverse physiological roles in vivo. Toward the end of replicating these roles in vitro, investigators have collected numerous epithelial cells, both primary and immortalized, to help advance our understanding of the physiological significance.

Gastrointestinal Epithelial Cell Cultures

Results from a three-dimensional organotypical tissue culture experiment from adult murine colon performed by Bareiss et al[17] showed that this model could be useful in studying epithelial cell–cell interactions, cellular signaling, and microbiological infections. Primary cell cultures of bovine colon epithelium can further be used in toxicological studies in vitro as a model for the colon epithelium.[18]

Hepatocyte Tissue Cultures

Hepatocytes play an integral role in the metabolism of numerous drugs, nutrients, and other constituents in vivo. In vitro culture of hepatocytes has facilitated extending our understanding of the physiological steps important in mediating these physiological functions. Additionally, in vitro culture has allowed investigators to understand the importance of viral infection, drug interactions, and other physiological signals that alter metabolism and normal cellular function.

Hepatocytes in Drug Metabolism

Studies performed by Nussler et al[19] concluded that the use of a collagen sandwich or three-dimensional membrane bioreactor can be used to study drug metabolism of human hepatocytes. Additionally, to study drug metabolism in human hepatocyte cells, Zeilinger et al[20] developed a three-dimensional bioreactor culture model of liver cells under continuous medium perfusion with decentralized mass exchange of integral oxygenation.

Hepatocellular Carcinoma

Zen et al,[21] using tissue culture methods, discovered support for the maintenance of the cancer cell hierarchy in human hepatocellular carcinoma. Utilizing a radial flow bioreactor system, Kosuge et al[22] prompted differentiation of human hepatocellular carcinoma cells in three-dimensional culture. The resulting differentiated cells could be useful for improvements in genetic or pharmacological reinforcement and for the monitoring of bioartificial livers.[22]

Nervous System Tissue Cultures

To potentially use reconstructed functional cortical-like tissues for drug screening, in detection of environmental toxins, and in neural cell therapy, Ma et al[23] cultured neural stem and progenitor cells on three-dimensional type I collagen gels and then transferred the cell-collagen constructs to three-dimensional bioreactors. Additionally, using three-dimensional collagen gel cultures, Ma et al[24] succeeded in using neural stem and progenitor cells to recapitulate the central nervous system stem cell development demonstrating a functional synapse and neuronal network formation in a three-dimensional matrix. A tissue-like primary cortical cell culture using Matrigel and serum free Start V medium was developed by Braun et al[25] to be used in the investigation of basic cell-cell interactions in vitro.

Lymphatic Tissue Culture

The complexity of the lymphatic system has made the engineering of functional tissues difficult, but progress has been made in the engineering of lymphatic capillaries and the development of a bioreactor designed to culture lymph nodes.[26]

Pulmonary Endothelial Cells

Pulmonary endothelial cells represent a unique type of endothelium because of their paradoxical responses to hypoxia. This uniqueness highlights the need to utilize cell culture models of pulmonary endothelium as opposed to other endothelia when interested in investigating their role(s) in pulmonary physiology. Several investigators have described the isolation and culture of pulmonary endothelial cells. Persistent pulmonary hypertension of the newborn, also known as neonatal pulmonary hypertension, is caused by a disorder of the pulmonary vasculature from fetal to neonatal circulation, culminating in hypoxemic respiratory failure and death. The inciting events that culminate in neonatal pulmonary hypertension are multifactorial. Despite this, decreased production of vasodilator molecules such as nitric oxide and prostaglandin I_2 in the pulmonary endothelium has been shown to be a critical component of disease progression.[27]

Infectious Diseases

Infectious diseases play a unique role in lung pathology in light of their roles as either important contributors or consequences of many lung diseases. For instance, certain lung diseases may predispose patients to infection: patients afflicted with obstructive lung diseases, as well as cystic fibrosis patients, commonly suffer from severe and recurrent bacterial infections. Additionally, patients may become superinfected following a viral respiratory infection. Systemic infections, such as gram-negative bacterial sepsis, may lead to lung diseases such as ARDS.

Human Type II Alveolar Pneumocytes and Acute Lung Injury/Acute Respiratory Distress Syndrome

Pulmonary alveolar type II cells are a unique cell subset that carries out highly specialized functions which include synthesis and secretion of surfactant, a unique composition of lipoproteins that act to reduce surface tension at the alveolar air–liquid interface.[28] Defining the molecular mechanisms leading to production of surfactant by type II pneumocytes is important in many disease processes.

The pathogenic sequence that results in ARDS, the most severe manifestation of alveolar lung injury, is generally thought to be initiated by a systemic inflammatory response.[29] Despite this knowledge, there still exist many questions about the initial triggers and pathological steps that occur in ARDS. Greater understanding of these steps may help to develop new treatment regimes. Currently, treatment of ARDS consists of mechanical ventilation, which helps to stabilize blood gases. However, mechanical ventilation itself may provoke further inflammation in the alveoli, thereby decreasing compliance and gas exchange in the alveoli.[30]

The cell type of particular interest in ARDS and diffuse alveolar damage is the type II pneumocyte.[31–35] Until recently, studies trying to decipher the pathological sequence in acute lung injury have had to rely on standard lung epithelial cell lines. Recently, however, human type II alveolar epithelial cells (pneumocytes) have been successfully isolated from fetal human lung tissue by collagenase digestion.[36] Briefly, fetal lung tissues were minced and incubated in a serum-free medium containing dibutyryl cyclic adenosine monophosphate for 5 days. The tissue explants were then treated with collagenase and incubated with DEAE-dextran to eliminate contaminating fibroblasts. Cells were then plated onto tissue culture dishes treated with extracellular matrix derived from MDCK cells and cultured overnight in Weymouth's medium containing 10% serum. These steps resulted in relatively pure populations of human type II pneumocytes that were then cultured at an air–liquid interface. Using these methods, Alcorn et al[36] were able to maintain a primary culture that retained the morphological and biochemical characteristics of type II pneumocytes for up to 2 weeks.

Three-Dimensional Biology

Conventional Bioreactors and Three-Dimensionality: The Origins of Three-Dimensional Culture

Carrel postulated that tissue development was linked to access to nutrient supply, noting that peripheral cells grew readily, and internal cells became necrotic, presumably based on their distance from the nutrient source. To circumvent this issue, Carrel implemented cultures on silk veils, preventing the plasma clots of the growth media from deforming or becoming spherical, thus facilitating the internal cell's ability to obtain nutrient replenishment. Many attempts were made in standard culture systems (bioreactors) and other culture apparatuses to escape the constraints of two-dimensional cell culture, with the intent of yielding high-fidelity human and mammalian tissues, and thus emphasizing the need for development of three-dimensional biology.[2,3]

Another famous researcher, Leighton, improved on Carrel's techniques in the 1950s and 1960s. Leighton's major contribution to three-dimensional culture technology was the introduction of the idea of a sponge matrix as a substrate on which to culture tissues.[37,38] Leighton first experimented on cellulose sponges surrounded by plasma clots resident within glass tubes. He devised a system to grow tissue explants 1–5 mm^2 in area on sponges, using small amounts of chick plasma and embryo extract. After the mixture solidified on the sponge, Leighton added the nutrient media and inserted the "histoculture" in a roller apparatus to facilitate nutrient mass transfer. He experimented with many sponge combinations, discovering that collagen-impregnated cellulose sponges were optimal for sustaining the growth of native tissue architecture.[4,39]

Leighton was successful in growing many different tissue types on the sponge-matrix cultures.[4,39] Leighton also found that C3HBA mouse mammary adenocarcinoma cells, when grown on sponge-matrix histoculture, aggregated "much like the original tumor, forming distinct structures within the tumors such as lumina and stromal elements, and glandular structures." An extremely important difference of this three-dimensional histoculture from the standard two-dimensional culture is the apparent quiescence of the stromal component and the balanced growth of these cells with regard to the overall culture. Leighton further advanced the concept of three-dimensional histoculture to histophysiological gradient cultures.[40] These cultures are conducted in chambers that allow metabolic exchange between "the pool of medium and the culture chamber by diffusion across a membrane." Histophysiological gradient cultures mimic, to some degree, diffusion in tissues.[39]

From the pioneering work of Carrel and Leighton, other methods of emulating three-dimensional cultures have been developed, such as embedding cells and tissues in collagenous gels of rat tail as per the techniques of Nandi and colleagues. Clusters of mammary cells were initiated into cultures, and Nandi and colleagues observed three-dimensional formation of ductile structures and sustained growth over a period of several weeks concordant with proliferation at the edge of the cell mass with minimal fibroblast contamination.[41] Thus, mammary tissue cultured in the collagen gels resembled, histologically, in vivo tissue, differing significantly from two-dimensional monolayer cultures. Nandi et al suggested an important criterion for successful growth of mammary tissue-like structures in collagen gels, versus the monolayer culture, was maintenance of mammary cell shape in three dimensions as opposed to conventional monolayer culture. Many of the advantages of three-dimensional cultures seen by Leighton, Nandi, and others may be attributed to permitting the cells to retain their normal shape and special associations.[4] This global concept will be important as we begin to understand and recall the physical and environmental characteristics of the rotating-wall vessel systems.

Other methods of three-dimensional culture encompass a technique known as organ culture or culture on a filter, a strategy developed by Strangeways[42] and Fell and Robison.[43] Tissue explants were grown on lens paper in a watch glass containing liquid culture medium. Browning and Trier[44] found "that for some tissues, it is critical to keep the cultures at the air-liquid interface," thus allowing the tissues to experience conditions similar to the in vivo environment.

Another strategy is the use of three-dimensional cultures known as *proto-tissues*, or aggregates of cells, used to form spheroids. This technique was popularized by Sutherland and colleagues more than 20 years ago when they manipulated aggregates of cells into a spherical configuration by spinning agitation of the cells in spinner flasks.[45] This technique produced pseudo-tissue-like organoids useful for research evaluations. Each of these methodologies will be of benefit as we continue to examine strategies for achieving three-dimensional lung tissue constructs.[4,39]

Bioreactors and three-dimensional systems have been and are used by medical researchers and by pharmaceutical production plants to create mimics of human, mammalian, microbial, and plant physiology. As reviewed, many approaches have been employed to create these synthetic physiologies; however, none of these has been entirely satisfactory for broad-based mammalian cell culture use.

The impact of molecular genetic engineering in the early 1980s was for the most part confined to pharmaceutical proteins such as interferon, growth hormones, and insulin molecules. During processing of these valuable, but low-yield, products, emphasis was placed on downstream recovery rather than on bioconversion or bioprocesses, let alone three-dimensional tissue development. Advances in modern biotechnological product recovery are expanding to medium- and high-volume enzyme products and limited forms of bulk products. In the world of high-volume, high-value products, cell culture bioreactor productivity contributes an important role in determining the economic influence of bioprocessing. Ideal bioprocessing bioreactors will in future have high reactor productivity with an in situ separation capability of products.[46]

Membrane bioreactors fit this description and already contribute remarkably to the bioprocessing industry. Membrane bioreactors are capable of retaining enzymes, organelles, and microbial, animal, and plant cells behind a membrane barrier, trapped in a matrix or adherent to the membrane surface. In 1963, Gallup and Gerhardt[47] initially used the membrane bioreactor for dialysis culture of *Serratia marcescens*. Immobilized enzyme microencapsulation was pioneered by Chang,[48] but Butterworth et al[49] first developed the enzyme membrane reactor to successfully accomplish starch hydrolysis with α-amylase. Similarly, for animal cell culturing, Knazek et al[50] cultured human choriocarcinoma cells on compacted bundles of Amicon fibers. Many reviews on the particular applications of hollow-fiber and immobilized bioreactant bioreactors for enzyme catalysts, microbial cells, and animal cell culture are available.[46,51–55]

Prolonged human residence in microgravity results in the atrophy of healthy skeletal muscle. Continued space exploration requires a better understanding of this phenomenon.[56] Muscle wasting is likely caused by changes at the system level marked by increased circulating glucocorticoids and decreased circulating growth hormone and at local levels as seen by decreased myofiber resting tensions. Differentiated skeletal muscle cultures have served as model systems for gaining a better understanding of the roles of exogenous and endogenous cytokines and of muscle fiber tension in regulating muscle cell growth. This effort has led to the beginning of tissue engineering studies by the space science community. Engineering these tissues into three-dimensional bioartificial muscle (BAM) constructs has extended their utility for space flight investigations as a prelude to developing countermeasures for the microgravity environment. Serendipitously, these bioreactor models in part mimic diseases of the musculature such as dystrophy.[57]

BAMs have been sustained in a viable condition for as much as a month in perfused hollow-fiber bioreactor cartridges.[58] Further, investigators have documented[57] that growth hormone and/or insulin-like growth factors are potential protein therapeutics, which may in part attenuate skeletal muscle degeneration in space as well as show promise for the treatment of muscle-wasting diseases of the terrestrial community.[57]

An associated application of space bioscience to an Earth-based disease is the investigation of bone degeneration. The microgravity effect on bone decalcification has been documented beginning with early space flights and subsequently found to mimic conditions analogous to osteoporosis frequently seen in aging females. Over the past three decades, the relationship between cellular morphology and metabolism and the role of physical stress in bone loss has been studied. Physiomechanical stresses induce shape changes in osteoblasts, possibly mediated by reorganization of focal contacts. Thus, gravity variations (Gz) have been studied to understand their influence on osteoblast adhesion of ROS 17/2.8 rat osteosarcoma cells during 15–30 parabolic flights. Nontrivial flight-induced cell shape changes consisting of decreased area concordant with contact plaque reordering have been observed; however identical periods of continuous mechanical stress induced by centrifugation (2 Gz) or clinostatic rotation (Gz randomization) had no discernible effect on cell adhesion. Synchronization of ROS 17/2.8 G2/M accomplished via treatment in nocodazole inhibited the flight-induced decreases in adhesion parameters. Investigators thus concluded that ROS 17/2.8 cells possess the ability to sense Gz triggers and that adaptation is related to cytoskeletal function.[59]

As presented previously, tissue engineering applications of three-dimensional function and structure are well known in medical science research.[60] In microgravity, three-dimensional aggregates form, facilitating the expression of differentiated organotypic assemblies. Investigations to determine the effect of composite matrices, spiked with esterified hyaluronic acid and gelatin, to augment osteochondral differentiation of cultured, bone marrow-derived mesenchymal progenitor cells, and the effects of the matrix on cellular differentiation have been examined in vitro and in vivo.[60]

Briefly, empty and populated matrices cultured for 28 days with and without transforming growth factor (TGF)-β1 demonstrated the following results. Cells implanted in the matrix produced a robust type II collagen extracellular matrix in vitro. Matrices placed in immunodeficient mice yielded no differentiation in empty constructs, osteochondral differentiation in loaded implants, and an enhanced level of differentiation in preimplantation in vitro-cultured matrices containing TGF-β1. These results demonstrate the utility of a three-dimensional matrix for presentation of bone mesenchymal progenitor cells in vivo for repair of cartilage and bone defects as well as indicate the efficacy for in vitro tissue engineering regimes.[60] These techniques lend themselves to microgravity and ground-based research tissue cultures alike.

Many earth-based laboratories are researching and developing hematopoietic bone marrow cultures of stem cell origin, and three-dimensional configurations are providing promising results, as illustrated by Schoeters and coworkers.[61] They report that murine bone marrow cells, cultured under long-term hematopoietic conditions, produce mineralized tissue and bone matrix proteins in vitro but only when precipitated by the presence of adherent bone stroma cells in three-dimensional collagen matrices. At a concentration of 8×10^6 stromal cells, mineralization occurs in 6 days. In contrast, two-dimensionally oriented marrow fragments at 1×10^7 cells require requires more than 10 days before mineralization can similarly be detected.[61]

Two-dimensional long-term marrow culture facilitates and enhances expansion of the stromal component and rudimentary differentiation of osteogenic-like cells in the adherent stromal layer, as verified by type I collagen or cells positive for alkaline phosphatase. Production of osteonectin and osteocalcin, a bone-specific protein, combined with calcification is observed only in three-dimensional cultures. These studies demonstrate the need for and benefit of three-dimensionality and the application to the microgravity environment.[61] As we can see, this further reinforces the quest for three-dimensionality and the potential of modeling the microgravity environment.

Three-Dimensional Models for Physiological Study

Investigations clearly show the need for the application of three-dimensional study techniques in pathophysiological studies. Interestingly, three-dimensional biology has facilitated full-scale investigations into most areas of tissue engineering, cell biology and physiology, immunology, and cancer research.

Several important factors are necessary when considering the successful recapitulation of a three-dimensional mammalian tissue. These factors include specific elements of differentiation which, when replicated accurately, result in the functionality of the original tissue being mimicked. Figure 18.1 graphically represents the stages of "assembly" and differentiation needed to result in a functionally accurate tissue. These stages are three-dimensional assembly, three-dimensional growth, cellular matrix formation [extracellular matrix (ECM) and basement membranes], differentiation (associated with cellular specialization), and vascular formation (or pseudo-vasculature).

To accomplish the successful recapitulation of an ex vivo tissue, a series of defined steps are necessary. The creation of a three-dimensional structure is accomplished by systematic tissue engineering of the desired tissue architectures. Figure 18.2 demonstrates the assembly process and illustrates some of the variables that may be employed in the engineering process.

The assembly of complex functional mammalian tissues in 1× gravity is problematic because of the effects of shear stress, turbulence, and inadequate oxygenation in conventional cell culture systems. This study describes the culture and three-dimensional assembly of baby hamster kidney (BHK) mammalian cells on microcarriers under controlled oxygenation, low shear stress, and turbulence in the NASA-designed integrated rotating-wall vessel (RWV).[62–64] Anchorage-dependent cells are widely cultured on microcarriers.[65] Studies show that for the purposes of improved surface-to-volume ratio and scale-up, the microcarrier suspension culture provides excellent potential for high-density cell growth.[66] In addition; microcarriers serve well as structural supports for three-dimensional assembly, the composite of which is the basis for three-dimensional tissue growth.[67]

Conventional culture systems for microcarrier cultures (i.e., bioreactors) use mechanical agitation to suspend microcarriers and thus induce impeller strikes as well as fluid shear and turbulence at the boundary layer between the wall and the fluid. Investigators have attempted to make a complete study of the most efficient bioreactor designs and agitation regimens.[68] They concluded that virtually all stirred-tank bioreactors operate in the turbulent regimen. It has been demonstrated that bead-to-bead bridging of cells is enhanced significantly at lower agitation rates in a stirred reactor.[69] Aggregates of as many as 12–15 microcarriers

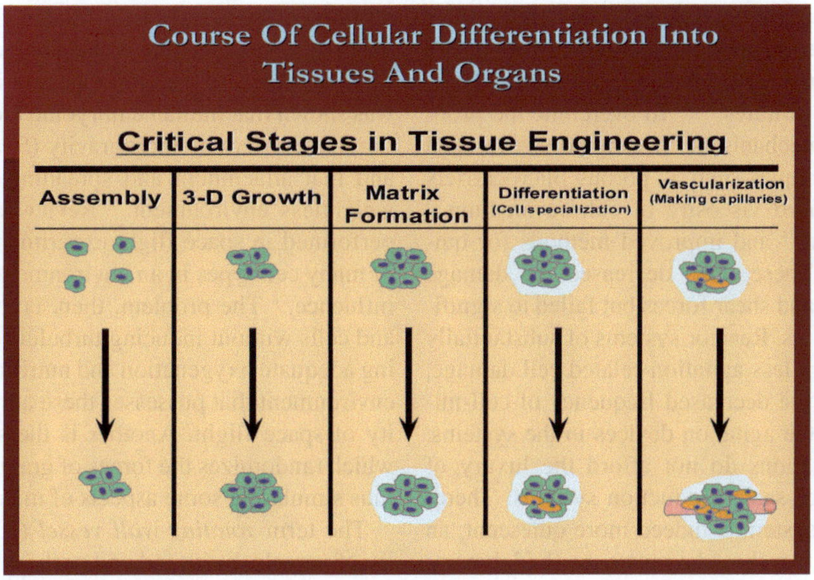

FIG. 18.1. Stages of cellular differentiation requisite for tissue functionality.

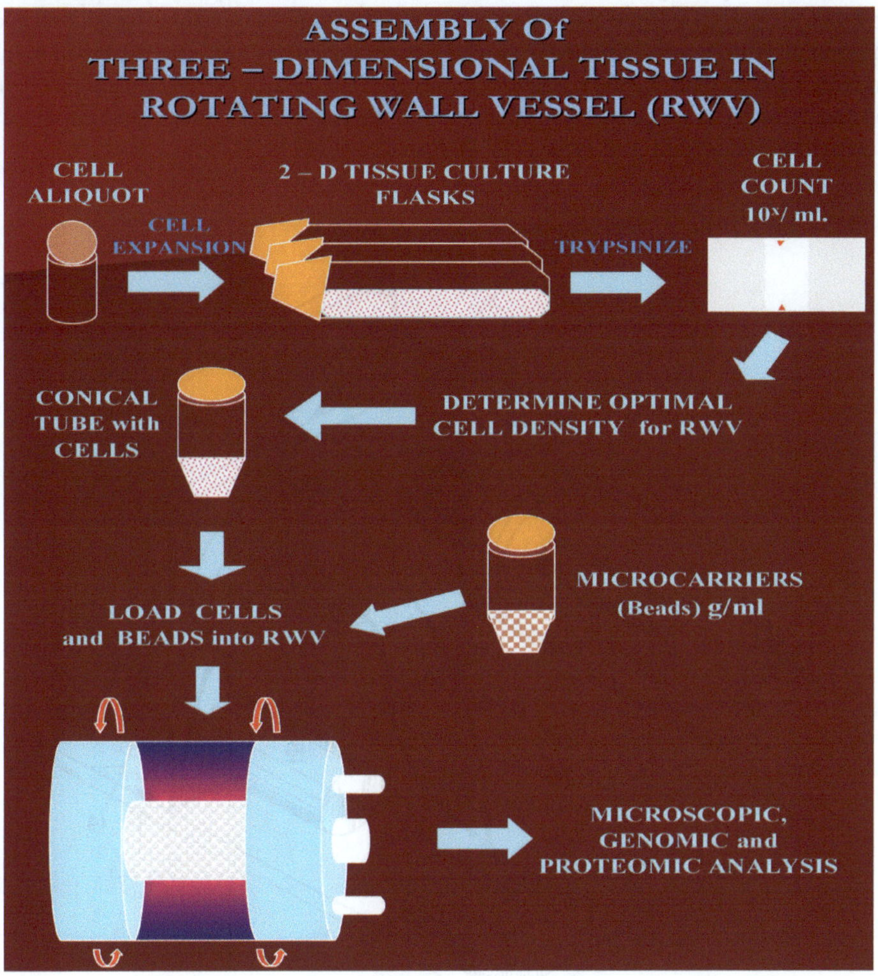

FIG. 18.2. Development and tissue assembly process for engineering of a three-dimensional tissue.

bridged with bovine embryonic kidney cells (BEK) have been reported. Excessive agitation from either stirring or gas bubble sparging has been documented as a cause of cell damage in microcarrier cell cultures.[70,71] To overcome the problems induced by these mechanisms, investigators developed alternative culture techniques such as porous microcarriers to entrap cells,[72] increased viscosity of culture medium,[73] bubble-free oxygenation,[74] and improved methods for quiescent inoculation.[75,76] These steps decreased the damage attributed to turbulence and shear forces but failed to significantly rectify the problems. Reactor systems of substantially increased volume exhibit less agitation-related cell damage, presumably because of the decreased frequency of cell-microcarrier contact with the agitation devices in the systems. Research-scale investigations do not afford the luxury of experimenting with large-scale production systems. Therefore, if a large-volume system is indeed more quiescent, an improved bioreactor system should emulate the fluid dynamics present in the upper regions of large-scale reactors in which cells and microcarriers reside with minimal agitation. Microcarriers and cells in large-volume reactors are semi-buoyant, rendering them practically weightless, a condition that we hypothesize may be analogous to the environment of microgravity. It is stated that "Unless a cell culture is growing in an environment free of gravitational forces, moderate levels of agitation are required to suspend microcarriers that are not neutrally buoyant."[73] The ability to grow cells in a rudimentary horizontally rotated chamber has been previously demonstrated and is intended to simulate in the laboratory the effects of weightlessness or microgravity on cells. It was shown that human embryonic kidney (HEK) cells attach to microcarriers in microgravity (Space Shuttle experiment) and that attachment and spreading may be enhanced in a weightless environment.[77] Review articles of cell biology performed in space flight experiments delineate the growth of many cell types in an environment devoid of gravitational influence.[78] The problem, then, is to suspend microcarriers and cells without inducing turbulence or shear while providing adequate oxygenation and nutritional replenishment. One environment that possesses these attributes is the microgravity of space flight. Another is the system referenced here, which randomizes the forces of gravity by classical methods, thus simulating some aspects of microgravity.

The term *rotating-wall vessel* (RWV) comprises a family of vessels, batch fed and perfused, that embody the same fluid dynamic operating principles. These principles are (1) solid body rotation about a horizontal axis that is characterized by (a) colocation of particles of different sedimentation rates, (b) extremely low fluid shear stress and turbulence, and (c) three-dimensional spatial freedom; and (2) oxygenation by active or passive diffusion to the exclusion of all but dissolved gasses from the reactor chamber, yielding a vessel devoid of gas bubbles and gas-fluid interface (zero head space) (Fig. 18.3).[79,80]

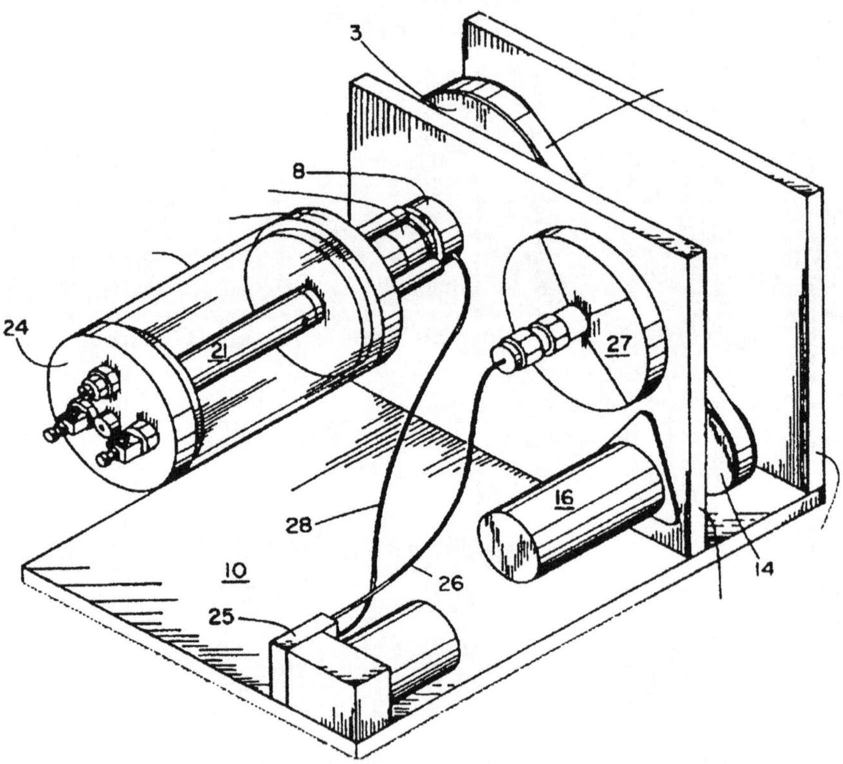

Fig. 18.3. NASA Rotating-Wall Bioreactor. United States Patent: Schwarz et al.[64] (Courtesy of NASA).

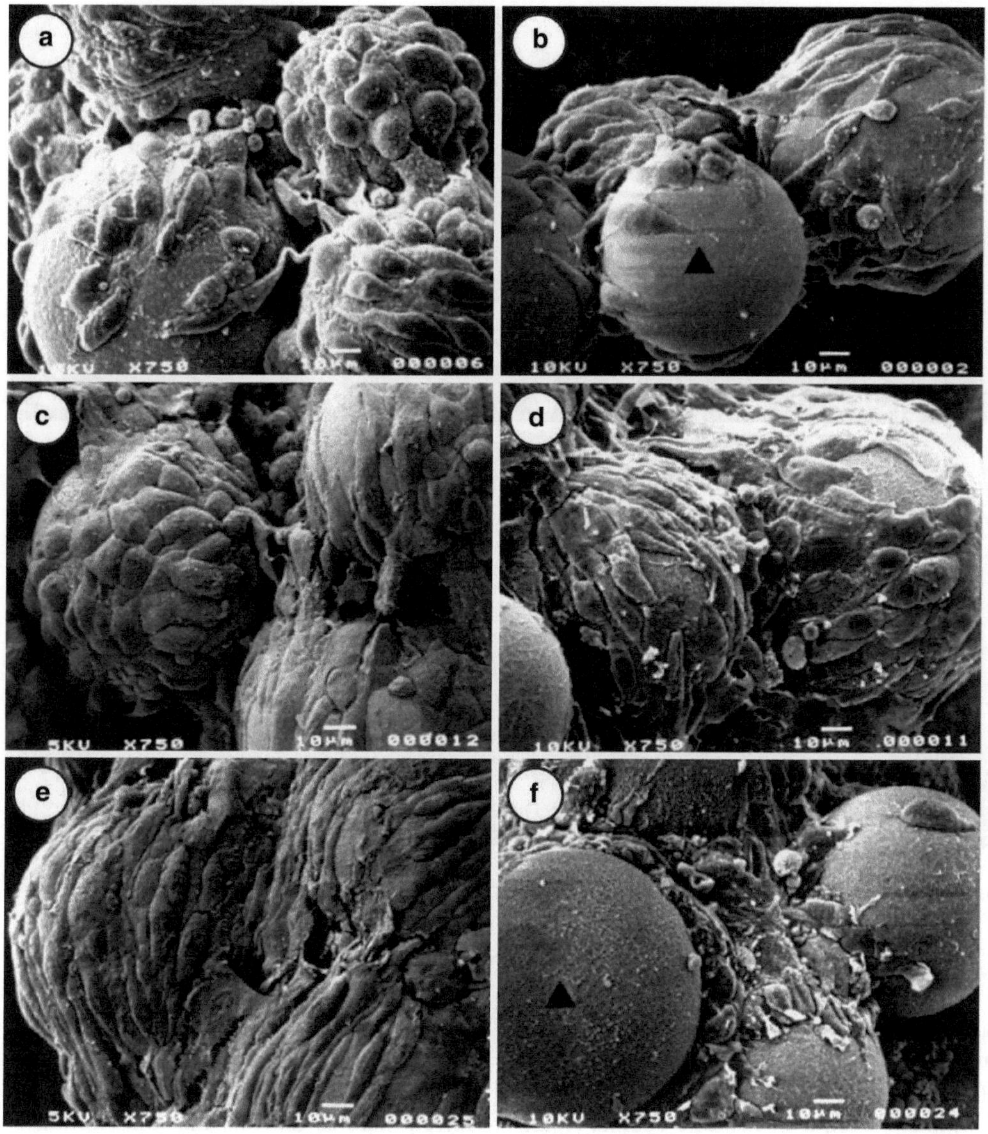

Fig. 18.4. Three-dimensional baby hamster kidney (BHK) tissues created in rotating-wall vessel (RWV) culture. (From Goodwin TJ, et al.[62] Reprinted by permission of Wiley-Liss, Inc., a subsidiary of John Wiley & Sons, Inc).

The baby hamster kidney (BHK) model (Fig. 18.4) described and illustrated here has been used extensively as a means of transfecting viral and bacterial DNA into cells to accomplish gene splicing protocols. These three-dimensional models are proving extremely valuable in viral, microbial, and genetic research.

Normal Tissue Models

As reviewed in several of the previous sections, investigators have cultured bone, cartilage (chondrocytes), muscle tissue, cells of the immune system, and others. Baker and Goodwin in 1997 employed the RWV to culture chondrocytes (Fig. 18.5) for 36 days to observe the influence of low-shear and quiescent culture conditions on the development of three-dimensional differentiation and extracellular matrix formation in bovine chondrocytes.

Chondrocytes from bovine cartilage were inoculated into the RWV with 5 mg/ml Cytodex-3 microcarriers in nonadherent petri dishes with identical conditions and were initiated with microcarriers as standard tissue culture controls. Differentiated chondrocytes were observed in all sections of RWV material through 36 days, while few were observed in the sections of petri dish material."[81] These results indicate that the unique conditions provided by the RWV afford access to cellular processes that signify the initiation of differentiation as well as production of normal matrix material. These experiments demonstrated the ability to create a synthetic model of cartilage or bone without leaving Earth's orbit.

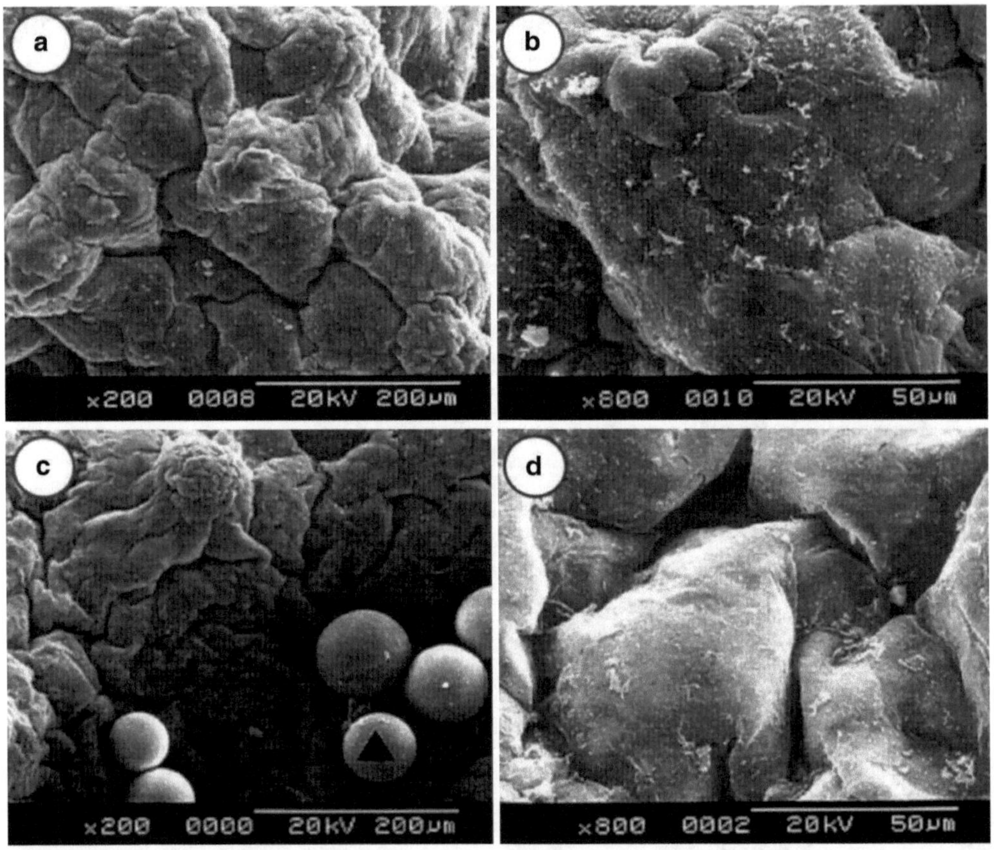

FIG. 18.5. Chondrocytes. (From Baker TL, Goodwin TJ.[81] Reprinted by permission of Springer Berlin/Heidelberg).

Figure 18.6 illustrates a 34-day time course of three-dimensional culture of normal human chondrocytes (HCH). Increased Actin (alexa 555 phalloidin; red) and nuclear staining (syto16; green) is observed over the time course, resulting in yellowish-red staining by day 34. Goodwin et al[63] describe the results of experiments where normal human small intestine epithelium and mesenchyma were cultured in rotating-wall vessel cultures. They found that the rotating-wall vessels are a major advance toward constructing a functionally accurate, large-scale, in vitro, three-dimensional (3-D) tissue model of the small intestine. Because of the presence of large tissue-like masses that express differentiated epithelial, mesenchymal, and developing endothelial cells, a multitude of possibilities are afforded for cell biological investigations, and further, functional epithelial cell brush borders with extracellular matrix and basal lamina components that represent ordering of tissue and cellular polarity were nurtured by the molecular conditions and physical orientations of the culture system. An important additional finding was that this 3-D model demonstrates a significantly diminished requirement of complex culture media, which suggests specific cell-cell interactions and the production of paracrine and autocrine factors essential to the growth and development of these fragile tissues.[63]

The role of basement membranes and extracellular matrix and their relationship to epithelial mesenchymal development and differentiation are the subjects of considerable research. Current studies indicate that the stromal component exerts a strong and driving influence over developing intestinal mucosa.[82–84] Stallmach et al[85] have shown that only organ-specific mesenchyme will produce differentiation in epithelium from a given organ site and that embryonic mesenchyme of the same age but from different organs was ineffective. Studies of rat small intestine development have detailed a shift in the membrane molecular components of crypt cells as measured by monoclonal antibody binding in fetal versus newborn rat epithelium. Quaroni[86] postulates that the presence of specific markers in the crypts indicates undifferentiated crypt cells, which may be able to perform specific intestinal functions similar to that of the villus cells. Additional information shows that a single fetal epithelial cell type in the last 2–3 days of gestation may express the function of more than one cell, thereby giving rise to DNA synthesis and proliferation.[87]

Cocultures of small intestine produced in the RWV were initiated with adult epithelium composed of dividing and terminally differentiated cells and predominantly adult mesenchymal cells with one exception, a 2-month-old female

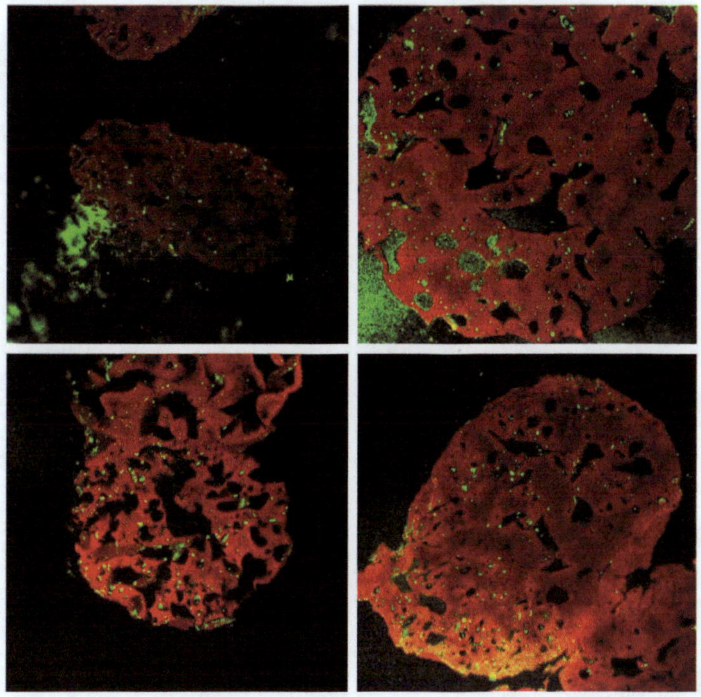

FIG. 18.6. Human articular chondrocytes stained for actin-binding protein (*red*) and nuclear protein (*green*).

donor. Levels of epithelial cell (keratin-positive) and mesenchymal cell-specific (vimentin-positive) immunostaining decreased slightly but consistently over the course of the RWV experiments, while total numbers of cells increased. Levels of endothelial cell (Factor VIII-positive) and epithelial cell (villin-positive) staining remained reasonably constant, while staining for angioblasts (endothelial precursor cells) developed toward the end of each experiment. Functional brush border markers such as sucrase were also present in this coculture system. Other brush border-specific monoclonal antibodies, kindly supplied by Dr. Andrea Quaroni (Cornell University, Ithaca, NY), have similarly proved positive in more recent studies (unpublished results). Collectively, these results indicate that the model, although not histotypically correct, exhibits many facets of functional normal small intestine. Growth conditions, however, may still be skewed slightly in favor of the mesenchymal versus the epithelial cell types. Work will continue on further definition of the correct nutritional and growth-matrix requirements directed toward attaining the best representation possible. Interestingly, all the RWV cocultures exhibited laminin, fibronectin, and type IV collagen production, as well as large amounts of proteoglycan. Hahn states that only undifferentiated, highly proliferative intestinal epithelial cells (usually fetal) synthesize the aforementioned proteins.[88] Of these, only laminin appeared to promote differentiation of intestinal epithelial cells. Type IV collagen and fibronectin had no effect.

The literature reviewed and the data obtained here would suggest that this model of human small intestine embodies many aspects of differentiation observed in other in vitro and in vivo cell and organ models. Primary distinctions would be (1) the overall scale of the model, (2) the ability to culture epithelium for long periods (in excess of 40 days) without loss of functional cell markers, and (3) the ability of the system to respond to extensive analyses and manipulations without the termination of a given experiment.

The significance of a complex three-dimensional in vitro culture system for the growth of normal small intestine should not be underestimated. The molecular basis and clinical treatment of diseases such as inflammatory bowel disease (Crohn's, ulcerative colitis), malabsorptive syndromes (short-gut syndrome), numerous infectious diseases, and tumors of the small bowel may be investigated with the advent of this new technology. For example, the recent demonstration that the human immunodeficiency virus can replicate in human small intestine and columnar epithelium may be impacted through studies of virus-cell interactions in the RWV culture system.[89] This hypothesis is currently being tested. Additionally, general application of this culture model may lead to advances in understanding growth and differentiation in developing organisms and the potential treatment of a myriad of clinical conditions as well as tissue renewal (Figs. 18.7 and 18.8).

The RWV systems have been demonstrated as useful for the development of sophisticated models, which emulate facets of

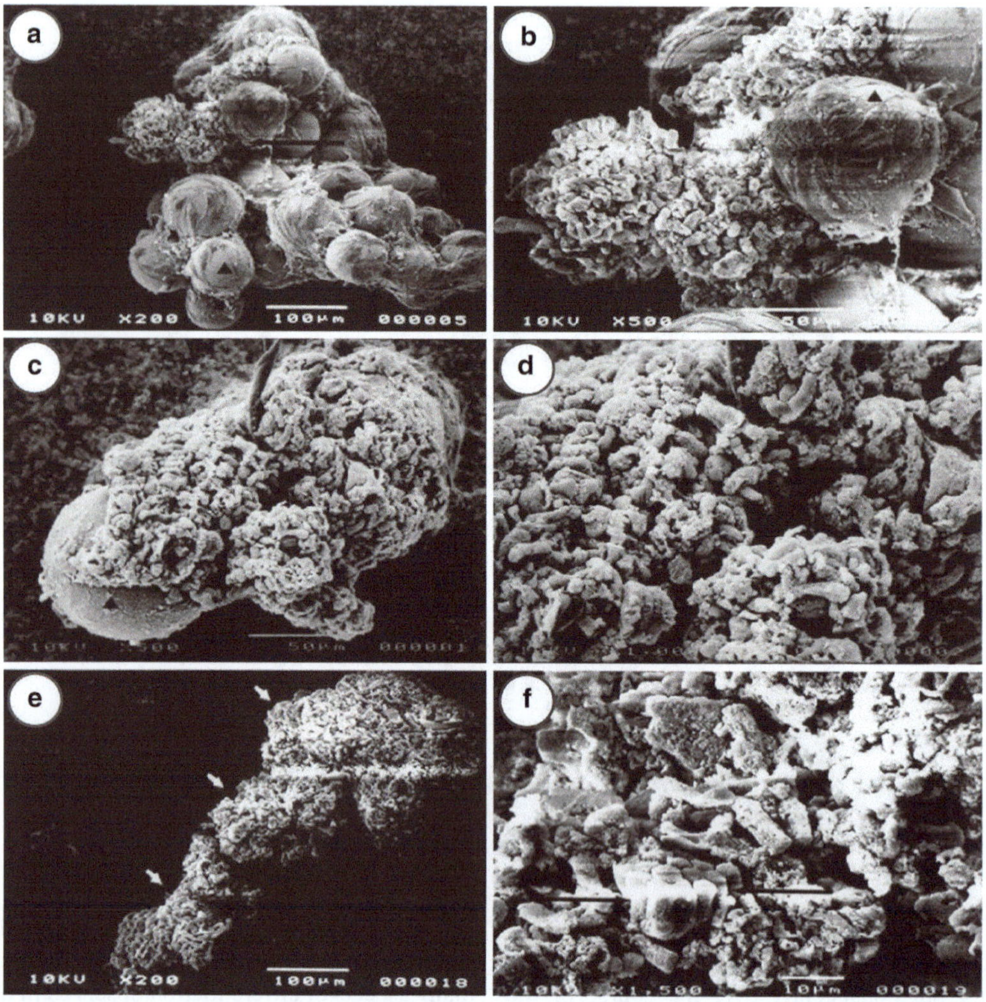

FIG. 18.7. Early-stage scanning electron micrographs of human small intestine RWV cocultures. (From Goodwin et al.[63] Reprinted by permission of the Society for Experimental Biology and Medicine).

microgravity and at the same moment present characteristics of in vivo tissues. We have examined the production of normal tissue models for research and experimentation; as we will see in the next section, the RWV has the ability to serve as a tremendous tool to construct synthetic tumor tissues, which allows an in-depth understanding of tumor biology and oncology.

Three-Dimensional Models of Lung Disease

Current cell culture models have shortcomings resulting in unreliable tumor growth, uncharacteristic tumor development, nonhuman tumors, and inadequate methods of detection. Cells propagated under traditional culture conditions differ widely in their expression of differentiated markers, adhesion receptors, and growth factor receptors compared with cells in situ or those grown as tissue-like structures.[90,91] This factor is of concern because the phenotypic changes leading to malignant transformation often stem from alterations in the balanced and multifaceted roles of growth factors, receptors, and cytokines (reviewed by Herlyn et al[90]. With increasing evidence of the importance of adhesive contacts, paracrine cross-talk between different cell types, and signaling cascades that link the cell with a complex substratum, there is now recognition that models must be developed that better simulate these complexities. There is still much to learn about the dynamic relationships among the different phenotypes found in the normal lung and in lung cancers. Until a cell culture system is developed that allows differentiation to occur,[62] it is difficult to make any conclusive statement about relating effects in cell culture to clinical practice. Tissue engineering is very embryonic in development and currently nearly universally focused on building replacement tissues. A new technology developed at the NASA Johnson Space Center used to study colon cancer has been adapted to three-dimensional in vitro lung tissue culture models but has not been reported on to date.

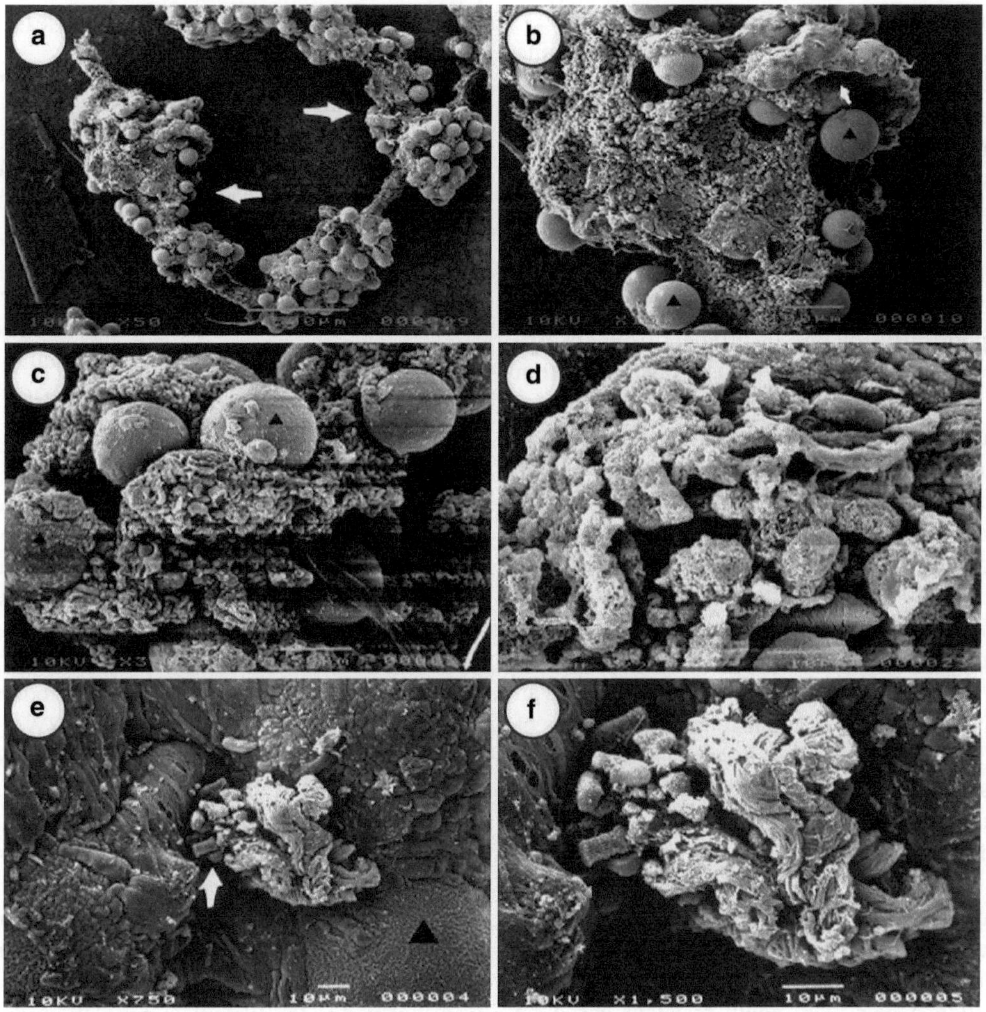

FIG. 18.8. Scanning electron micrographs of mid- and late-stage human small intestine RWV cocultures. (From Goodwin et al.[63] Reprinted by permission of the Society for Experimental Biology and Medicine).

Rotating-wall vessels are horizontally rotating cylindrical tissue culture vessels that provide controlled supplies of oxygen and nutrients with minimal turbulence and extremely low shear.[79] These vessels suspend cells and microcarriers homogeneously in a nutrient-rich environment, which allows the three-dimensional assembly of cells to tissue. Before seeding rotating-wall vessels (Synthecon, Houston, TX, USA), cells were cultured in standard T flasks (Corning, Corning, NY, USA) in GTSF-2 medium (1993 PSEBM) in a humidified 37°C, 5% CO_2 incubator. The rotating-wall vessels were seeded with 1–2 mg/ml Cultispher-GL microcarriers (Hyclone Laboratories, Logan, UT, USA) followed by BEAS2-B or BZR-T33 cells (ATCC, Baltimore, MD, USA) at a density of 2×10^5 cells/ml. Cultures were grown in the rotating-wall vessels for 14–21 days for formation of 3- to 5-mm-diameter tumor masses. Rotating-wall vessel rotation was initiated at 25 rpm and increased as aggregate size became larger. Stationary control cultures were initiated under the same conditions using FEP Teflon bags (American Fluoroseal, Columbia, MD, USA). At 24-h intervals pH, dissolved CO_2, and dissolved O_2 were determined using a Corning 238 model clinical blood gas analyzer. Glucose concentration was determined using a clinical glucose analyzer. Cell samples were harvested every 48 h and fixed for immunohistochemistry or for scanning electron microscopy.

Cancer models already developed by NASA investigators include growth and differentiation of an ovarian tumor cell line,[62,92,93] growth of colon carcinoma lines,[62] and three-dimensional aggregate and microvillus formation in a human bladder carcinoma cell line.[93] In support as an appropriate model for cancer, even the most rudimentary three-dimensional cellular structures exhibit different phenotypes than cell lines cultured under two-dimensional conditions. Properties such as responses to TGF-β, drug resistance to cisplatin or cyclophosphamide, and resistance to apoptosis are all altered in various types of cell aggregates.[94]

Many investigations sustain consistent evidence that cells growing in three-dimensional arrays appear more resistant to cytotoxic chemoagents than cells in monolayer culture.[39] Li et al[95] found that spheroids were more resistant to cytosine arabinoside by 11-fold and methotrexate by 125-fold when compared with single-cell suspensions. Further monolayer cultures of colon carcinoma cells were sensitive to piericidin C in contrast to responses within in vivo colon tumors or three-dimensional slices of tumors grown in vitro.[96] Numerous other investigations have revealed increased levels of drug resistance of spheroids compared with single-cell monolayers.[4,39]

Questions of poor diffusion and insufficient drug absorption within spheroids and a relatively frequent high proportion of resting cells have clouded differences in drug resistance, which could be the result of nutrient deprivation and hypoxia. Heppner and colleagues executed precise experiments that confirmed three-dimensional structure and function as the causative agent which was responsible for drug resistance rather than simple inaccessibility to nutrients or the drug concentration. Heppner embedded tumor specimens or cell aggregates in collagen gels, exposed the culture to various cytotoxic drugs, and compared the drug responses of the same cells in monolayers. These experiments revealed an increased resistance in the three-dimensional tumor arrays of a remarkable 1,000-fold greater than in monolayer cultures, and a similar result was seen in three-dimensional histocultures in collagen. The tumor cells grew in the presence of drug concentrations that rendered monolayers to a viability less than 0.1% of control cultures. Amazingly, Heppner observed that the cells became sensitive again when replated as monolayers, and finally showed that even when exposed to melphalan and 5-fluorouracil in monolayer culture, cells transferred to collagen gels were again resistant based on three-dimensional architecture. Thus, the cells were exposed to the drugs as monolayers, facilitating access to the drugs, and, once the cells were transferred after drug exposure to a three-dimensional structure, high resistance to the drugs was sustained.[39,97–100]

Based on the caliber of data referenced above, Teicher et al[101] serially passaged through multiple (ten) transfers EMT-6 tumors in mice that were treated with thiotepa, cisplatin, and cyclophosphamide over a prolonged 6-month period, thus producing extremely drug-resistant tumors in vivo. When these tumors were grown as monolayer cultures, they were as drug sensitive as the parental cells. Kobayashi and colleagues[102] grew the same in vivo drug-resistant tumor cell lines as spheroids in three-dimensional arrays, and resistance was almost 5,000 times that of the parent line with selected drugs, an example being the active form of cyclophosphamide used in vitro. Similarly extreme resistance was also observed to cisplatin and thiotepa. This resistance was not seen in monolayer cultures, even when the monolayers were cultured on traditional extracellular matrix substrates. These experiments reconfirmed that cells in a three-dimensional array are more drug resistant than monolayer cells in vitro and demonstrated that three-dimensional cellular configurations can and do become resistant to super-pharmacological doses of drugs by forming compact structures.[39]

Rotating-Wall Vessel Tumor Models

Several important human tumor models have been created in rotating-wall vessel cultures, specifically, lung, prostate, colon, breast carcinoma, and ovarian.[15,67,92,103] Many of these models involve cancers that are leading killers in our society. We present two such examples in this section, colon and prostate carcinoma. As previously reviewed, the literature indicates the remarkable difference between chemotherapeutic cytotoxicity in two-dimensional and three-dimensional cellular constructs, which may be predicated on a number of criteria. Therefore, a three-dimensional tumor model that emulates differentiated in vivo-like characteristics would provide unique insights into tumor biology.

Colon Cancer

Goodwin et al[67] detailed the first construction of a complex three-dimensional ex vivo tumor in rotating-wall vessel (RWV) culture composed of a normal mesenchymal base layer (as would be seen in vivo), and either of two established human colon adenocarcinoma cell lines, HT-29, an undifferentiated line, and HT-29KM a stable, moderately differentiated subline of HT-29, were grown in RWV culture. RWVs were used in conjunction with multicellular cocultivation to develop a unique in vitro tissue modeling system. Cells were cultivated on Cytodex-3 microcarrier beads, with and without mixed normal human colonic fibroblasts, which served as the mesenchymal layer. Culture of the tumor lines in the absence of fibroblasts produced spheroid-like growth and minimal differentiation. In contrast, when tumor lines were cocultivated with normal colonic fibroblasts, initial growth was confined to the fibroblast population until the microcarriers were covered. The tumor cells then commenced proliferation at an accelerated rate, organizing themselves into three-dimensional tissue masses that achieved 1.0- to 1.5-cm diameters. Each of these engineered tumor tissues produced tissue-like aggregates (TLAs) with glandular structures, apical and internal glandular microvilli, tight intercellular junctions, desmosomes, cellular polarity, sinusoid development, internalized mucin, and structural organization akin to normal colon crypt development. Necrosis was minimal throughout the tissue masses up to 60 days of culture while achieving >1.0 cm in diameter. Other notable results included enhanced growth of neoplastic colonic epithelium in the presence of mixed normal human colonic mesenchyme. These results mimic the cellular differentiation seen in vivo and are similar to results obtained with other tumor types.

Prostate carcinoma has also been modeled in the RWV system by several investigators.[104–106] One of the most comprehensive descriptions of these engineered tissues is detailed by Wang et al[107] In that review, the authors describe the ability of the RWV system to recapitulate human prostate carcinoma (LNCaP) and bone stroma (MG63) to illuminate the evolution of prostate tumorigenesis to the metastatic condition. In particular, the LNCaP and ARCaP models represented in the review are known to be lethal in the human, being androgen independent and metastatic. Rotating-wall vessel TLA engineering also allowed in-depth study of epithelial and stromal interactions, which are the facilitating elements of the continuance of LNCaP prostate-specific antigen production in vitro. When LNCaP was cultured in three dimensions without stroma, production of prostate-specific antigen ceased, and metastatic markers were not observed. The authors outline the process of malignant transformation, demonstrating that these metastatic models are only possible in three-dimensional TLAs and are achieved by specific geometric relationships in three-dimensional configuration. Furthermore, they show through direct comparison with other culture systems the advantages of the RWV system to allow synergistic relationships to study this disease state.[107]

The final two carcinoma models to be reviewed here are the subject of original ongoing research, breast carcinoma in Dr. Goodwin's lab and lung carcinoma in Dr. Vertrees' lab. As we have seen with the previous data, the modeling of three-dimensional tissues is not only achievable, but also desirable to create an improvement in the fidelity of the tumor model for research and diagnostic purposes. Construction of three-dimensional tumor models using RWVs has revealed striking similarities to tissues grown in vivo.[63,67]

As a demonstration of the capability of these advanced tumor models to produce unique results, we have developed three-dimensional breast carcinoma models and subjected them to assault by lymphokine-activated killer (LAK) cells. Tumor-infiltrating lymphocyte (TIL) and lymphokine-activated killer (LAK) cell research has primarily dealt with tumor cells in single suspension cultured in the presence of LAK cells in close proximity using 96-well plates. Although a substantial amount of literature exists in this field, no literature could be found investigating three-dimensional tumor models and the kill response elicited in the presence of LAK cells. We hypothesized that not only would the three-dimensional models more closely resemble in vivo tissue histologically, but they also could demonstrate a useful application of using three-dimensional models. Three breast carcinoma cell lines were chosen for this investigation: BT-20, an undifferentiated estrogen receptor (ER)– cell line; MCF-7, a moderately differentiated and metastatic ER+ cell line; and T-47D, an ER+ cell line. Each cell line was cultured in the RWV for a minimum of 21 days. Mononuclear cells were isolated from peripheral blood lymphocytes and placed in T-flasks at a cell density of 1.5×10^6 cells/ml in the presence of rhIL-2. A modified procedure based on the standard ^{51}Cr release assay was developed to quantitate the activity of LAK cell invasion of three-dimensional (3-D) RWV-grown tumor aggregates. Cells from T-flasks (2-D controls) or RWV-grown aggregates (3-D) were labeled with ^{51}Cr and seeded into 12-well plates with LAK cells at a 10:1 ratio to the tumor cells. We chose the lowest ratio of effector cells to target cells compared to standard suspension cell protocols of using 10:1, 100:1, or 1,000:1. Spontaneous release controls were initiated identically without LAK cells. Supernatant samples were taken at 24-h and 48-h intervals and counted using a gamma counter. Maximum release data were obtained by lysing the entire cell population using Triton X-100. Cytokine analysis of the supernatant and flow cytometry analysis of the cytokine receptors were performed from samples taken at the same time points from a duplicate nonradiolabeled assay.

Comparative data accumulated on the two estrogen receptor (ER)-positive and one ER-negative cell lines indicated that, in contrast to generally accepted data obtained from standard two-dimensional LAK assay, three-dimensional tumor aggregates present significantly different kill rates for all the breast carcinoma lines than do their two-dimensional analogues. Additionally, these altered kill ratios are associated with dramatic response changes in both the receptor and soluble molecular expression for the cytokines interleukin (IL)-8, tumor necrosis factor (TNF)-α, transforming growth factor (TGF)-β, and epidermal growth factor (EGF).

In Fig. 18.9, a well-developed histology section of MCF-7 is seen with organizational structure before LAK invasion. The scanning electron microscopy (SEM) panel in Fig. 18.10 demonstrates the adherence and invasion of LAK cells at 24 and 48 h post inoculation onto a mature, three-dimensional culture of breast carcinoma. Concomitantly, at 48 h LAK cells are seen digesting the surrounding tumor cells (Fig. 18.10).

Vertrees et al[108] reported on a study that compared various characteristics between an immortalized (not malignant) and its malignant transfected counterpart grown both as 2-D and 3-D cultures (Figs. 18.11 and 18.12). Electron microscopy identified significant mitochondrial and granular endoplasmic reticular pathology in the 2-D cells not seen in the 3-D cells.[108] The degree of differentiation determined by immunohistochemistry shows that ultrastructure and antibody expressions were more representative of control tissue when cells (both immortalized and transfected) were grown in 3-D culture than when grown as a 2-D culture. Electron microscopy identified the presence of lipid inclusion and lamellar (surfactant) bodies indicative of type II pneumocytes, a differentiation not seen in monolayer cultures. Fluorescence microscopy allowed for determination of individual migrating cells (unpublished data). The coculture experiment (simultaneous cultures of normal and transformed lung cells) in the three-dimensional environment demonstrates the ability of the transformed cell to migrate through the preexisting normal cells.

In contrast to two-dimensional models, these rotating-wall vessel tumor tissues were devoid of metabolic and nutrient deficiencies and demonstrated in vivo-like architecture. These data suggest that the rotating-wall vessel affords a new model for investigation and isolation of growth, regulatory, and structural processes within neoplastic and normal tissues.

Rotating-Wall Vessel Normal Human Tissue Models as Disease Targets

In this section, we explore the utility of rotating-wall vessel tissue-like aggregates (TLA)s as targets for microbial infection and disease. Several studies have been conducted

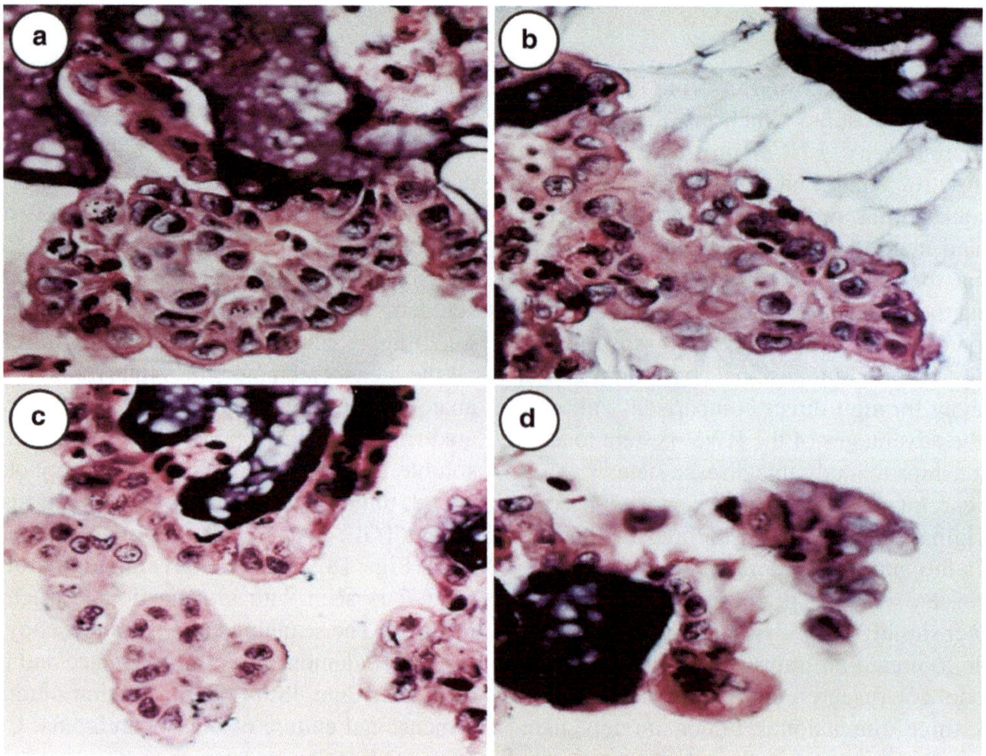

Fig. 18.9. Hematoxylin and eosin (H&E)-stained sections (×400) demonstrating the superior organization of the RWV-grown cells at days 14 and 21 (**a** and **c**) versus the static control cultures at days 14 and 21 (**b** and **d**). (From Goodwin, 2001: original work; supported by NASA grant #962-23-0132).

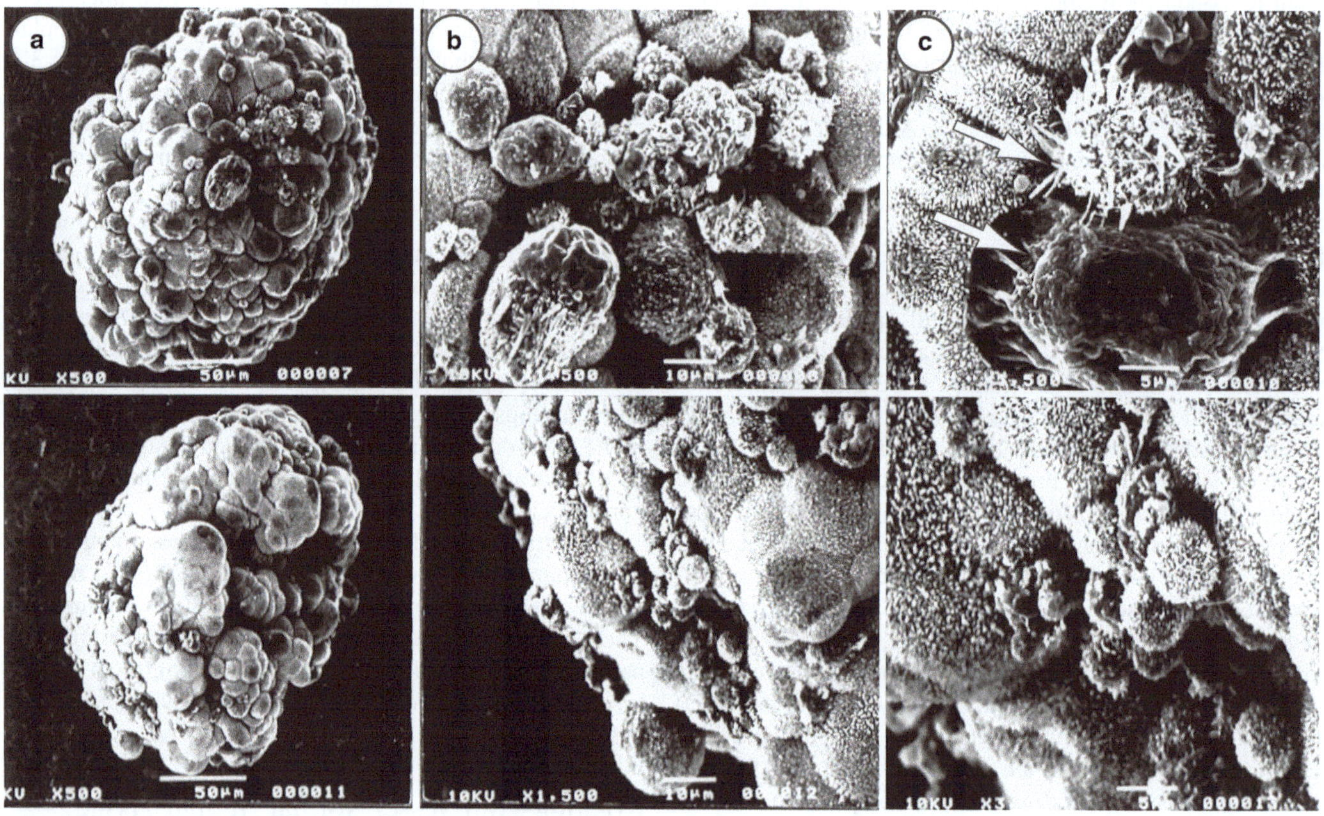

Fig.18.10. Scanning electron microscopy at 48 h post-LAK invasion. Notice the increased invasion potential of the LAK cells exposed to the RWV-grown cells (**a**) as compared to the static control cultures (**b**). (From Goodwin et al., 2001: original work; supported by NASA grant #962-23-0132).

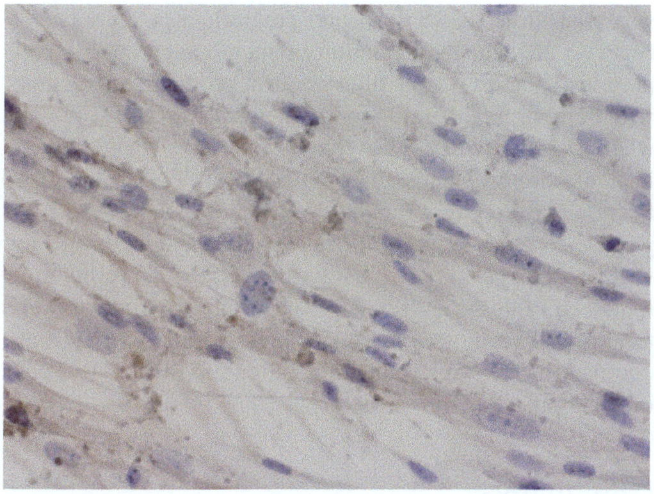

FIG. 18.11. Micrograph of a human lung cancer grown in two-dimensional (2-D) cultures. Note the lack of any architecture resembling that of lung.

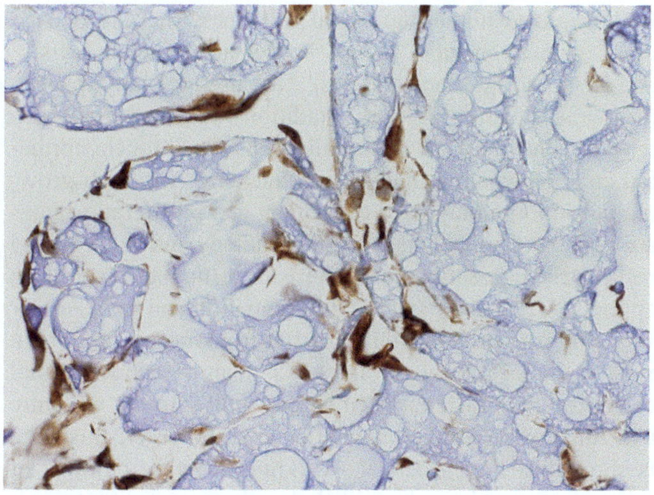

FIG. 18.12. Micrograph of a human lung cancer in three dimensions (3-D). Note the overall spherical geometry and lung architecture

recently that indicate that three-dimensional tissues respond to infective agents with greater fidelity and with a more in vivo-like response than traditional two-dimensional cultures. Nickerson et al[109] describe the development of a three-dimensional TLA engineered from INT-407 cells of the human small intestine, which were used as targets for the study of *Salmonella typhimurium*. In this study, three-dimensional TLAs were used to study the attachment, invasion, and infectivity of *Salmonella* into human intestinal epithelium. Immunocytochemical characterization and scanning and transmission electron microscopic analyses of the three-dimensional TLAs revealed that the TLAs modeled human in vivo differentiated tissues more accurately than did two-dimensional cultures. The level of differentiation in the INT-407 TLAs was analogous to that found in previously discussed small intestine TLAs[62] and from other organ tissues reconstructed in rotating-wall vessels. Analysis of the infectivity studies revealed *Salmonella* attached and infected in a manner significantly different from that in control two-dimensional cultures. During an identical exposure period of infection with *Salmonella*, the three-dimensional TLAs displayed a minor loss of structural integrity when compared with the two-dimensional INT-407 cultures. Furthermore, *Salmonella* demonstrated a greatly reduced ability to adhere, invade, and induce the apoptotic event in these INT-407 three-dimensional TLAs than in two-dimensional cultures. This result is not unlike the in vivo human response. Two-dimensional cultures were significantly damaged within several hours of contact with the bacteria; conversely, although "pot marks" could be seen on the surfaces of the three-dimensional TLAs, they remained structurally sound.

Cytokine analysis and expression postinfection of three-dimensional TLAs and two-dimensional cultures with *Salmonella* exhibited remarkable differences in expressed levels of interleukin (IL)-1α, IL-1β, IL-6, IL-Ra, and TNF-α mRNAs. Additionally, noninfected three-dimensional TLAs constitutively demonstrated elevated levels of TGF-β1 mRNA and prostaglandin E_2 compared with noninfected two-dimensional cultures of INT-407.[109]

As previously stated, traditional two-dimensional cell monolayers lack adequate fidelity to emulate the infection dynamics of in vivo microbial adhesion and invasion. The respiratory epithelium is of critical importance in protecting humans from disease. Exposed to the environment, the respiratory epithelium acts as a barrier to invading microbes present in the air, defending the host through a multilayered complex system.[110] The three major layers of the human respiratory epithelium are pseudo-stratified epithelial cells, a basement membrane, and underlying mesenchymal cells. Ciliated, secretory, and basal epithelial cells are connected by intercellular junctions and anchored to the basement membrane through desmosomal interactions. Together with tight junctions and the mucociliary layer, the basement membrane maintains the polarity of the epithelium and provides a physical barrier between the mesenchymal layer and the airway.[111,112] Infiltrating inflammatory and immune cells move freely between the epithelial and subepithelial compartments.

Airway epithelial cells play a vital role in host defense[110] by blocking paracellular permeability and modulating airway function through cellular interactions. Ciliated epithelial cells block invasion of countless inhaled microorganisms by transporting them away from the airways.[113] As regulators of the innate immune response, epithelial cells induce potent immunomodulatory and inflammatory mediators such as cytokines and chemokines that recruit phagocytic and inflammatory cells that remove microbes and enhance protection.[110,112,114,115]

Ideally, cell-based models should reproduce the structural organization, multicellular complexity, differentiation state, and function of the human respiratory epithelium.

Immor-talized human epithelial cell lines, such as BEAS-2B,[116] primary normal human bronchial epithelial cells,[117] and air-liquid interface cultures,[118] are used to study respiratory virus infections in vitro. Traditional monolayer cultures (two-dimensional) of immortalized human bronchoepithelial cells represent homogeneous lineages. Although growing cells in monolayers is convenient and proliferation rates are high, such models lack the morphology and cell-cell and cell-matrix interactions characteristic of human respiratory epithelia. Thus, their state of differentiation and intracellular signaling pathways most likely differ from those of epithelial cells in vivo. Primary cell lines of human bronchoepithelial cells provide a differentiated model similar to the structure and function of epithelial cells in vivo; however, this state is short lived in vitro.[117,119] Air-liquid interface cultures of primary human bronchoepithelial cells (or submerged cultures of human adenoid epithelial cells) are grown on collagen-coated filters in wells on top of a permeable filter. These cells receive nutrients basolaterally, and their apical side is exposed to humidified air. The result is a culture of well-differentiated heterogeneous (ciliated, secretory, basal) epithelial cells essentially identical to airway epithelium in situ.[118,120] Although this model shows fidelity to the human respiratory epithelium in structure and function, maintenance of consistent cultures is not only difficult and time consuming but also limited to small-scale production and thus limits industrial research capability.

True cellular differentiation involves sustained complex cellular interactions[121–123] in which cell membrane junctions, extracellular matrices (e.g., basement membrane and ground substances), and soluble signals (endocrine, autocrine, and paracrine) play important roles.[124–127] This process is also influenced by the spatial relationships of cells to each other. Each epithelial cell has three membrane surfaces: a free apical surface, a lateral surface that connects neighboring cells, and a basal surface which interacts with mesenchymal cells.[128]

Viral studies by Goodwin et al[129] and Suderman et al[130] were conducted with RWV-engineered TLA models of normal human lung. This model is composed of a coculture of in vitro three-dimensional human bronchoepithelial TLAs engineered using a rotating-wall vessel to mimic the characteristics of in vivo tissue and to provide a tool to study human respiratory viruses and host–pathogen cell interactions. The TLAs were bioengineered onto collagen-coated cyclodextran beads using primary human mesenchymal bronchial tracheal cells as the foundation matrix and an adult human bronchial epithelial immortalized cell line (BEAS-2B) as the overlying component. The resulting TLAs share significant characteristics with in vivo human respiratory epithelium, including polarization, tight junctions, desmosomes, and microvilli. The presence of tissue-like differentiation markers, including villin, keratins, and specific lung epithelium markers, as well as the production of tissue mucin, further confirm these TLAs differentiated into tissues functionally similar to in vivo tissues. Increasing virus titers for human respiratory syncytial virus (*wt*RSVA2) and parainfluenza virus type 3 (*wt*PIV3 JS) and the detection of membrane-bound glycoproteins (F and G) over time confirm productive infections with both viruses. Viral growth kinetics up to day 21 pi with *wt*RSVA2 and *wt*PIV3 JS were as follows: *wt*PIV3 JS replicated more efficiently than *wt*RSVA2 in TLAs. Peak replication was on day 7 for *wt*PIV3 JS [approximately 7 $\log_{10}$ particle-forming units (pfu) per milliliter] and on day 10 for wtRSVA2 (approximately 6 $\log_{10}$ pfu/ml). Viral proliferation remained high through day 21 when the experiments were terminated. Viral titers for severe acute respiratory syndrome coronavirus were approximately 2 $\log_{10}$ pfu/ml at 2 days post inoculation.

Conclusion

Although great strides have been made in recent years in the understanding of the biology of cancer, the role that cellular differentiation and three-dimensional structural organization play in metastasis and malignancy is still unclear. A model that more closely links the expression of specific antibodies in normal lung cells, transformed lung cells, and cancerous lung cells to cellular biology may provide new information about cancer. The development of three-dimensional cell cultures in bioreactors may ultimately provide a model that facilitates discovery and interpretation of more relevant information for the expression of an antibody and its role in cellular pathobiology of the lung.

Human lung tissue-like aggregations mimic aspects of the human respiratory epithelium wall and provide a unique opportunity to study the host-pathogen interaction of respiratory viruses and their primary human target tissue independent of the host immune system, as there can be no secondary response without the necessary immune cells. These rotating-wall vessel-engineered tissues represent a valuable tool in the quest to develop models that allow analysis and investigation of cancers and infectious disease in models engineered with human cells alone.

We have explored the creation of three-dimensional tissue-like aggregates for normal and neoplastic studies and finally as targets for microbial infections. Perhaps Carrel and Leighton would be fascinated to know that from their early experiments in three-dimensional modeling and the contributions they made has sprung the inventive spirit to discover a truly space age method for cellular recapitulation.

References

1. Clements JA, King RJ. Composition of surface active material. In: Crystal RG, ed. *The Biochemical Basis of Pulmonary Function*. New York: Marcel Dekker; 1976:363–387.
2. Carrel A. On the permanent life of tissues outside the organism. *J Exp Med.* 1912;15:516–528.
3. Huber HL, Koessler KK. The pathology of bronchial asthma. *Arch Intern Med.* 1922;30:689–760.

4. Hoffman RM. Three-dimensional histoculture: origins and applications in cancer research. *Cancer Cells.* 1991;3:86–92.
5. Sherwin RP, Richters A, Yellin AE, Donovan AJ. Histoculture of human breast cancers. *J Surg Oncol.* 1980;13:9–20.
6. Scherer WF, Syverton JT, Gey GO. Studies on the propagation in vitro of poliomyelitis viruses. IV. Viral multiplication in a stable strain of human malignant epithelial cells (strain HeLa) derived from an epidermoid carcinoma of the cervix. *J Exp Med.* 1953;97:695–710.
7. Look DC, Walter MJ, Williamson MR, et al. Effects of paramyxoviral infection on airway epithelial cell Foxj1 expression, ciliogenesis, and mucociliary function. *Am J Pathol.* 2001;159:2055–2069.
8. Freshney RI. *Culture of Animal Cells: A Manual of Basic Technique*. 3rd ed. New York: Wiley-Liss; 1994.
9. Tsao M, Viallet J. Preclinical models of lung cancer: cultured cells and organ culture. In: Lenfant C, ed. *Biology of Lung Cancer.* New York: Marcel Dekker; 1998:215–246.
10. Walboomers JM, Meijer CJ, Steenbergen RD, van Duin M, Helmerhorst TJ, Snijders PJ. Human papillomavirus and the development of cervical cancer: concept of carcinogenesis. *Ned Tijdschr Geneeskd.* 2000;144:1671–1674.
11. Adamson IY, Young L, Bowden DH. Relationship of alveolar epithelial injury and repair to the induction of pulmonary fibrosis. *Am J Pathol.* 1988;130:377–383.
12. Krump-Konvalinkova V, Bittinger F, Unger RE, Peters K, Lehr HA, Kirkpatrick CJ. Generation of human pulmonary microvascular endothelial cell lines. *Lab Invest.* 2001;81:1717–727.
13. Rep M, van Dijl JM, Suda K, Schatz G, Grivell LA, Suzuki CK. Promotion of mitochondrial membrane complex assembly by a proteolytically inactive yeast Lon. *Science.* 1996;274:103–106.
14. Gruenert DC, Finkbeiner WE, Widdicombe JH. Culture and transformation of human airway epithelial cells. *Am J Physiol.* 1995;268:L347–L360.
15. Vertrees RA, Das GC, Coscio AM, Xie J, Zwischenberger JB, Boor PJ. A mechanism of hyperthermia-induced apoptosis in ras-transformed lung cells. *Mol Carcinog.* 2005;44:111–121.
16. Vertrees RA, Nason R, Hold MD, et al. Smoke/burn injury-induced respiratory failure elicits apoptosis in ovine lungs and cultured lung cells, ameliorated with arteriovenous CO2 removal. *Chest.* 2004;125:1472–1782.
17. Bareiss PM, Metzger M, Sohn K, et al. Organotypical tissue cultures from adult murine colon as an in vitro model of intestinal mucosa. *Histochem Cell Biol.* 2008;129:795–804.
18. Follmann W, Weber S, Birkner S. Primary cell cultures of bovine colon epithelium: isolation and cell culture of colonocytes. *Toxicol In Vitro.* 2000;14:435–445.
19. Nussler AK, Wang A, Neuhaus P, et al. The suitability of hepatocyte culture models to study various aspects of drug metabolism. *Altex.* 2001;18:91–101.
20. Zeilinger K, Sauer IM, Pless G, et al. Three-dimensional co-culture of primary human liver cells in bioreactors for in vitro drug studies: effects of the initial cell quality on the long-term maintenance of hepatocyte-specific functions. *Altern Lab Anim.* 2002;30:525–538.
21. Zen Y, Fujii T, Yoshikawa S, et al. Histological and culture studies with respect to ABCG2 expression support the existence of a cancer cell hierarchy in human hepatocellular carcinoma. *Am J Pathol.* 2007;170:1750–1762.
22. Kosuge M, Takizawa H, Maehashi H, Matsuura T, Matsufuji S. A comprehensive gene expression analysis of human hepatocellular carcinoma cell lines as components of a bioartificial liver using a radial flow bioreactor. *Liver Int.* 2007;27: 101–108.
23. Ma W, Tavakoli T, Chen S, et al. Reconstruction of functional cortical-like tissues from neural stem and progenitor cells. *Tissue Eng Part A.* 2008;14:1673–1686.
24. Ma W, Fitzgerald W, Liu QY, et al. CNS stem and progenitor cell differentiation into functional neuronal circuits in three-dimensional collagen gels. *Exp Neurol.* 2004;190:276–288.
25. Braun H, Buhnemann C, Neumann J, Reymann KG. Preparation of a tissue-like cortical primary culture from embryonic rats using Matrigel and serum free Start V Medium. *J Neurosci Methods.* 2006;157:32–38.
26. Hitchcock T, Niklason L. Lymphatic tissue engineering: progress and prospects. *Ann N Y Acad Sci.* 2008;1131:44–49.
27. Shaul PW. Regulation of vasodilator synthesis during lung development. *Early Hum Dev.* 1999;54:271–294.
28. Gikas EG, King RJ, Mescher EJ, et al. Radioimmunoassay of pulmonary surface-active material in the tracheal fluid of the fetal lamb. *Am Rev Respir Dis.* 1977;115:587–593.
29. Goodwin MN Jr. Selected anatomic burn pathology review for clinicians and pathologists. *Aviat Space Environ Med.* 1989;60 :B39–B43.
30. Lionetti V, Recchia FA, Ranieri VM. Overview of ventilator-induced lung injury mechanisms. *Curr Opin Crit Care.* 2005;11:82–86.
31. Balamugesh T, Kaur S, Majumdar S, Behera D. Surfactant protein-A levels in patients with acute respiratory distress syndrome. *Indian J Med Res.* 2003;117:129–133.
32. Herrera MT, Toledo C, Valladares F, et al. Positive end-expiratory pressure modulates local and systemic inflammatory responses in a sepsis-induced lung injury model. *Intensive Care Med.* 2003;29: 1345–1353.
33. Imai Y, Parodo J, Kajikawa O, et al. Injurious mechanical ventilation and end-organ epithelial cell apoptosis and organ dysfunction in an experimental model of acute respiratory distress syndrome. *JAMA.* 2003;289:2104–2112.
34. Schiller HJ, Steinberg J, Halter J, et al. Alveolar inflation during generation of a quasi-static pressure/volume curve in the acutely injured lung. *Crit Care Med.* 2003;31:1126–1133.
35. Song ZF, Yu KL, Shan HW, Ma J. Evaluation of mechanical ventilation for patients with acute respiratory distress syndrome as a result of interstitial pneumonia after renal transplantation. *Zhongguo Wei Zhong Bing Ji Jiu Yi Xue.* 2003;15:358–361.
36. Alcorn JL, Smith ME, Smith JF, Margraf LR, Mendelson CR. Primary cell culture of human type II pneumonocytes: maintenance of a differentiated phenotype and transfection with recombinant adenoviruses. *Am J Respir Cell Mol Biol.* 1997;17:672–682.
37. Leighton J, Justh G, Esper M, Kronenthal RL. Collagen-coated cellulose sponge: three dimensional matrix for tissue culture of Walker tumor 256. *Science.* 1967;155:1259–1261.
38. Leighton J, Tchao R, Stein R, Abaza N. Histophysiologic gradient culture of stratified epithelium. *Methods Cell Biol.* 1980;21B: 287–307.
39. Hoffman RM. The three-dimensional question: can clinically relevant tumor drug resistance be measured in vitro? *Cancer Metastasis Rev.* 1994;13:169–173.
40. Leighton J. Structural biology of epithelial tissue in histophysiologic gradient culture. *In Vitro Cell Dev Biol.* 1992;28A: 482–492.

41. Yang J, Richards J, Bowman P, et al. Sustained growth and three-dimensional organization of primary mammary tumor epithelial cells embedded in collagen gels. *Proc Natl Acad Sci U S A.* 1979;76:3401–3405.
42. Strangeways T. *Tissue Culture in Relation to Growth and Differentiation.* Cambridge, England: Heffer and Sons; 1924.
43. Fell HB, Robison R. The growth, development and phosphatase activity of embryonic avian femora and limb-buds cultivated in vitro. *Biochem J.* 1929;23:767–845.
44. Browning TH, Trier JS. Organ culture of mucosal biopsies of human small intestine. *J Clin Invest.* 1969;48:1423–1432.
45. Inch WR, McCredie JA, Sutherland RM. Growth of nodular carcinomas in rodents compared with multi-cell spheroids in tissue culture. *Growth.* 1970;34:271–282.
46. Chang HN, Furusaki S. Membrane bioreactors: present and prospects. *Adv Biochem Eng Biotechnol.* 1991;44:27–64.
47. Gallup DM, Gerhardt P. Dialysis fermentor systems for concentrated culture of microorganisms. *Appl Microbiol.* 1963; 11:506–512.
48. Chang TM. Semipermeable microcapsules. *Science.* 1964;146: 524–525.
49. Butterworth TA, Wang DI, Sinskey AJ. Application of ultrafiltration for enzyme retention during continuous enzymatic reaction. *Biotechnol Bioeng.* 1970;12:615–631.
50. Knazek RA, Gullino PM, Kohler PO, Dedrick RL. Cell culture on artificial capillaries: an approach to tissue growth in vitro. *Science.* 1972;178:65–66.
51. Belfort G. Membranes and bioreactors: a technical challenge in biotechnology. *Biotechnol Bioeng.* 1989;33:1047–1066.
52. Chang HN. Membrane bioreactors: engineering aspects. *Biotechnol Adv.* 1987;5:129–145.
53. Gekas VC. Artificial membrane as carrier for the immobilization of biocatalysts. *Enzyme Microb Technol.* 1986;8:450–460.
54. Hopkinson J. Hollow fiber cell culture in industry. In: Mattiasson B, ed. *Immobilized Cells and Organelles.* Boca Raton, FL: CRC Press; 1983.
55. Kitano H, Ise N. Hollow fiber enzyme reactors. *Trends Biotechnol.* 1984;2:5–7.
56. Vandenburgh HH, Shansky J, Del Tatto M, Lee P, Meir J. *Tissue Engineering Organs for Space Biology Research.* First Biennial Space Biomedical Investigators Workshop; 1997.
57. Vandenburgh H, Del Tatto M, Shansky J, et al. Attenuation of skeletal muscle wasting with recombinant human growth hormone secreted from a tissue-engineered bioartificial muscle. *Hum Gene Ther.* 1998;9:2555–2564.
58. Chromiak J, Shansky J, Perrone CE, Vandenburgh HH. Bioreactor perfusion system for the long term maintenance of tissue-engineered skeletal muscle organoids. *In Vitro Cell Dev Biol.* 1998;34:695–703.
59. Guignandon A, Usson Y, Laroche N, et al. Effects of intermittent or continuous gravitational stresses on cell-matrix adhesion: quantitative analysis of focal contacts in osteoblastic ROS 17/2.8 cells. *Exp Cell Res.* 1997;236:66–75.
60. Angele P, Kujat R, Nerlich M, Yoo J, Goldberg V, Johnstone B. Engineering of osteochondral tissue with bone marrow mesenchymal progenitor cells in a derivatized hyaluronan-gelatin composite sponge. *Tissue Eng.* 1999;5:545–554.
61. Schoeters G, Leppens H, Van Gorp U, Van Den Heuvel R. Haemopoietic long-term bone marrow cultures from adult mice show osteogenic capacity in vitro on 3-dimensional collagen sponges. *Cell Prolif.* 1992;25:587–603.
62. Goodwin TJ, Prewett TL, Wolf DA, Spaulding GF. Reduced shear stress: a major component in the ability of mammalian tissues to form three-dimensional assemblies in simulated microgravity. *J Cell Biochem.* 1993;51:301–311.
63. Goodwin TJ, Schroeder WF, Wolf DA, Moyer MP. Rotating-wall vessel coculture of small intestine as a prelude to tissue modeling: aspects of simulated microgravity. *Proc Soc Exp Biol Med.* 1993;202:181–192.
64. Schwarz RP, Wolf DA, Trinh TT. Horizontally rotated cell culture system with a coaxial tubular oxygenator. US patent no 5,026,650; June 25, 1991.
65. Van Wezel AL. Microcarrier/cultures of animal cells. In: Paterson MK, ed. *Tissue Culture: Methods and Applications.* New York: Academic Press; 1973.
66. Glacken MW, Fleischaker RJ, Sinskey AJ. Mammalian cell culture: engineering principles and scale-up. *Trends Biotechnol.* 1983;1:102–108.
67. Goodwin TJ, Jessup JM, Wolf DA. Morphologic differentiation of colon carcinoma cell lines HT-29 and HT-29KM in rotating-wall vessels. *In Vitro Cell Dev Biol.* 1992;28A:47–60.
68. Croughan MS, Hamel JF, Wang DI. Hydrodynamic effects on animal cells grown in microcarrier cultures. *Biotechnol Bioeng.* 1987;29:130–141.
69. Cherry RS, Papoutsakis ET. Physical mechanisms of cell damage in microcarrier cell culture bioreactors. *Biotechnol Bioeng.* 1988;32:1001–1004.
70. Cherry RS, Hulle CT. Cell death in the thin films of bursting bubbles. *Biotechnol Prog.* 1992;8:11–18.
71. Croughan MS, Wang DI. Growth and death in over agitated microcarrier cell cultures. *Biotechnol Bioeng.* 1989;33:731–744.
72. Nilsson K, Buzsaky F, Mosbach K. Growth of anchorage-dependent cells on macroporous microcarriers. *Biotechnology.* 1986;4:989–990.
73. Croughan MS, Sayre SS, Wang DI. Viscous reduction of turbulent damage in animal cell culture. *Biotechnol Bioeng.* 1989;33: 862–872.
74. Thalmann E. Biological experiences in bubble-free aeration system. *Acta Biotechnol.* 1989;39:511–516.
75. Clark J, Hirstenstein H, Gebb C. Critical parameters in the microcarrier culture of animal cells. *Dev Biol Stand.* 1980;46: 117–124.
76. Feder J, Tolbert WR. The large-scale cultivation of mammalian cells. *Sci Am.* 1983;248:36–43.
77. Tschopp A, Cogoli A, Lewis ML, Morrison DR. Bioprocessing in space: human cells attach to beads in microgravity. *J Biotechnol.* 1984;1:287–293.
78. Gmunder FK, Cogoli A. Cultivation of single cells in space. *Appl Micrograv Technol.* 1988;3:115–122.
79. Schwarz RP, Goodwin TJ, Wolf DA. Cell culture for three-dimensional modeling in rotating-wall vessels: an application of simulated microgravity. *J Tissue Cult Methods.* 1992;14: 51–57.
80. Wolf DA, Schwarz RP. Analysis of Gravity-Induced Particle Motion and Fluid Perfusion Flow in the NASA-Designed Rotating Zero-Head-Space Tissue Culture Vessel. *NASA Technical Paper* 3143; 1991.

81. Baker TL, Goodwin TJ. Three-dimensional culture of bovine chondrocytes in rotating-wall vessels. *In Vitro Cell Dev Biol Anim.* 1997;33:358–365.
82. Haffen K, Kedinger M, Simon-Assmann P. Mesenchyme-dependent differentiation of epithelial progenitor cells in the gut. *J Pediatr Gastroenterol Nutr.* 1987;6:14–23.
83. Kedinger M, Haffen K, Simon-Assmann P. Intestinal tissue and cell cultures. *Differentiation.* 1987;36:71–85.
84. Kedinger M, Simon-Assmann PM, Lacroix B, Marxer A, Hauri HP, Haffen K. Fetal gut mesenchyme induces differentiation of cultured intestinal endodermal and crypt cells. *Dev Biol.* 1986;113: 474–483.
85. Stallmach A, Hahn U, Merker HJ, Hahn EG, Riecken EO. Differentiation of rat intestinal epithelial cells is induced by organotypic mesenchymal cells in vitro. *Gut.* 1989;30:959–970.
86. Quaroni A. Crypt cell development in newborn rat small intestine. *J Cell Biol.* 1985;100:1601–1610.
87. Quaroni A. Development of fetal rat intestine in organ and monolayer culture. *J Cell Biol.* 1985;100:1611–1622.
88. Hahn U. Extracellular matrix proteins in small-intestinal cell cultures. *Scand J Gastroenterol Suppl.* 1988;151:70–78.
89. Moyer MP, Huot RI, Ramirez A Jr, Joe S, Meltzer MS, Gendelman HE. Infection of human gastrointestinal cells by HIV-1. *AIDS Res Hum Retroviruses.* 1990;6:1409–1415.
90. Herlyn M, Kath R, Williams N, Valyi-Nagy I, Rodeck U. Growth-regulatory factors for normal, premalignant, and malignant human cells in vitro. *Adv Cancer Res.* 1990;54:213–234.
91. Shih IM, Herlyn M. Autocrine and paracrine roles for growth factors in melanoma. *In Vivo.* 1994;8:113–123.
92. Goodwin TJ, Prewett TL, Spaulding GF, Becker JL. Three-dimensional culture of a mixed mullerian tumor of the ovary: expression of in vivo characteristics. In Vitro Cell Dev Biol Anim. 1997;33:366–374.
93. Prewett TL, Goodwin TJ, Spaulding GF. Three-dimensional modeling of T-24 human bladder carcinoma cell line: a new simulated microgravity vessel. J Tissue Cult Methods. 1993;15: 29–36.
94. Kerbel RS, Rak J, Kobayashi H, Man MS, St Croix B, Graham CH. Multicellular resistance: a new paradigm to explain aspects of acquired drug resistance of solid tumors. Cold Spring Harbor Symp Quant Biol. 1994;59:661–672.
95. Li LH, Bhuyan BK, Wallace TL. Comparison of cytotoxicity of agents on monolayer and spheroid systems. Proc Am Assoc Cancer *Res.* 1989;30:2435A.
96. Smith KS, Badiner GJ, Adams FG, Wilson DK, Li LH, Bhuyan BK. Modified 2-tumour (L1210, colon 38) assay to screen for solid tumor selective agents. In: *Proceedings of the Symposium on Anticancer Drug Discovery and Development.* Detroit, MI; 2006.
97. Lawler EM, Miller FR, Heppner GH. Significance of three-dimensional growth patterns of mammary tissues in collagen gels. *In Vitro.* 1983;19:600–610.
98. Miller BE, Miller FR, Heppner GH. Assessing tumor drug sensitivity by a new in vitro assay which preserves tumor heterogeneity and subpopulation interactions. *J Cell Physiol Suppl.* 1984;3:105–116.
99. Miller BE, Miller FR, Heppner GH. Factors affecting growth and drug sensitivity of mouse mammary tumor lines in collagen gel cultures. *Cancer Res.* 1985;45:4200–4205.
100. Simon-Assmann P, Bouziges F, Vigny M, Kedinger M. Origin and deposition of basement membrane heparan sulfate proteoglycan in the developing intestine. *J Cell Biol.* 1989;109: 1837–1848.
101. Teicher BA, Herman TS, Holden SA, et al. Tumor resistance to alkylating agents conferred by mechanisms operative only in vivo. *Science.* 1990;247:1457–1461.
102. Kobayashi H, Man S, Graham CH, Kapitain SJ, Teicher BA, Kerbel RS. Acquired multicellular-mediated resistance to alkylating agents in cancer. *Proc Natl Acad Sci U S A.* 1993;90: 3294–298.
103. Margolis J, Li ZP. Heat sterilisation to inactivate AIDS virus in lyophilised factor VIII [letter]. *Aust N Z J Med.* 1986;16:413.
104. Ingram M, Techy GB, Saroufeem R, et al. Three-dimensional growth patterns of various human tumor cell lines in simulated microgravity of a NASA bioreactor. *In Vitro Cell Dev Biol Anim.* 1997;33:459–466.
105. Margolis L, Hatfill S, Chuaqui R, et al. Long term organ culture of human prostate tissue in a NASA-designed rotating wall bioreactor. *J Urol.* 1999;161:290–297.
106. Zhau HE, Goodwin TJ, Chang SM, Baker TL, Chung LW. Establishment of a three-dimensional human prostate organoid coculture under microgravity-simulated conditions: evaluation of androgen-induced growth and PSA expression. *In Vitro Cell Dev Biol Anim.* 1997;33:375–380.
107. Wang R, Xu J, Juliette L, et al. Three-dimensional co-culture models to study prostate cancer growth, progression, and metastasis to bone. *Semin Cancer Biol.* 2005;15:353–364.
108. Vertrees RA, Zwischenberger JB, Boor PJ, et al. Cellular differentiation in three-dimensional lung cell cultures. *Cancer Biol Ther.* 2008;7:404–412.
109. Nickerson CA, Goodwin TJ, Terlonge J, et al. Three-dimensional tissue assemblies: novel models for the study of *Salmonella enterica* serovar *Typhimurium* pathogenesis. *Infect Immun.* 2001;69:7106–7120.
110. Hiemstra PS, Bals R. Series introduction: innate host defense of the respiratory epithelium. *J Leukoc Biol.* 2004;75:3–4.
111. Gibson MC, Perrimon N. Apicobasal polarization: epithelial form and function. *Curr Opin Cell Biol.* 2003;15:747–752.
112. Knight DA, Holgate ST. The airway epithelium: structural and functional properties in health and disease. *Respirology.* 2003;8:432–446.
113. Cotran R, Kumar V, Collins T. *Robbins Infectious Diseases.* 6th ed. Philadelphia: Saunders; 1999.
114. Garofalo RP, Haeberle H. Epithelial regulation of innate immunity to respiratory syncytial virus. *Am J Respir Cell Mol Biol.* 2000;23:581–585.
115. Polito AJ, Proud D. Epithelia cells as regulators of airway inflammation. *J Allergy Clin Immunol.* 1998;102:714–718.
116. Ke Y, Reddel RR, Gerwin BI, et al. Human bronchial epithelial cells with integrated SV40 virus T antigen genes retain the ability to undergo squamous differentiation. *Differentiation.* 1988;38: 60–66.
117. Gray GD, Wickstrom E. Evaluation of anchorage-independent proliferation in tumorigenic cells using the redox dye alamar Blue. *Biotechniques.* 1996;21:780, 782.
118. Wu R, Sato GH, Whitcutt MJ. Developing differentiated epithelial cell cultures: airway epithelial cells. *Fundam Appl Toxicol.* 1986;6:580–590.

119. Stoner GD, Katoh Y, Foidart JM, Myers GA, Harris CC. Identification and culture of human bronchial epithelial cells. *Methods Cell Biol.* 1980;21A:15–35.
120. Adler KB, Li Y. Airway epithelium and mucus: intracellular signaling pathways for gene expression and secretion. *Am J Respir Cell Mol Biol.* 2001;25:397–400.
121. Fukamachi H. Disorganization of stroma alters epithelial differentiation of the glandular stomach in adult mice. *Cell Tissue Res.* 1986;243:65–68.
122. Sutherland AE, Calarco PG, Damsky CH. Expression and function of cell surface extracellular matrix receptors in mouse blastocyst attachment and outgrowth. *J Cell Biol.* 1988;106: 1331–1348.
123. Wiens D, Park CS, Stockdale FE. Milk protein expression and ductal morphogenesis in the mammary gland in vitro: hormone-dependent and -independent phases of adipocyte-mammary epithelial cell interaction. *Dev Biol.* 1987;120:245–258.
124. Buset M, Winawer S, Friedman E. Defining conditions to promote the attachment of adult human colonic epithelial cells. *In Vitro Cell Dev Biol.* 1987;23:403–412.
125. Daneker GW Jr, Mercurio AM, Guerra L, et al. Laminin expression in colorectal carcinomas varying in degree of differentiation. *Arch Surg.* 1987;122:1470–1474.
126. Durban EM, Knepper JE, Medina D, Butel JS. Influence of mammary cell differentiation on the expression of proteins encoded by endogenous BALB/c mouse mammary tumor virus genes. *Virus Res.* 1990;16:307–323.
127. Kaye GI, Siegel LF, Pascal RR. Cell replication of mesenchymal elements in adult tissues. I. The replication and migration of mesenchymal cells in the adult rabbit dermis. *Anat Rec.* 1971;169:593–611.
128. O'Brien LE, Zegers MM, Mostov KE. Opinion: Building epithelial architecture: insights from three-dimensional culture models. *Nat Rev Mol Cell Biol.* 2002;3:531–537.
129. Goodwin TJ, Deatly AM, Suderman MT. Three-dimensional engineered high fidelity normal human lung tissue-like assemblies (TLA) as targets for human respiratory virus infection. American Society for Virology 22nd Annual Meeting; 2003 (abstract).
130. Suderman MT, Mossel E, Watts DM, et al. Severe acute respiratory syndrome (SARS)-CoV infection in a three-dimensional human bronchial-tracheal (HBTE) tissue-like assembly. American Society for Virology 23rd Annual Meeting; 2004 (abstract).

Index

MIX
Papier aus verantwortungsvollen Quellen
Paper from responsible sources
FSC® C105338

If you have any concerns about our products,
you can contact us on
ProductSafety@springernature.com

In case Publisher is established outside the EU,
the EU authorized representative is:
Springer Nature Customer Service Center GmbH
Europaplatz 3, 69115 Heidelberg, Germany

Printed by Libri Plureos GmbH
in Hamburg, Germany